In Quest of Great Lakes Ice Age Vertebrates

In Quest of Great Lakes Ice Age Vertebrates

J. Alan Holman

Michigan State University Press
East Lansing

Michigan State University Press
East Lansing, Michigan 48823–5202

Printed and bound in the United States of America.

07 06 05 04 03 02 01 1 2 3 4 5 6 7 8 9 10

LIBRARY OF CONGRESS CATALOGING-IN-PUBLICATION DATA

Holman, J. Alan, 1931-
  In quest of Great Lakes Ice Age vertebrates / J. Alan Holman.
      p. cm.
Includes bibliographical references and index.
  ISBN 0-87013-591-0 (cloth : alk. paper)
 1. Vertebrates, Fossil—Great Lakes Region. 2.
Paleontology—Pleistocene. 3. Animals, Fossil—Great Lakes
Region. I. Title.
  QE841 .H65 02001
  566'.0977—dc21
                              2001003708

Cover design by Heidi Dailey
Book design by Sans Serif Inc., Saline, Michigan

Visit Michigan State University Press on the World Wide Web at:
www.msupress.msu.edu

Contents

Preface

For me, collecting fossils, holding these ancient bones in my hands, studying to find out what they are, how the creatures they represent lived, has enabled me to reach back in time, to touch the past.

—Amateur paleontologist
Joan Wiffen, New Zealand

The last Ice Age (Pleistocene epoch) ended only about 140 human generations ago. It was characterized by gigantic, moving ice sheets that changed the face of the earth and a massive, worldwide extinction of large mammals. This book details the Pleistocene vertebrates: fishes, amphibians, reptiles, birds, and mammals of Ontario, Michigan, Ohio, Indiana, Illinois, and Wisconsin (fig. 1). The spectacular extinct mammals as well as more familiar vertebrates that survived into modern time are featured, and unanswered questions about the sudden extinction of the many large mammalian species at the end of the epoch are addressed. The intended audience includes the general public, students and teachers, and professionals with biological or geological interests.

The book begins with a definition of concepts and terms for the general audience and a discussion of the Pleistocene and how it affected the physical and biological world. A general account of the Pleistocene in the Great Lakes region is given. Next, the methods employed and tools used in collecting vertebrate fossils are explained, and the ethics and protocol involved in maintaining a proper collection of vertebrate fossils are discussed.

This is followed by a species-by-species account of the Pleistocene vertebrates of the region. Here the structure, habits, and habitats of these animals are discussed. Some important fossils and species represented by the fossils are illustrated. A site-by-site description of the major Pleistocene vertebrate faunas of the Great Lakes region is presented next, which includes a list of all of the vertebrate species found in each site, as well as ecological and climatic interpretations for each locality.

The final portion of the book examines the compelling problems of the Pleistocene relative to faunal interpretations in the region. Major topics will be vertebrate range adjustments that occurred in the region, how the great Pleistocene extinction affected the animals of the region, and the aftermath of the Ice Age.

I have attempted to write this book in a way that reflects the main objective of the book, which is to introduce the reader to the fascinating vertebrate life of the Ice Age of the Great Lakes region.

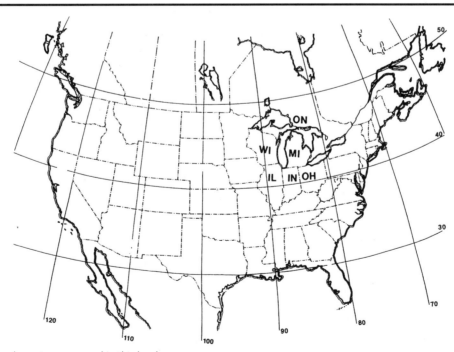

FIGURE 1. States and provinces covered in this book.

Acknowledgments

I am especially grateful to the people who have shared their special knowledge and enthusiasm about Pleistocene vertebrates of the Great Lakes region with me—most of them in person and all of them through their publications and/or correspondence. Without their help this book could not have been written. These people are: Laura Abraczinskas, Andrea Bair, Charles (Rufus) Churcher, Charles Cleland, the late John Dorr, James Farlow, Daniel Fisher, Kenneth Ford, Russell Graham, Michael Hansen, James Harding, the late Claude Hibbard, Richard Harington, Margaret Holman, the late Ronald Kapp, Paul Karrow, Graham Larsen, Bradley Lepper, Gregory McDonald, Paul Martin, Ronald Mason, William Monaghan, Patrick Munson, David Overstreet, Paul Parmalee, Ronald Richards, Jeffery Saunders, Kevin Seymour, Jeheskel Shoshani, William Wakefield, David Westjohn, Richard Wilson, and John-Paul Zonneveld.

I thank Martha Bates, Julie L. Loehr, and the other staff members of Michigan State University Press for their efforts in the production of this book. Drs. David D. Gillette and C. R. Harington, as well as an anonymous reviewer provided helpful corrections and comments on the manuscript and Bob Burchfield edited the entire work—I am grateful to all. The National Geographic Society has kindly provided grants that have funded some of my work on Michigan fossil vertebrates. The Michigan State University Museum has helpfully provided office and research space for my studies. For years, the annual meeting of the Michigan Academy of Science, Arts, and Letters has been a very special place to exchange ideas about the Pleistocene flora and fauna of Michigan, and I thank that organization for providing a forum for these exchanges.

Introduction

From the human standpoint, the Ice Age (Pleistocene epoch) is the most important unit of geological time, for changes wrought during this short epoch still strongly influence almost every aspect of human life. The Pleistocene consisted of cold stages when massive ice sheets moved down into the Great Lakes region from the north, smothering the landscape, and warmer stages in when the ice retreated. The power of the glaciers etched a new topography on the land and carried vast amounts of sedimentary material (including huge boulders) great distances. In what is now the Canadian prairies, gigantic masses of Cretaceous bedrock called megablocks that weighed millions of tons and measured up to 3 miles long were sometimes transported more than 200 miles by the ice sheet.

A discomforting thought is that there could be another glacial advance in the future, which, among other disruptions, would cover Chicago, Milwaukee, Cleveland, Detroit, Indianapolis, and Toronto with a mile or two of ice! The last interglacial age, the Sangamonian (Ipswichian in Britain) is now thought to have lasted about 40,000 years, and the last glacial age, the Wisconsinan (Devensian in Britain) is thought to have lasted about 100,000 years. Since there is no evidence that the underlying cause for the cyclic events of the last 1.9 million years has subsided and since it has been about 10,000 years since the end of the Wisconsinan

cold interval, we may be due for another ice age 30,000 years or so down the road.

However, in this book, rather than nervously speculating about the future of our ancestors forty generations from now, I will examine the fascinating vertebrate life (fishes, amphibians, reptiles, birds, and mammals) of the past in the Great Lakes region. We shall find that a much different vertebrate fauna existed in the region during the Ice Age, a fauna dominated by elephant-like mastodonts (preferred spelling) and mammoths as well as other large mammalian herbivores and the carnivores that ate them. I will not only examine the large mammals of the epoch but will discuss the smaller vertebrate animals that formed the "supporting cast" of the Pleistocene panorama of life.

Some may be surprised that many familiar present-day vertebrates existed alongside the extinct Pleistocene giants in the region. The book details these extant animals and attempts to reflect upon their ecological role in the Ice Age world. Why these animals survived while so many large mammals became extinct is an important question that will be discussed later in the book.

The Great Lakes region of the Pleistocene may be divided into northern and southern subregions based on geological and vertebrate paleontological evidence. The problem of the range adjustments of extant species and the massive extinction of large mammals in the region will be analyzed, and the possible interactions between all of these forms and humans will be detailed. Finally, the aftermath of the Ice Age in the region will be addressed.

Before we begin the major parts of the book, basic concepts and terms for the general reader are discussed and defined.

EVOLUTION

The theory of evolution has united the scientific disciplines of the world like no other concept before or since. This is especially true of the biological and geological sciences. It is often said that paleontology, the study of the life of the past and the subject of this book, illustrates the patterns of evolution better than any other scientific discipline.

Evolution is a firmly established theory that has been tested in many different ways by thousands of highly trained scientists. Evolutionary theory explains that life originated from simple groups of large organic molecules and that these early living things evolved into more and more complex units over millions of years.

Charles Darwin is the person most clearly associated with bringing the theory of evolution sharply before the scientific community with his concept of natural selection as put forth in his *Origin of Species*, published in 1859. Essentially, he pointed out that in nature, more offspring are produced than will survive and that forces in the biotic and physical environment will select out those that are most adapted for existing conditions. Since the best-adapted organisms are most likely to survive to reproductive age, following generations should become more and more adapted to existing conditions. On the other hand, if the environment changes, it may produce further organic changes by selecting for those organisms best adapted to those changes.

The genetic basis of evolution was unknown to Darwin (even though the contemporary Gregor Mendel was beginning the foundation of genetics by simple experiments with peas), but shortly after the turn of the nineteenth century, several scientists began to understand the genetic basis of inheritance. Modern scientists now understand a great deal about the molecular nature of the genetic material, and we know now that for evolution to occur these molecular units (genes) must either physically change or become adapted in some way themselves to produce different results.

A few evolutionary terms are basic for understanding some concepts in this book. Adaptive radiation, convergence, and parallelism are such terms and are discussed below. Large landmasses usually support a variety of organisms (living things). These organisms tend to fill the various ecological niches, or "modes of life," that are available to them. When a new landmass, such as a large volcanic island, becomes available for colonization by living things, many ecological niches are usually available. But typically, based on chance, only a few species reach these new landmasses. These few taxa may evolve rapidly into species that fill the available ecological niches. This situation produces one type of adaptive radiation.

Another type of adaptive radiation occurs when a group of organisms evolves a mechanism that allows the group to exploit a major ecological situation. The evolution of wings in birds in the age of dinosaurs set the stage for an adaptive radiation of birds early in the age of mammals.

When an adaptive radiation occurs in one group of organisms, some species may take on the appearance of totally unrelated forms. For instance, the adaptive radiation of a small group of very ancient amphibians called microsaurs produced some species that looked very much like small reptiles. It seems that the both microsaurs and the very earliest reptiles were adapting to a terrestrial, insect-eating niche. This evolutionary process is called convergence. In nature, striking examples of convergence occur, such as those between some types of sharks (fishes) and porpoises (mammals).

When rather closely related animals that live in different geographic areas tend to evolve resemblances to one another because they occupy similar ecological niches, the situation is termed parallelism. Certain North American frogs of the family Ranidae behaviorally and structurally tend to resemble certain South American frogs of the family Leptodactylidae because they have been evolving in parallel fashion in different continents.

Two other evolutionary terms, *homology* and *analogy*, are essential for understanding paleontology. The concept of homology relates to structures being inherited from a common ancestor. Thus the bones of human forelimbs and hind limbs are homologous to the bones of the forelimbs and hind limbs of ancestral four-legged vertebrates. The forelimbs of apes and humans are homologous to the wings of birds, as both have a like origin, even if they have different functions. The term *analogy* relates to structures having similar functions. The wings of birds and butterflies are analogous as they have similar functions, but they are not homologous, as they have different ancestral origins.

GEOLOGY

Geology is the scientific study of the earth and earth processes. A discussion of some geological concepts and terms will be helpful in understanding vertebrate paleontology.

The Major Classes of Rocks

There are three major classes of rocks: igneous, sedimentary, and metamorphic. All of these occur in the Great Lakes region (table 1). Igneous rocks form by the solidification of gaseous or molten material. Early scientists called them plutonic rocks because they either form by cooling at great depths in the earth, flow out of the earth as volcanic lava, or are blasted into the air from volcanoes as gas or fragments of molten material. Common igneous rocks are basalt, obsidian, granite, and volcanic ash.

Sedimentary rocks are formed from older rocks by the action of wind, water, and ice (clastic rocks) or by chemical or organic means. Common clastic sedimentary rocks are sandstones, mudstones, shales, and conglomerates (cemented, unsorted accumulations of other clastic rocks). Chemically or organically produced sedimentary rocks include salt, gypsum, limestones, chalk, and coal. Most fossils in the world, including those in the Great Lakes region, are found in sedimentary rocks.

Metamorphic rocks are formed when igneous or sedimentary rocks are changed in structure by intense pressure, by great heat, or by infiltration of other material at great depths. Metamorphic rocks are either flaky like slate, schist, or gneiss or nonflaky like marble (metamorphosed limestone). Fossils are occasionally found in the nonflaky metamorphic rocks, but the fossils are often distorted and difficult to identify.

The Geological Cycle

The geological cycle (fig. 2) consists of the processes of uplift, erosion, and deposition, which have been repeated again and again throughout geological time. If we begin by viewing the tilted uplifted rock strata of the cycle, we note that erosion has removed the top part of the layers (fig. 2). In nature, erosional processes move material from the uplifted highlands or mountain ranges to lowlands or basins. Deposition then takes place in the low places until processes within the earth cause uplift to occur once again.

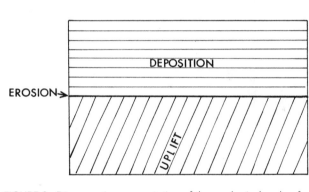

FIGURE 2. Diagramatic representation of the geological cycle of uplift, erosion, and deposition of sediments.

In nature, breaks may occur between rock layers because of erosional intervals. These breaks are called disconformities. The heavy line in figure 2 represents a disconformity caused by an erosional interval. If a long erosional interval has occurred between two layers of rocks, the fossils in the two layers are often quite different. In much of the Great Lakes region, a vast erosional interval termed the great lost interval occurred between the Pleistocene sediments and the very ancient Paleozoic rocks that they lie directly upon. As will be explained later, this is why there are no dinosaur remains in the area.

TABLE I. A few common rocks of the Great Lakes region.

Igneous	Sedimentary	Metamorphic
Andesite	Clastics:	Gneiss
Basalt	Clay and shale	Marble
Diorite	Sand and sandstone	Phyllite
Felsite	Silt and siltstone	Schist
Gabbro	Chemical/biologicals:	Slate
Granite	Chert and flint	Quartzite
	Coal	
	Gypsum	
	Limestone and dolomite	
	Salt	

Uniformitarianism

Uniformitarianism is a doctrine that holds that events of the past may often be interpreted by observing processes going on in the present. Uniformitiarian principles were very important in establishing the chronology and relationships of rock strata in the Great Lakes region. Obviously, there have been violent changes in the Earth's crust in the past that have never been observed in the present (for example, the violent volcanic ash storms that occurred during the uplift of the Rocky Mountains), but this does not mean that the study of physical processes that are working in the earth today cannot be useful in explaining the great crustal disturbances of the past.

Very often the principle of uniformitarianism is used in interpreting the lives of modern vertebrates that lived in the region during the Ice Age in that we assume that they had essentially the same needs and habits in the Pleistocene that they do today.

The Geological Formation

The geological formation is the basic rock unit that the geologist maps; thus it is often an important part of the locality data for Pleistocene vertebrates, especially in Ontario where several geological formations of Pleistocene age have been named. A formation is a natural rock unit that has characteristics by which it may be traced from place to place and distinguished from other formations. Often, a formation depicts a single sedimentary event of a rather large magnitude, such as the filling of a very large depressed area or basin.

Formations are mainly composed of sedimentary rocks, and some formations may transcend established units of time. More often, formations represent a depositional sequence that took place within or even contiguous with one of the established time periods. Formations are named after the geographic locality where they are characteristically exposed and where they were first described and defined.

When a formation consists of a single type of rock category, the rock category is included in its name. The Late Pleistocene Halton Till, for instance, was named at Halton, Ontario, and is composed of glacial till, sedimentary material deposited by the ice. If a formation consists of different assemblages of rock types, it is merely designated as a formation. The Early Pleistocene Scarborough Formation, for instance, was named at Scarborough, Ontario. It is formed of different kinds of rock types and thus is called the Scarborough Formation.

Geological Time

A geological time scale (fig. 3) was set up worldwide in the nineteenth century on the basis of what is termed relative chronology and later was supplemented by absolute chronological studies. Both relative chronology and absolute chronology are important in interpreting the vertebrate fossils of the region, and Pleistocene absolute chronology will be discussed in detail later. The relative chronology of rock strata is based on the relationship of the strata to one another and the fact that fossil assemblages change from older to younger beds. When younger beds lie upon older beds, it is referred to as superposition, and faunal succession occurs when fossil assemblages change from older strata to younger ones. A schematic representation of faunal succession in the glaciated part of the Great Lakes region is shown in figure 4.

TIME IN MILLIONS OF YEARS BEFORE PRESENT	ERAS	PERIODS	EPOCHS
.1	CENOZOIC	QUATERNARY	HOLOCENE
1.9			PLEISTOCENE
5		TERTIARY	PLIOCENE
25			MIOCENE
35			OLIGOCENE
55			EOCENE
65			PALEOCENE
140	MESOZOIC	CRETACEOUS	
210		JURASSIC	
250		TRIASSIC	
290	PALEOZOIC	PERMIAN	
320		PENNSYLVANIAN	
360		MISSISSIPPIAN	
410		DEVONIAN	
440		SILURIAN	
500		ORDOVICIAN	
550		CAMBRIAN	
4550	PRECAMBRIAN		

FIGURE 3. The geological time scale.

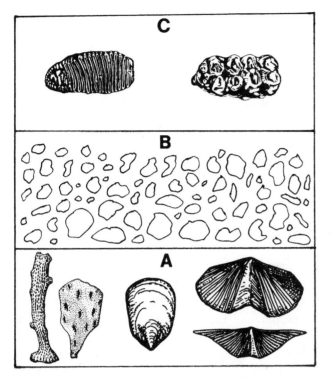

FIGURE 4. A schematic representation of faunal succession in the Great Lakes region. A, bryozoans (the two figures on the left) and brachiopods (the three figures on the right) from Paleozoic beds. B, disconformable clastic rocks left by the Wisconsinan glaciation. C, mammoth tooth (left) and mastodont tooth (right) from the postglacial Wisconsinan.

Paleontology

Paleontology is the study of ancient life and is based on the interpretation of fossils. As a working definition, a fossil is any recognizable and interpretable evidence of prehistoric life. Vertebrate paleontology, the subject of this book, is the study of fossil animals with backbones. Vertebrate fossils consist mainly of bones and teeth. Subdisciplines of paleontology that supplement the interpretation of vertebrate fossils are invertebrate paleontology, the study of fossil animals without backbones (mainly the external shells of animals; see fig. 4); paleobotany, the study of fossil plants; micropaleontology, the study of very small fossils that must be studied under a microscope; and palynology, the microscopic study of fossil pollen and spores.

Invertebrate paleontologists tend to be associated with geology departments because the science was basically founded by William "Strata" Smith (1769–1839), an English geologist. Smith used invertebrate fossils to identify the layers of rocks that bore them. Vertebrate paleontologists often receive their degrees from zoology departments because vertebrate paleontology was essentially founded by the French comparative zoologist Georges Cuvier (1769–1832).

WHAT ARE VERTEBRATES?

Vertebrates are animals with a spinal column composed of vertebrae. Figure 5 is a schematic representation of a generalized vertebrate body. Vertebrates are a subphylum of a phylum of animals called chordates, which are distinguished from all other animal groups by the presence of a long, supple rod called the notochord. The notochord is not only soft enough to allow the body to move by lateral and occasionally horizontal undulations, but this rod stiffens enough along its axis that the thrust generated by the undulations pushes primitive chordates such as *Amphioxus* (lancelets) through the water effectively. Although the notochord occurs in embryonic vertebrates, it is usually supplanted by the vertebral column. Some of the major vertebrate characteristics are as follow.

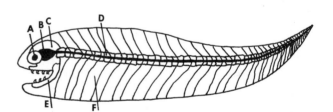

FIGURE 5. Some essential higher vertebrate (gnathostome) structures. A, well-developed, paired eyes; B, well-developed brain at anterior end of nerve cord; C, well-developed chondrocranium or skull; D, vertebral column surrounding nerve cord; E, lower jaw; F, segmental muscles.

Bilateral Symmetry, Cephalization, and Regionalization

The vertebrate body plan relates directly to the fact that vertebrates are actively moving animals. Thus bilateral symmetry, where half of the body is a mirror image of the other, and cephalization, where the brain and sense organs are concentrated in the head region at the front of the body, are of primary importance. Other forms of life may be asymmetrical, as in the amoeba, or have spherical bodies, as in some other simple life forms. Radial symmetry, whereby the parts of the body are arranged around a central oral (mouth) and aboral (away from the mouth) axis, like the spokes of a bicycle wheel, is found in such animals as jellyfish and sea anemones.

But in truly active animals (for example, marine worms, insects, and vertebrates), there tends to be a leading end where the sense organs are concentrated (head) and a body that follows. The head directs the rest of the animal toward food, shelter, and reproductive partners, whereas the tail tends to propel the animal forward in fishlike vertebrates or to act as a balance or grasping organ in land vertebrates. In essence, an active animal tends to move toward the essential things in front of it (hence cephalization), with nonessential things on either side of it (hence bilateral symmetry). Regional differences in vertebrates also go from front to back. Regional units between the head and the tail consist of a neck (in vertebrates above fishes) and a trunk that contains the major body organs. In fishes there is no neck region, as the region between the head and trunk is an immovable gill-bearing area.

Skeletal and Hard Tissues

The framework of the vertebrate body is a skeleton of cartilage or bone. Without a skeleton the vertebrate body would exist as a collapsed sack of skin filled with soft tissue. Invertebrate animals tend to have external skeletons (for example, clam shells) that are heavy and break easily, whereas vertebrates have internal skeletons that are light and supple. The vertebrate skeleton usually consists of two units, the axial and the appendicular skeleton. The axial skeleton consists of the head and vertebral column, ribs, and sternum. The appendicular skeleton consists of either fins or limbs and their supporting structures.

Bone. Bone is wonderful stuff. Because of its lightness and suppleness, it allows vertebrates to be more active than any other group of animals (with the exception of some insects). It has been shown that bone acts as a so-called two-phase substance, with an organic matrix of fibrous collagen (a complex, stringy protein) onto which inorganic tiny crystals of the mineral hydroxyapatite are packed. This combination of organic and inorganic parts gives bone its remarkable structure.

The inorganic part of bone (hydroxyapatite) is full of tiny cracks. When compression >———< is applied, these cracks close to take up the stress. But when tension <———> is applied, these cracks tend to spread open and fracture may occur. In other words, the inorganic part of the bone is stronger under compression than under tension. On the other hand, the organic material, collagen, is not rigid and tends to deform under compression. But when collagen is placed under tension, its fibers are able to stretch like a rubber band before they finally break. Thus the organic part of the bone is stronger under tension than compression. In summary, the inorganic hydroxyapatite and the organic collagen produce a tissue with a high compressive and tensile strength.

Cartilage. Cartilage is an elastic structure that supplements the bony part of the skeleton in most higher vertebrates and forms a large part of all the skeleton in some fishes. Cartilage is composed of a matrix called chondromucsin that is secreted by cartilage cells and that is laced through with collagen fibers. In many vertebrates, cartilage is replaced by bone in the adult, especially in the appendicular skeleton and outer part of the skull.

Hard tissues. Hard tissues of the vertebrate body other than bone include the very hard substance enamel and enamel-like structures found as the outer parts of teeth and fish scales; dentine, a less hard tissue found in teeth, tusks, and fish scales; and cementum, a component of vertebrate teeth. Obviously, bones and hard tissues are the parts of the vertebrate body that are most likely to fossilize.

Muscular Tissue

The muscles of the vertebrate body may form as much as half of the body weight. Three distinct types of muscle tissue are found in vertebrates: smooth, skeletal, and cardiac. The individual cells (fibers) of smooth muscle tissue have a single, central nucleus and lack crossbands. These muscles are not controlled by conscious processes. Smooth muscles line the digestive tract, ducts of various glands, bladder, trachea, bronchial tubes, circulatory vessels, and sex organs.

Skeletal muscle fibers have multiple peripheral nuclei and prominent crossbands, and they are controlled by conscious processes. These are the muscles that move the vertebrate skeleton. Cardiac muscles occur as an interconnected meshwork of fibers. There are multiple, superficially located nuclei in the fibers of this mass, and both thick and thin crossbands are present.

Cardiac muscles are not controlled by conscious processes and form the muscles of the heart.

Segmentation

Some highly developed invertebrates, the annelids and the arthropods, have bodies that are completely composed of serially arranged segments that are nearly or totally identical. Vertebrates are also segmented, but rather than being completely segmented from the outside as in invertebrates, vertebrates have an internal segmentation that is in the form of the segmented nature of the vertebral column and the associated musculature of the trunk and the spinal nerves.

Nervous System

In vertebrates, a single, dorsal, hollow, fluid-filled nerve cord (spinal cord) is surrounded and protected by the vertebral column. At the front end of the nerve cord is an enlargement called the brain. Large nerves branching from the brain (cranial nerves) and the spinal cord (spinal nerves) serve other parts of the head and body.

Celom

Many invertebrate groups have their internal organs embedded within the body tissues. In other groups of invertebrates and in vertebrates, most of the body organs are suspended within cavities called celomic cavities. In vertebrates, one celomic cavity, the abdominal cavity, contains most of the digestive organs, the organs of the urinary system, and the sex organs. Another celomic cavity contains the heart and lungs. In mammals, there are separate cavities for the heart and lungs.

Circulatory System

The circulatory system of vertebrates is a closed system where the blood is always enclosed within the heart or blood vessels. In some invertebrates, the blood oozes through the body tissues on its way back to the heart. Vertebrate blood contains an iron compound called hemoglobin, which is the blood's oxygen carrier. Hemoglobin in vertebrates is not free in the bloodstream but is carried in red blood cells. In most invertebrates, whether hemoglobin is present or not, the oxygen-carrying materials are free in the bloodstream and not confined to any particular type of cell.

Digestive System

Many organisms above the single-celled level of organization have a digestive cavity with an entrance to it and an exist from it. In bilaterally symmetrical animals with cephalization, the entrance is usually near the head and is called the mouth, whereas the exit is usually near the tail and called the anus. In the invertebrate annelids and arthropods, the anus is at the terminal end of the body, but in vertebrates there is a postanal tail.

In most vertebrates, the digestive tube (gut) is subdivided into a series of regions running between the mouth and the anus. These regions are the pharynx (throat), esophagus, stomach, intestine, and cloaca. The cloaca is a unit of the gut that is a common chamber for the products of the digestive system, urinary system, and reproductive system and is absent in most mammals.

The chief digestive organs of vertebrates are the liver, where food materials are converted into usable substances and stored, and the pancreas, which has the primary function of secreting digestive enzymes.

Kidneys

Vertebrates have paired, dorsal kidneys that function to rid the body of nitrogenous wastes (which would build up to toxic levels without these organs) and also to maintain the proper fluid balance within the tissues. In animals below the vertebrate level of organization, the tubules that serve as excretory organs are not structured as discrete kidneys.

Breathing Tissues

The process of breathing occurs in vertebrates when oxygen from the air or water diffuses into a network of tiny blood vessels called capillaries, thence to the general circulatory system, and finally to all of the tissues of the body. The complex of capillaries that traps the oxygen from the air or water may be in the form of external gills, gills in the throat region, lungs within the body, specialized areas in the mouth or cloacal cavity, or networks throughout the entire skin, as in the amphibians.

Sex Organs

The male and female organs of reproduction are always separate in vertebrates, although there are some all-female species in fishes, amphibians, and reptiles. The male sex organs are paired structures called testes; the female sex organs are paired structures called ovaries. Many invertebrate groups, including some of the most complex ones, have both sexes in the same individual.

PRESERVATION OF VERTEBRATE FOSSILS

The two most important criteria for the preservation of vertebrate fossils are the possession of hard parts and immediate burial. Soft tissues usually decay too rapidly for fossilization to occur; thus most vertebrate fossils are composed of skeletal parts and teeth. In vertebrates, teeth are especially common fossils because they are harder than bone. If immediate burial does not occur, the animal will be torn apart by scavengers such as vultures or hyenas. Then bacterial and/or fungal decay will do away with the soft parts. Next, the hard parts weather to dust. But if the dead vertebrate animal is quickly buried under sediments, scavengers are generally avoided and the skeleton remains more or less associated; oxygen is largely excluded and decay is greatly reduced; finally, weathering is prevented and permineralization can occur

Permineralization involves minerals, such as calcium carbonate (lime) or silica, filling in the spaces that naturally occur in bone. A general rule is, the more ancient the bone, the more it tends to be permineralized, but this is not always the case. Most vertebrate fossils in the Great Lakes region are preserved as the result of the permineralization process.

Actual preservation of soft parts of vertebrates sometimes occurs in asphalt pits ("tar pits"), in oil seeps, in frozen arctic soil, or in dry caves in the southwestern part of the United States. Fossil vertebrate soft parts, however, are rare or nonexistent in the Great Lakes region. Molds, casts, and imprints of vertebrate fossils occur in other areas but are rare or absent in the region. Coprolites, fossilized fecal masses or pellets, are unusual in the Great Lakes region, as are stomach contents, although these have been rarely reported in mastodonts in the region.

Taphonomy is a very important term that relates to the preservation of fossils in all geographic regions. The term *taphonomy* may be confusing to people since another term, *taxonomy*, is commonly used in paleontological studies. Taxonomy deals with the procedure of giving organisms scientific names. Taphonomy, in essence, is the study of the death burial and preservation of organisms that become fossils. Taphonomic studies demand that fossils be recorded exactly as they are located in the sediments so that the exact orientation of the fossils with respect to the pattern of sediments around them as well as to other fossils in the deposit can be determined. Information yielded by such studies can indicate the way the animal died, how it became entrapped in the sediments, and how the animal became preserved as a fossil. In the anthropological sense, taphonomic studies yield important information about the relationship of vertebrate fossils to the humans that utilized them for various purposes.

PALEOECOLOGY

Early studies in vertebrate paleontology centered on the phylogeny (evolutionary history of a group or lineage) of such taxa as pigs, horses, and rhino. The study of Pleistocene communities involves an ecological approach to paleontology. Ecology may be defined as the study of the relationship of organisms or groups of organisms to their environment. The environment may be divided into the physical environment and the biotic environment.

We know that higher organisms are composed of ascending levels of the organization of cells (cells–tissues–organs–organ systems–individuals), so it can be said that ecology mainly deals with levels of organization above the individual level. A central unit for ecological and paleoecological study is the community. A community may be defined as all of the populations of organisms living in a specific area—in other words, all of the organisms living in a pond form a pond community. The organisms in a community usually interact together in a more or less harmonious way. The ecosystem concept reflects the interaction between the community and the physical environment.

A community has a definite structure relative to the energy that flows through it. All of the original energy for any biological community ultimately comes from the sun. Thus a community without primary solar energy, such as a cave, must depend upon energy brought into the system. In Pleistocene and modern cave communities, bats have been important in bringing energy into caves in the form of guano.

There are several trophic levels as this energy flows through the community. Producers are the photosynthetic organisms (such as plants) that bind the energy from the sun into food sources for the other living members of the community. Producers are eaten by primary consumers, but there may be secondary, tertiary, and even quaternary consumers in a community. For instance, the algae (producer) in a Pleistocene pond may have been eaten by a frog tadpole (primary consumer), which may have been eaten by a small fish (secondary consumer), which may have been eaten by a northern pike (tertiary consumer), which may have been eaten by a Paleo-Indian (quaternary consumer).

Decomposers, mainly bacteria and fungi, break down dead bodies and waste materials excreted by the organisms of the various trophic levels into nutrients that may be reused by the community.

The energy flow through modern and Pleistocene communities may be considered in different ways. A so-called pyramid of energy illustrates the concept that energy is lost to the system as it flows through the trophic levels. Thus there are usually many more herbivores (producers) than carnivores (consumers) in any biotic community. This loss occurs in a variety of ways but mainly through the inefficient use of materials from one trophic level to the other.

The concept of the food web indicates the weblike interdependence between the living things in a community. Food webs in a Pleistocene pond in the Great Lakes region were simple compared to those that exist in tropical ponds where many more kinds of consumers exist.

The study of Pleistocene communities has its difficulties. A fossil assemblage is a thanatocoenosis (an assemblage of organisms brought together after death) that never really represents the entire assemblage of organisms that were present in the original community. For instance, in the Great Lakes region, for taphonomic reasons, we often tend to get assemblages of herbivorous mammals, but carnivores tend to be missing, so that only one trophic level is depicted. On the other hand, if several carnivores were present in association with only one kind of herbivore, we would get an unnatural, topsy-turvy pyramid of energy (reversed energy pyramid). This situation rarely occurs in Pleistocene fossil assemblages in the region.

CLASSIFICATION

In this book, I shall use the Linnaean system of classification. Worked out in the middle of the eighteenth century by the Swedish naturalist Carolus Linnaeus, this is the current worldwide method of classifying both fossil and living organisms. I will most often use the terms phylum, subphylum, class, order, family, genus, and species to refer to the vertebrates that are featured in the book. This system tends to put organisms in larger and larger groups as their resemblances become less and less. Each group is called a taxon (plural taxa). The names for these taxa are usually derived from the Latin or Greek language. The system works like this for the human species *(Homo sapiens)*:

Phylum Chordata (animals with a notochord)
  Subphylum Vertebrata (animals with a vertebral column)
    Class Mammalia (animals with hair, milk glands, etc.)
      Order Primates (monkeys, apes, humans, etc.)
        Family Hominidae (humanlike primates)
          Genus *Homo* (the human genus)
            Species *Homo sapiens* (modern humans)

Modern species are classified on the basis of many biological criteria, including skeletal parts, soft parts, and molecular characteristics. Modern taxa are usually described on the basis of large numbers of individuals. Vertebrate fossils, however, are described mainly on the basis of individual bones and/or teeth. Moreover, vertebrate fossil species are often named, by necessity, on the basis of a few specimens. Thus some paleontological species are, at best, the result of educated guesses. Paleontological genera, on the other hand, are usually based on more definitive characters, and suites of fossil species are sometimes available to represent each genus. Paleontological genera, therefore, often tend to be viewed with more credibility by the scientific community than are paleontological species.

The Pleistocene Ice Age

The Ice Age that I shall discuss in this book is confined to the Pleistocene epoch of the Quaternary period (see fig. 3). Other ice ages have occurred far back in time including the Permian and Ordovician periods and even in the Precambrian, where readily discernable glacial features may be observed in the rocks. Ancient striated rocks as well as those with many other glacial features have even been discovered near the equator. The Pleistocene began about 1.9 million B.P. (1.9 million years Before Present) and ended about 10,000 B.P. with the worldwide extinction of many large mammals. The time since the end of the Pleistocene is termed the Holocene (see fig. 3) in North America. The term Holocene is generally equivalent to the term Recent.

Portions of time that include glacial advances are called glacial stages and long-term temperate periods are called interglacial stages. These stages have names that differ in different parts of the world (table 2). For instance, the Wisconsinan, which is the last glacial stage of the Pleistocene in North America, is called the Weichselian in northern Europe, the Devensian in the British Isles, the Tali in China, and the Gamblian in East Africa.

With the advent of radiocarbon dating in the 1950s, it became possible to obtain dates for the terrestrial Pleistocene based on the analysis of fossil material. Correlation of terrestrial and marine deposits was not possible until sedimentary cores from the sea bottom were collected that represented continuous deposition over long periods of time. These cores showed that alternating biotas of warm-water and cold-water organisms, mainly foraminifera, existed. Coupled with studies of oxygen isotopes and magnetic reversals, these cores provided an accurate time scale for the expansion and contraction of continental ice sheets.

At present, the sequence of cold and warm cycles shown by evidence from the sea bottom provides the best information about what was probably happening on land. Good correlations between marine and terrestrial Pleistocene events have been established for the later parts of the Pleistocene throughout the world.

It has been recently demonstrated that many glacial advances and withdrawals (fig. 6) as well as climatic fluctuations occurred within the classic Pleistocene stages shown in table 2. Nevertheless, these classic terms are still used internationally to depict generally cold (glacial) and temperate (interglacial) stages.

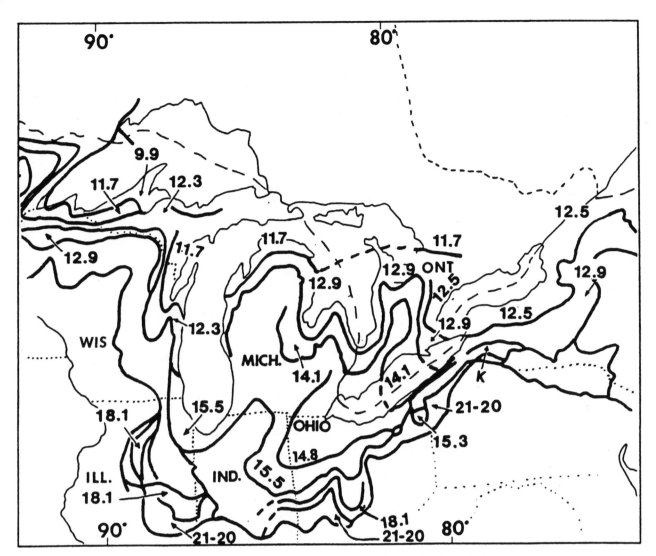

FIGURE 6. Positions of the Pleistocene ice margin in the Great Lakes region in thousands of years B.P.

TABLE 2. Some Pleistocene Stage Names (not strictly equal in time in many cases).

N. EUROPE	ALPS	BRITISH ISLES	NORTH AMERICA
Weichselian	Würm	Devensian	Wisconsinan
GLACIAL	GLACIAL	GLACIAL	GLACIAL
Eemian	Riss-Würm	Ipswichian	Sangamonian
Interglacial	Interglacial	Interglacial	Interglacial
Saalian	Riss	Wolstonian	Illinoian
GLACIAL	GLACIAL	GLACIAL	GLACIAL
Holsteinian	Mindel-Riss	Hoxnian	Yarmouthian
Interglacial	Interglacial	Interglacial	Interglacial
Elsterian	Mindel	Anglian	Kansan
GLACIAL	GLACIAL	GLACIAL	GLACIAL
Cromerian	Günz-Mindel	Cromerian	Aftonian
Interglacial	Interglacial	Interglacial	Interglacial
	Günz	Beestonian	Nebraskan
	GLACIAL	GLACIAL	GLACIAL

An explanation for the cause of glacial ages has been searched for ever since the learned Louis Agassiz expounded in the middle of the nineteenth century on the "Great Ice Age." Nevertheless, a widely accepted theory (such as natural selection to explain evolution or plate tectonics to explain continental drift) has not emerged to address Pleistocene or earlier glacial events. Such a theory could involve cyclic activities in the atmosphere due to changes in solar output, irregularities in Earth's orbit and rotation, volcanic periodicity, or even variations in Earth's magnetic field.

In Europe, special terms are used for human cultural industries of the Quaternary. These names are mainly based on human artifacts. Hand axes and spear points were common in the Pleistocene, whereas more sophisticated artifacts occurred in the Holocene. Palaeolithic (Old Stone Age) industries occurred in the Pleistocene, whereas Mesolithic (Middle Stone Age), Neolithic (New Stone Age) and Bronze Age and subsequent cultures occurred in the Holocene (table 3). Since humans did not arrive in North America until the late part of the Late Pleistocene, the European Paleolithic subdivisions do not apply in the Americas.

TABLE 3. Cultural Industries in Europe

	Bronze Age and Later Industries
HOLOCENE	Neolithic (New Stone Age)
	Mesolithic (Middle Stone Age)
PLEISTOCENE	Palaeolithic (Old Stone Age)

Ice sheets of the past were originally recognized on the basis of evidence from ice formations of the present. Today, these ice formations are restricted to high latitudes and high altitudes. Giant continental ice sheets still cover Greenland and Antarctica, and the Arctic Ocean is frozen over permanently. Moreover, extensive glaciers occur in the southern Andes, the Rockies, the Alps, and the Himalayas.

Both modern and ancient ice sheets leave undeniable evidence of their movements. This evidence includes scoured and polished bedrock, U-shaped valleys formed by erosion, and relocated foreign rocks called erratics that were plucked from their original location by the moving ice. Beyond the extent of the glaciers, one finds sands and gravels from glacial meltwaters as well as landscape features that will be discussed in more detail later.

Studies of this kind of evidence have established a detailed picture of where the ice was during the Late Wisconsinan as well as equivalent ages such as the Late Devensian and Weichselian. All of the modern ice sheets and glaciers expanded tremendously during this time. Giant ice sheets covered the northern part of Europe, including most of the British Isles and all of Finland and Scandinavia. In North America, the ice extended as far south as southern Illinois, Indiana, and Ohio (fig. 7). Oddly, Alaska and part of the Yukon and Northwest Territories remained mainly unglaciated. In the Southern Hemisphere the ice cover increased in the Andes, there was a significant increase in the amount of the sea ice in Antarctica, the glaciers in Africa extended lower, and even Tasmania, which has no ice cover today, had an ice cap.

FIGURE 7. Maximum extent of the Illinoian and Wisconsinan glaciation in the Great Lakes region. This is the farthest south the ice penetrated in North America in the Pleistocene.

THE PLEISTOCENE IN NORTH AMERICA

Two giant ice masses existed in the North American Pleistocene. The largest of those was the Laurentide Ice Sheet, which extended from Nova Scotia and the northeastern United States across the continent to western

Canada. The Laurentide Ice Sheet impacted the Great Lakes region in the Illinoian and Wisconsinan stages of the Pleistocene. A smaller mass, the Cordilleran Ice Sheet, covered the mountain ranges of the Northwest from Montana and Washington up to the Aleutian Islands.

In North America, the ice sheet penetrated farthest in the central Great Lakes region. The most southern penetration took place during the Illinoian glacial stage when the Laurentide Ice Sheet extended to southern Illinois, Indiana, and Ohio (see fig. 7). During the Wisconsinan, the maximum penetration of the ice did not extend quite as far south in these states (see fig. 7).

EFFECTS OF THE ICE SHEET

The general effects of the Laurentide and Cordilleran Ice Sheets in North America are summarized below.

Community Destruction

When ice sheets advance, vast areas of habitat are destroyed. Some simple organisms, plants, and insects can live on or within the ice, but major life forms are obliterated from areas lying under the masses of ice. For instance, in the central Great Lakes region, in the Late Wisconsinan, about 20,000 B.P., the ice sheet blanketed huge areas of habitat and changed the nature of habitats in the southern, unglaciated portions of Illinois, Indiana, and Ohio.

Additional advances occurred at intervals of about 1,000 years before a final advance extended to northern Indiana and Ohio about 15,000 B.P. About 14,800 B.P. the ice started its final withdrawal in the Midwest and exposed the land for recolonization by animals and plants.

Topographic Changes

The Pleistocene ice sheets drastically changed the landscapes over which they passed. The thickness of the North American ice sheets varied from place to place. It has been estimated the average thickness was about 1.25 miles and that in places it was 2 miles thick or more. Valleys, ridges, various types of small hills, lakes (including the Great Lakes), streams, swamps, and bogs were all produced by ice sheet activity. All of these features influenced the ecology of Pleistocene vertebrates.

Climatic Changes

Certainly the movements of the ice sheets produced dramatic climatic changes. Nevertheless, the classic idea of alternating cold glacial and warm interglacial climates is now considered to be oversimplified. Modern evidence in North America indicates that climates were cold in areas near the ice sheet borders, but in the central and southern United States the climate is believed actually to have been more equable than it is today, with warmer winters and cooler summers. This theory, termed the Pleistocene climatic equability model, will be discussed in detail in later chapters.

Vegetational Changes

The advancing and retreating ice sheets of the Pleistocene altered vegetational communities. The classic idea was that major vegetational associations were caused to withdraw southward in bandlike units by the advancing ice and that these units were caused to move northward in the same way by ice sheet retreats (fig. 8). It was thought that in the eastern United States during glacial times, a barren tundra association existed in a deep band south of the glacial front and that this was followed by a deep band of coniferous forest that graded into temperate deciduous forest, which penetrated far into the Southeast. This classic concept is often referred to as the stripe hypothesis.

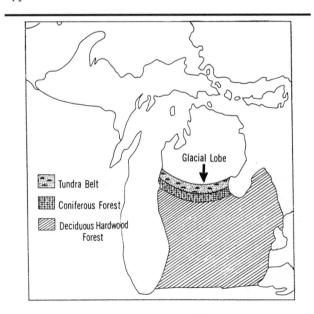

FIGURE 8. "Stripe hypothesis" illustrated. Biological communities in an orderly advance ahead of the ice sheet during the Pleistocene in Michigan.

The modern theory, however, is that during a large portion of the Pleistocene, a cold climate and tundra or coniferous vegetation existed in areas rather near the edge of the glacier but that the vegetational communities in the central and southern United States existed as a mixture of the original plants of the area coexisting with invading northern forms. This theory has arisen from the idea that plant and animal species reacted individually rather than as groups to Ice Age changes and that the mixed communities in the region would have been able to coexist in the equable climates of the time. The modern theory is often referred to as the plaid hypothesis.

Sea and Lake Level Changes

During Pleistocene glacial times, so much of the world's water was bound up in the ice sheets that sea levels fell. For example, in Florida, the peninsula enlarged greatly as the sea withdrew during glacial times. On the other hand, during interglacial times, sea and lake levels rose considerably. In the Great Lakes region, the Great Lakes themselves rose and fell, as did large rivers such as the St. Lawrence.

Reorganization of Biotic Communities

When the great ice sheets melted, they left a mass of virtually sterile mud, silt, sand, and gravel in their wake. This new material had to be recolonized by pioneer species during the period of ecological succession that must have followed. Certainly there must have been a considerable lag time between the retreat of the Pleistocene ice sheets and the development of fully developed and stable plant communities.

DIVISIONS OF THE NORTH AMERICAN PLEISTOCENE

Two systems of dividing the North American Pleistocene into temporal units exist. The older system, the Pleistocene Glacial and Interglacial Age system, discussed in part earlier, relies principally on the glacial and interglacial sedimentary record. The second system, the Land Mammal Age (LMA) system, relies on the biochronology (dating of biological events using biostratigraphic or objective paleontological data) of land mammals.

Pleistocene Glacial and Interglacial Ages

Before the 1840s, North American scientists attributed what we now know are glacial deposits to flooding. Then, Louis Agassiz's seminal work showed that ancient glacial landforms and deposits existed far south of existing glaciers and that these features indicated former cold climates. Later, when stratigraphic studies were made, it was found that some layers contained weathered zones of organic soils and plant remains between other layers of glacial sediments such as sands and gravels. Early scientists suggested that organic layers formed in an unglaciated environment and that the glaciers must have advanced and retreated several times.

Shortly after the turn of the nineteenth century, four major drift sheets were identified, each of which was separated from the others on the basis of organic layers and/or fossils that indicated an interglacial environment. The classic glacial and interglacial sequence in North America (oldest at the bottom) is:

Wisconsinan Glacial Age
Sangamonian Interglacial Age
Illinoian Glacial Age
Yarmouthian Interglacial Age
Kansan Glacial Age
Aftonian Interglacial Age
Nebraskan Glacial Age

Since the most deeply penetrating borders of all of the glacial deposits are rather close to one another, it has been suggested that the drift sheets were formed under about equally cold climates.

But at present, the classic glacial and interglacial ages before the Late Illinoian are considered to be poorly defined and highly questionable. At best, "Nebraskan" and "Kansan" glacial strata are difficult to identify because they have been exposed to very long periods of weathering and erosion and to the scouring effects of later glacial movements, especially those of the Wisconsinan. Since it is the youngest glacial stage, the Wisconsinan contains the most detailed record of Pleistocene events.

It has often been suggested that Pleistocene cold stages, worldwide, lasted much longer than the warm or temperate stages. It has recently been proposed that the Sangamonian interglacial (Ipswichian of the British Isles) lasted about 40,000 years but that the Wisconsinan glacial (Devensian of the British Isles) was about 100,000 years long. Unfortunately, estimates of the ages of earlier glacial and interglacial events are obscure.

Wisconsinan Interstadials

The Wisconsinan is the most well documented Pleistocene stage in the Great Lakes region and in North America as well. Several warmer periods within the generally cold Wisconsinan led to the temporary withdrawal of the ice sheet. These warmer periods are called interstadials and have been given names in certain regions. For instance, in the Toronto, Ontario, region, two interstadials are recognized in the Wisconsinan: a Port Talbot interstadial that has yielded a biota (biological assemblage) including vertebrates somewhat younger than 54,000 to 45,000 B.P. and a Plum Point interstadial that has yielded a biota that existed from about 34,000 to 23,000 B.P.

Land Mammal Ages in the Pleistocene

The North American Land Mammal Ages are based on biochronology, namely the absolute dates that are associated with the changes that occur in the North American mammalian fossil faunas based on the evolution, extinction, and dispersal of mammalian genera and species from one area to another. In the Pleistocene, the dispersal of mammalian taxa from one area to another is considered to be especially important in these determinations. Both large and small mammals have been considered, but recently the evolution and dispersal of small mammals, especially microtine rodents (small, mouselike rodents) have been very important. The two North American Pleistocene Land Mammal Ages are termed the Irvingtonian (oldest) and the Rancholabrean (youngest).

The Irvingtonian Land Mammal Age was originally defined based on a mammalian fauna from a gravel pit southeast of Irvington, California. Based on a recent study on the dispersal of microtine rodents from Eurasia into North America, the Irvingtonian Land Mammal Age is considered to have begun about 1.9 million B.P. and to have lasted until about 150,000 B.P. Three Irvingtonian subunits, Irvingtonian I, II, and III, are also based on microtine rodent dispersal studies. Irvingtonian I is considered to date from about 1.9 million years B.P. to 850,000 B.P.; Irvingtonian II from about 850,000 B.P. to 400,000 B.P.; and Irvingtonian III from about 400,000 B.P. to 150,000 B.P.

The Rancholabrean Land Mammal Age was originally defined on the basis of the famous Rancho La Brea faunal assemblage in Los Angeles. Based on recent microtine rodent dispersal studies, the Rancholabrean is considered to have begun about 150,000 B.P. and to have lasted until about 10,000 B.P., when the Pleistocene is considered to have ended. The Rancholabrean has not been differentiated into subunits. The great majority of Great Lakes region vertebrate sites fit within the time frame of the Rancholabrean Land Mammal Age.

Now that we have examined the Pleistocene in general, we are ready to move into a more specific consideration of the Pleistocene in the Great Lakes region.

3

The Pleistocene in the Great Lakes Region

The Great Lakes region may divided into two subregions. Subregion I is the very large northern area presently mainly covered by glacially derived Wisconsinan sediments and dominated by Wisconsinan glacial topography. Subregion II is in a much smaller southern area beyond the limits of the Wisconsinan ice (fig. 9). Although areas in Subregion II have been previously overridden by ice predating the Wisconsinan, it is generally free of glacially derived surficial sediments, and outcrops of bedrock are much more common than in Subregion I.

The modern topography, flora, and fauna of these subregions are considerably different, as are their Pleistocene vertebrate faunas. The geological dynamics of these subregions will be discussed separately.

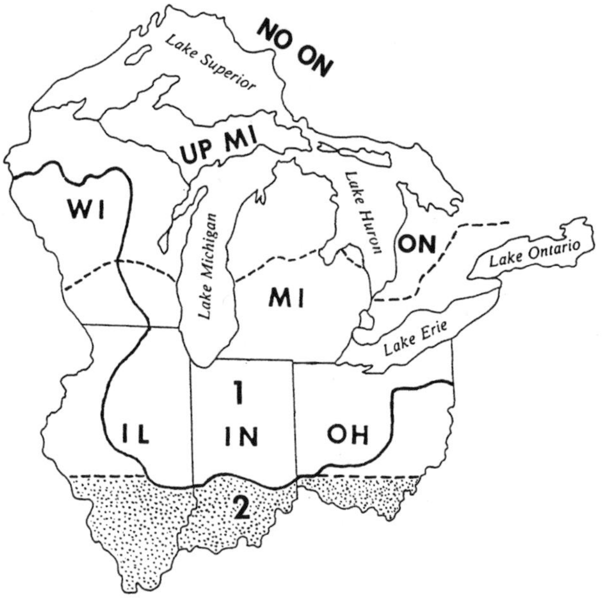

SUBREGION I

Glacially derived sediments have been collectively called drift since Charles Lyell published the first volume of his classic *Principles of Geology* in 1830, a work essentially based on the doctrine of uniformitarianism. Subregion I was generally covered with Wisconsinan drift, and its modern landscape is dominated by a topography that reflects the dynamics of the Wisconsinan ice sheet.

FIGURE 9. Pleistocene Subregion I is the area between the upper broken line in Ontario, Michigan, and Wisconsin and the stippled area in Ohio, Indiana, and Wisconsin, which denotes Pleistocene Subregion II.

Glacial Features

Glacial features dominate the topography of much of Subregion I. Matter transported by the Wisconsinan Laurentide Ice Sheet was deposited whenever and wherever melting occurred. These deposits (drift) are classified either as till or as outwash deposits. Till is deposited directly from the ice and consists of a mix of particles of all sizes and shapes. Till forms structures called moraines. Outwash deposits form as aprons of sediments that typically develop in front of moraines.

Moraines may be subdivided into end moraines, lateral moraines, and ground moraines. End moraines form at the end of the ice lobe, lateral moraines form at the side of the ice, and ground moraines form when the ice moves across the land rather rapidly. Most of the big hills in the northern part of the Great Lakes region are portions of end moraines.

Terminal moraines mark the end points of major glacial advances. As an example, a very important terminal moraine is the Shelbyville Moraine in central Indiana, which marks the end of the deepest penetration of the Wisconsinan ice about 20,000 years ago. Recessional moraines mark the various stops of the ice sheet during retreats.

In many areas of Subregion I, one finds typical glacial topographic features called drumlins, eskers, kettle holes, and kames. Drumlins are attractive, elliptical hills, some of which may be quite large, that are composed of glacial till left behind by the glacier. Drumlins have their long axes pointing in the direction of the ice flow. Eskers are narrow ridges with a snakelike form that originated from till laid down in tunnels in or under the ice. Kettle holes form from rounded ice blocks that melted after the glacial retreat. Kames are rounded hills of glacial outwash sand and gravel.

Origin of the Great Lakes

The Great Lakes were of utmost importance for all of the Pleistocene life of the region, as they provided habitats for aquatic organisms and a water supply for terrestrial plants and animals. These giant lakes, since their presence, have had a modifying effect on the climate of the surrounding areas. The Great Lakes are the result of bedrock topography that was produced during the Paleozoic and the erosion of these rocks over a vast period of time, ending with the erosional events of the Pleistocene. In terms of geological time, they are infant bodies of water, but in due time, like all lakes, even the Great Lakes will fill in with sediments.

Today, however, they are exceedingly important to plants and animals, as they hold one-sixth of the world's freshwater supply. The glacial basins containing the Great Lakes are such large features that they are said to be the only glacially produced feature that can be seen from the moon. These lakes occupy 95,000 square miles and have over 8,000 miles of shoreline.

Before the Pleistocene, the basins that contain the modern Great Lakes were stream valleys in the bedrock. When the ice sheets arrived, they tended to move along these valleys, following the path of the least resistant rocks. Finally, glacial meltwater filled in the gouged-out basins. The modern lakes all drain in an easterly direction. Lake Superior drains into Lake Huron at Sault Ste. Marie; Lake Michigan drains into Lake Huron through the Straits of Mackinac; Lake Huron drains south along the St. Clair River into Lake Erie; and Lake Erie drains into Lake Ontario. Finally, Lake Ontario drains into the Atlantic Ocean through the St. Lawrence River.

Lake Erie is the shallowest and warmest of the Great Lakes and Lake Huron the second shallowest. Lake Superior is the coldest and deepest lake. Except for Erie and Huron, the deepest parts of all the lakes are more than 300 feet below sea level.

The Great Lakes began to fill up in earnest in the Late Wisconsinan sometime before 13,300 B.P., and the different lake stages have been given different names, even though they are all part of the same process. Glacial Lake Chicago, which occupied the lower part of the present Lake Michigan Basin, and Glacial Lake Maumee, which filled in the lower part of the Lake Erie Basin sometime before 13,300 B.P., were the earliest of these ancient glacial lakes. By the time of the so-called Nipissing stage, which occurred about 4,000 B.P., the Great Lakes were about like they are today. Between 13,300 B.P. and 4,000 B.P. the lakes had various shapes caused by glacial processes.

Reestablishment of Plant Communities

Studies of the patterns of the reestablishment of plant communities in previously glaciated regions are especially important, as these patterns relate to re-colonization patterns in animals. There are few studies of the earlier glacial ages because of the removal of sediments by succeeding glacial events. Wisconsinan postglacial vegetational events, however, have been relatively well studied in Michigan, Indiana, and Ohio, primarily on the basis of palynology.

When the Wisconsinan ice sheet retreated, it left a mainly sterile blanket of gravel, sand, and mud. Thus the establishment of well-developed, stable plant communities in these areas must have taken quite a bit of time, especially since the proglacial climate was probably quite cold.

In Michigan, during a temporary withdrawal of the ice sheet about 40,000 B.P. (Cherry Tree substage) in Kalkaska County and about 24,000 B.P. (Plum Point substage) in Muskegon County, a boreal or subboreal climate was indicated by paleobotanical studies. Open forests dominated by spruce and pine were present, and tamarack and cedar were prominent in the swamps. Sedges and cattails were found in marshy areas, and disturbed-ground herbaceous plants were found in the better-drained situations.

During the period when the ice sheet began its final withdrawal about 14,800 B.P., the fossil record in southern Michigan indicates evidence of marshes and muskegs in the lowlands. By about 13,000 B.P. almost half of the Lower Peninsula of Michigan was ice free. At this time, most of the well-drained areas contained scattered stands of pioneer trees such as juniper, aspen, ash, and spruce; as well as sun-tolerant shrubs such as crowberry, silverberry, and willow. But whether this early landscape was essentially a treeless tundra or an open forest of spruce, tamarack, and mixed deciduous trees is not known.

From about 12,500 to about 11,800 B.P., southern Michigan had developed a boreal forest that was dominated by spruce, but there were other areas of boreal parkland and open woodland. In the northern part of this area, boreal parkland dominated, and tundra vegetation occurred in open situations along the ice front and on exposed slopes and hills.

Changes in the pollen record occurred between 11,000 and 9,900 B.P. About 10,600 B.P. jack pine and red pine began to replace spruce trees in southwestern Lower Michigan, and by 10,000 B.P. white pine entered the picture. Hardwood trees such as birch, blue beech, and elm became abundant during this time.

Thus by 10,000 B.P., the time when the Pleistocene is considered to have ended, the forest vegetation in southern Michigan had become relatively diverse, with mixed forests of white and red pine, yellow and paper birch, aspen, oak, white ash, red and white elm, and blue beech. The paleobotanical and palynological studies in this area of the Great Lakes region, especially studies of the period between 10,600 and 9,500 B.P., have helped scientists to define the end of the Pleistocene.

Pollen records from Indiana and Ohio indicate that the time between 14,000 and 9,000 B.P. was also characterized by major vegetational changes. Beginning about 13,000 B.P., during a time span of about 1,000 years, spruce forest communities were largely replaced by mixed coniferous and deciduous forest communities. Between about 12,000 and 11,000 B.P. the climate was stable, but between 11,000 and 10,000 B.P. a series of marked changes led to the development of complex vegetation patterns that are different than any presently found in the area. Shortly after 10,000 B.P. oak-dominated forests that resemble modern ones were established.

The Mason-Quimby Line

The paleobotany, palynology, vertebrate paleontology, and archaeology of the postglacial Pleistocene has been especially well documented in Michigan in past decades. These studies have supplemented numerous discussions about an imaginary line that divides the state into northern and southern portions (fig. 10). The "Mason-Quimby Line" (often abbreviated MQ) reflects the early writings of two University of Michigan archaeologists, R. J. Mason and G. I. Quimby, who were interested in the hunting activities of the Paleo-Indians of the postglacial Pleistocene. Thus the Mason-Quimby Line was originally set up to depict the northernmost occurrences of Paleo-Indian artifacts and proboscidean (mastodont and mammoth) distribution in the state. The most important of these artifacts are fluted spear points (fig. 11), which are very similar to those used by Paleo-Indian mammoth hunters in the West.

Today, the Mason-Quimby Line depicts the northernmost records of most, if not all, of the state's Pleistocene vertebrates. The best explanation for this distribution is believed to relate to the positions of the ice sheet in Michigan during its final retreat (see fig. 6). About 14,000 B.P. the end of the ice sheet was about 35 miles south of the Mason-Quimby Line, but by 13,000 B.P. it had retreated to about 70 miles north of the line.

Retreating ice leaves a mass of essentially sterile gravel, sand, and mud in its path that must be recolonized by plants in order for stable biological communities to develop. Certainly it would take considerable time for such communities to develop in this sterile material, especially considering the cold climate that existed. Authentic records of postglacial Pleistocene vertebrates in Michigan span a relatively short time span of about 12,500 to 10,000 B.P. Thus it seems highly possible that during this interval, plant communities north of the Mason-Quimby were not stable enough to sustain the number of vertebrate species necessary to contribute significantly to the fossil record.

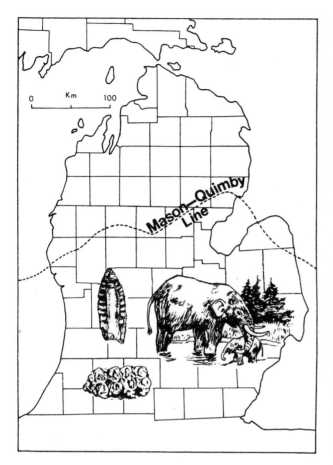

FIGURE 10. The broken line represents the Mason-Quimby Line. There is no undisputed evidence of Pleistocene vertebrates and very little evidence of Paleo-Indian activity above this line in Michigan.

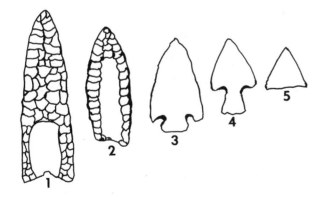

FIGURE 11. 1 and 2, fluted spear points of the type used by Paleo-Indians; 3, 4, and 5, points that occurred much later in time and are commonly termed arrowheads.

SUBREGION II

Pleistocene Subregion II of this book lies exclusively south of the maximum penetration of the Wisconsinan Laurentide Ice Sheet. It is much smaller than Subregion I (see fig. 9) but is extremely important in that it has yielded several Pleistocene vertebrates with southern affinities as well as species yielded by a rich cave system. In general, the subregion composes about the southern fourth of Ohio and about the southern third of Indiana and Illinois.

The Ice Cover

Evidence exists that Subregion II was, in sequence, covered by earlier Pleistocene and then Illinoian portions of the Laurentide Ice Sheet. However, much of the evidence of these earlier ice encroachments has been lost because of the erosional dynamics of the final ice movements in the Wisconsinan. We know that Illinoian ice extended farther southward than did the Wisconsinan ice (see fig. 7). In fact, the Illinoian ice reached its most southerly penetration in the United States in Illinois, where it covered all but the southern tip of the state. In Indiana, Illinoian ice penetrated more deeply than in Early Pleistocene times, especially in southwestern Indiana. In Ohio, where there is little evidence of earlier Pleistocene drift, the Illinoian ice made its deepest penetration in the southwestern portion of the state.

Topography

Glacial topography is rare in Subregion II compared to Subregion I. This is due to the fact that about 150,000 years have elapsed since the earlier Pleistocene and Illinoian glaciers passed over Subregion II, so that the features that the earlier glaciers left have been mainly eroded away. One might speculate whether these features were ever as prominent as those left behind by the Wisconsinan ice, but it is quite possible that they were even more pronounced.

Many features in Subregion II are due to the erosion of ancient plateaus composed mainly of limestones and shales, which has produced a topography of high relief and/or knobby hills. Most of the other topographic features have been produced by the area's important river and stream systems. These features include river bluffs and their valleys. Although some significant Pleistocene vertebrate fossils occur in stream sediments, the most important ones have come from caves that have developed in limestone areas.

4

Where to Find Pleistocene Vertebrate Fossils

I n the Great Lakes region, the best places to find Pleistocene vertebrate fossils in Subregion I are kettles and other small, glacially derived basins. Caves are by far the most important sources of vertebrate fossils in Subregion II.

GLACIALLY DERIVED KETTLES AND SMALL BASINS

The Wisconsinan ice left countless thousands of kettles and other small, glacially derived basins in its wake. These features filled with sterile glacial meltwater and with sufficient passage of time developed into pond communities that supported a thriving biota of bacteria, protists (single-celled organisms), fungi, plants, and animals. Eventually, these ponds filled in and disappeared.

Such infilled Pleistocene basins possess a sedimentary record that depicts the various successional stages from the birth to the death of the pond. These features not only contain fossil remains of the small biota that lived in the pond but sometimes include fossils of Pleistocene vertebrates. Since these former ponds represent essentially still bodies of water where little or no transport of fossils occurred, they are natural laboratories for paleoecological studies.

The stratigraphic layers (fig. 12) in the features

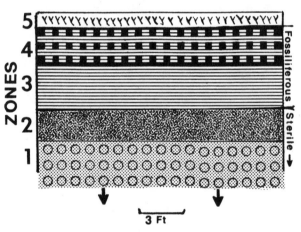

FIGURE 12. The numbered zones refer to the stratigraphic zones discussed in the text. Zones 1 (glacial sediments) and 2 (derived clay) are biologically sterile; zones 3 (shelly marl) and 4 (muck and peat) are fossiliferous; and zone 5 is topsoil.

that depict the birth and the death of a Pleistocene pond can be described from the bottom to the top.

Zone 1

Zone 1 is the bottom zone of the kettle or basin and consists of glacially derived clastic sediments such as sands, gravels, cobbles, and small boulders. This zone does not contain fossils and, depending on the locality, can be very thick.

Zone 2

Zone 2 is produced chemically and physically from the parent glacial material in zone 1. It may be 2 or 3 feet thick and often consists of a relatively pure bluish gray clay. Zone 2 represents an early, biologically sterile time in the filling in of the kettle or basin and does not typically contain fossils.

Zone 3

Zone 3 is almost always a very fossiliferous zone. It represents the pond stage of the infilling process, had open water, was well aerated, and was biologically very diverse. This zone usually consists of a grayish to grayish brown colloidal (sticky) shelly marl that may be from about 2 to 5 feet thick. The fossils usually found in zone 3 are composed of the detrital and skeletal remains of the biota that accumulated in shallow, oxygenated situations. A square yard of this material may contain many hundreds of freshwater shells. Fossils commonly found in this zone are pollen grains and spores, plant fibers, cones, stems and twigs, nuts, seeds, leaves, roots, logs, branches, bark, beaver-chewed wood, ostracods, clams, snails, beetle wings, and sometimes vertebrate bones. Vertebrate bones mainly include frogs, turtles, and small and large mammals.

Zone 4

Zone 4 is composed of dark, organic peat or muck and is also an intensively biotic zone. It represents the successional stage of the pond where mats of aquatic vegetation formed over the surface of the feature. This zone may occur in the form of commercial-grade peat, which may be sold in the Great Lakes region or other places. Zone 4 may range from about 2 to 8 feet in thickness. Pleistocene fossils from zone 4 may include pollen and spores, plant fibers, cones, stems, twigs, nuts, seeds, branches, many fewer clams and snails than in zone 3, beetle wings, and sometimes vertebrate bones.

Zone 5

Zone 5 is composed of recently derived humus and topsoil and often has been modified by humans in the region. The modern plants that typically grow in this zone compose a willow-sedge community.

How Were so Many Mastodonts and Mammoths Trapped in Kettles and Basins?

Most mastodont and many mammoth bones in the Wisconsinan drift-covered area of the Great Lakes region were recovered from kettle bogs and other shallow depressions. A clue to the reason for these occurrences lies in the composition of zone 4, where the dark, organic peat and muck represent the last biotic stage of the ancient pond. In this stage the pond is filled with floating mats of vegetation, or quaking bogs, which may have acted as traps for the mastodonts or mammoths that fell into them, especially during early winter. Smaller animals or even humans might have been able to walk across these mats without falling through, but it has been reasoned that large mammals such as mastodonts or mammoths often met their deaths when they ventured out onto such places.

One might imagine the violent struggles and trumpeting of terror of these animals as their feet failed to gain purchase on the slippery edges (fig. 13). It has been suggested that these animals merely drowned in such bogs. On the other hand, there is evidence that, even if only their feet were stuck in the sticky marl below the mats of vegetation, the mastodonts and mammoths would be unable to extract themselves from the natural trap. This would mean that they would either slowly starve to death or be very vulnerable to human hunters with spears and boulders. It has also been suggested that these animals usually became trapped when thin layers of snow and ice masked these features in the early fall or spring.

The large mammals preserved as fossils by sinking into the muck and marl of glacially derived kettle bogs and shallow basins were obviously somewhat younger than the ancient sediments that surrounded them. This is the opposite of most other fossil sites, where fossil bones are somewhat older than the sediments that blanket them. This also means that the snails, clams, and other small inhabitants of Pleistocene ponds should be expected to have somewhat older radiocarbon dates than the mastodonts and mammoths that sank into them. These differences in age depend on the amount of time that it took for the pond to change from an open, oxygen-rich structure that supported clams and snails to the quaking bog stage. If both the

FIGURE 13. A *mastodont (*Mammut americanum) struggles in a quaking bog in southern Michigan during the Late Wisconsinan (*denotes extinct animals).

"mollusk layer" and the mastodont or mammoth bones were in stratigraphic context and both were dated by radiocarbon methods, one should expect the mollusk layer to yield an older date than the proboscidean bones. If this is not the case, one might question the accuracy of the radiocarbon dates.

GLACIAL DEPOSITS

In certain areas in the Great Lakes region, notably in Wisconsin, vertebrate remains have been found in glacial deposits. These fossils usually consist of parts of mastodonts, mammoths, or large, hoofed herbivores. Some of this material consists of bones that were picked up by moving glaciers and transported some distance from their original location. These fossils occur mainly as small, abraded fragments of animals. Other vertebrate fossils consist of disarticulated bones of animals that lived on broad outwash plains and were transported by glacial meltwater. In at least one such instance, a mastodont tooth was recovered at Fond du Lac, Wisconsin. Because fossils in glacial deposits normally lack stratigraphic context, they usually provide only general distributional records.

CAVES

Caves and other solution features that contain Pleistocene vertebrates are characteristic of Subregion II, but there are some notable exceptions in Ontario, in the driftless area of Wisconsin, and in northwest Ohio. In temperate areas throughout the world, including the Great Lakes region, vertebrate fossils found in caves tend to have the same kinds of origins. Bones of small mammals, such as mice and shrews, tend to occur in the regurgitated pellets of the Pleistocene owls that were roosting in the cave. Bones of bats that used the cave as a roosting place are also common. Fragments of large bones are often present, and they are usually derived from large predators, such as bears, that brought their prey into the cave to eat. Finally, in pitfall caves, one finds bones of vertebrates that accidentally fell into the feature.

As in the case of other vertebrate deposits, vertebrates in caves must also be immediately buried by sediments for fossilization to occur. Some caves have such slow rates of sedimentation that few, if any, vertebrate remains have been preserved as fossils. On the other hand, caves subject to periodic flooding often have large accumulations of fossil bones.

How Are Caves Formed?

Most caves in the Great Lakes region are formed by the solution of limestone or dolomite by groundwater. The process occurs because rainwater contains free oxygen that combines with carbon dioxide in the atmosphere to form a weak solution of carbonic acid. This dilute acid is capable of dissolving limestone and dolomite. Caves usually begin to form when water from the water table seeps down into limestone or dolomite to create small, unconnected cavities. These cavities enlarge and become interconnected over the years.

In areas where the limestone is very near the surface or where erosion has removed cap rock layers over the limestone beds, rather large entrances to the cave system may form at the surface. Periodic floods tend to enlarge these cave openings. Sometimes, especially when the water level drops and the roofing material is thin, large collapses may suddenly occur and form quite large entrances to the system. If these entrances are vertical, they may act as natural traps for animals that are unable to fly or crawl out of the cave. If entrances are more accessible, the caves may provide dens and other shelters for vertebrates. In southern Indiana, sinks and caves are so abundant that they form a topography of small basins, springs, and other solution features. This type of topography is called karst topography.

Unfortunately, establishing the time relationships of vertebrate bones in caves may be difficult. The problem is that post-Pleistocene stream action might leave young sediments on the cave floor and the Pleistocene sediments might occur as remnants on the roof, on walls, and especially on ledges in the cave. Therefore, the older deposits in a cave may actually be higher than the younger ones, an inverted time sequence.

It is not unusual, then, for the sediments on the floor of a Pleistocene cave to contain the skeletal remains of bats and the other modern vertebrates that live in the cave. Chimney caves, whose only openings are vertical shafts, act as traps. Skeletons of farm animals, dogs, and even humans may be mixed together with those of other modern animals in the sediments below such openings. Sometimes flooding may wash

Pleistocene vertebrates out of the walls and ledges of caves, so that extinct vertebrates are mixed in with bones of modern animals and soft-drink cans.

Fortunately, deposits in caves sometimes form as identifiable stratigraphic layers. Such a process may take place if the cave roof opens and closes several times during the cave's history. If these layers form on a wide ledge, the sediments coming in from the outside may build up normally, that is with the oldest sediments at the bottom of the ledge and the younger ones on top. Changes in vegetation and soil through time outside the cave may produce sediments of different textures and colors that may be identified as distinctive layers within the cave. A time-stratigraphic sequence in the cave may be developed if bones and wood in these layers are radiocarbon dated.

Intrusive Vertebrates in Cave Faunas

As pointed out, modern vertebrate bones sometimes become mixed with Pleistocene bones in caves. This can happen various ways, and intrusive modern species from almost all vertebrate groups may be present. Unfortunately, some true Pleistocene fossils have been almost routinely considered to be intrusive elements and have been ignored.

Amphibian and reptile bones in caves, especially those associated with mammalian species that presently live in boreal or tundra situations, have sometimes been considered to be modern intrusive species in Pleistocene faunas and have often been passed over, even by experienced vertebrate paleontologists. Now we realize that most amphibian and reptile bones are part of the Pleistocene fauna.

Actually, it is not hard to recognize intrusive amphibians and reptiles or other vertebrate intrusive species in Pleistocene sediments in a cave. True intrusive forms are almost always burrowers, such as mole salamanders or burrowing toads, and the intrusive forms usually consist of suspiciously complete skeletons that are usually of a different color, texture, and density than the deposit's true fossils.

LAKE, STREAM, AND ANCIENT SEAWAY DEPOSITS

Lake deposits are considered here to be larger bodies of water than the glacial kettle holes and shallow basins previously discussed. In the proglacial areas of the Great Lakes Basin, glacial lake sediments reflect changes that occurred in the levels of the ancient Great Lakes. Vertebrate remains, including duck and toad bones, have been found below glacial lake sediments in water well diggings in Michigan, but other vertebrate finds in these situations have been rare and sometimes controversial.

Vertebrate bones that find their way into lakes or streams tend to be transported by wind currents (in lakes) or gravity currents (in rivers or streams) until they reach a quiet place (low-energy situation) where they tend to accumulate. Fossilization may occur if these bones at rest become buried by sediments. In prairie states such as Nebraska, shallow, braided streams with almost constantly changing channels and many quiet water oxbows have yielded rich fossil vertebrate deposits. Braided streams, however, do not generally occur in the Great Lakes region; thus stream-related fossil deposits are not nearly as abundant.

Lakes and streams may sometimes act together to form vertebrate fossil deposits. The Prairie Creek Pleistocene faunas in southwestern Indiana were originally deposited in and around a feature called Lake Prairie Creek. At one stage in its development, the Lake Prairie Creek outlet was captured by an adjacent valley system, flushing the fossil-rich sediments into the streambed of a feature called Prairie Creek. This fossil deposit will be discussed in detail in chapter 8.

The only known Pleistocene seaway fossil deposit known in the Great Lakes region formed in the St. Lawrence Valley in Ontario and Quebec in the Late Wisconsinan in what is called the Champlain Sea. Not only did this deposit yield an assemblage of marine plants and invertebrates, but it also produced an assortment of marine fishes, seals, walruses, and whales. This remarkable paleogeographic feature will be addressed later.

Collecting the Fossils

Ice Age vertebrate fossils are nonrenewable resources and invaluable scientific objects. Studies of Pleistocene vertebrates are significant in many ways. Not only are they necessary for the correlation of Pleistocene stages and land mammal ages, but they often reflect the patterns and processes of evolution. In some cases, vertebrate fossils indicate life patterns and food habits of prehistoric humans. Therefore, it is important to say something about the ethics of fossil collecting.

To the scientist, vertebrate fossils are never important as individual objects but only in the way they relate to the layers and sediments in which they were found and to the other fossils that occur with them. Moreover, as we often see in the media, a single fossil from a critical point in time can change geological or evolutionary thinking. For that reason, before one decides to make a private collection of vertebrate fossils, one should realize that certain ethical considerations are necessary. Although rock shops sometimes sell vertebrate fossils, many museum stores consider vertebrate fossils to be such valuable scientific resources that they will not sell them.

If one decides to establish a private collection of Pleistocene fossils, it is best to concentrate on invertebrate rather than on vertebrate fossils because invertebrate fossils are very abundant in some situations, whereas vertebrate fossils are usually rare. Pleistocene invertebrate fossils are relatively easy to collect in an undamaged state and are of local, cultural, and recreational value. They are also useful in teaching and in giving students field and hands-on experience with scientific objects.

Pleistocene vertebrate fossils, on the other hand, are fragile and usually require special collecting techniques. They also tend to disintegrate after they have been collected, and this is especially true of the teeth and especially the tusks of Pleistocene mastodonts and mammoths. Collectors need to know that it is important that Pleistocene vertebrate fossil remains find their way to institutions where they may be maintained properly for the good of the public as well as for scientific studies.

If one does decide to make a private collection of fossils, there are certain rules to follow. (1) Always get written permission to collect fossils, whether they are to be taken from federal, state, provincial, or private lands. It is illegal to collect on many public lands, and, of course, it is against the law to trespass on private lands. (2) Carefully record and photograph all of the stratigraphic and locality data possible with each fossil you collect. (3) Always take unusual or important fossils, invertebrates or vertebrates, to professional paleontologists for evaluation.

Most states and provinces have laws that govern the collection of fossils, and there are also laws for federal lands. Ethical guidelines for paleontological collecting in general are found in *Paleontological Collecting,* a large report published by the National Academy of Sciences in 1987 and issued to most museums and universities with paleontological collections or programs. State regulations are discussed in two publications:

R. W. West, State Regulation of Geological, Paleontological, and Archaeological Collecting, *Curator* 32(1989): 281–319.

R. W. West, State Regulation of Geological, Paleontological, and Archaeological Collecting, *Curator* 34(1991): 199–209.

Regarding vertebrate paleontology specifically, collectors should consider the Statement of Ethics issued by the Society of Vertebrate Paleontology (SVP). The SVP sponsors an annual meeting and produces the quarterly *Journal of Vertebrate Paleontology,* the quarterly *News Bulletin,* and other services for its members. Persons interested in membership in the SVP may visit its website (http://www.vertpaleo.org/) and click on membership.

- It is the responsibility of vertebrate paleontologists to strive to ensure that vertebrate fossils are collected in a professional manner, which includes the detailed recording of pertinent contextual data (e.g., geographic, stratigraphic, sedimentologic, taphonomic).
- It is the responsibility of vertebrate paleontologists to assist government agencies in the development of management policies and regulations pertinent to the collection of vertebrate fossils, and to comply with those policies and regulations during and after collection. Necessary permits on all lands administered by federal, state, and local governments, whether domestic or foreign, must be obtained from the appropriate agency or agencies before fossils are collected. Collecting fossils on private lands must only be done with the landowner's consent.
- Fossil vertebrate specimens should be prepared by, or under the supervision of, trained personnel.
- Scientifically significant fossil vertebrate specimens, along with ancillary data, should

be curated and accessioned in the collections of repositories charged in perpetuity with conserving fossil vertebrates for scientific study and education (e.g. accredited museums, universities, colleges, and other educational institutions).
- Information about vertebrate fossils and their accompanying data should be disseminated expeditiously to both scientific community and interested general public.
- The barter, sale, or purchase of scientifically significant vertebrate fossils is not condoned unless it brings them into, or keeps them within, a public trust. Any other trade or commerce in scientifically significant vertebrate fossils is inconsistent with the foregoing, in that it deprives both the public and professionals of important specimens, which are part of our natural heritage.

Turning now to some simple directions for collecting fossils in the region, some major types of collecting sites follow.

GLACIALLY DERIVED KETTLE BOGS AND SHALLOW BASINS

As stated, vertebrate bones, especially those of mastodonts and mammoths, are often found in kettle bogs and shallow basins that filled in the postglacial Pleistocene. Bones in these sites usually may be found in the peat and/or the marl layers or sometimes in a position somewhere between these layers. Most fossil vertebrates found in these situations are located by accident. People digging canals to access lakes, digging tile lines to drain low spaces, or plowing in the lowest parts of fields during dry years are most likely to unearth vertebrate bones. In Subregion I, most vertebrate bones are found in very low areas, usually near the surface. When one visits such a site or explores low areas for vertebrate bones, certain basic equipment is needed.

A long metal probe for pushing down through the peat to locate large bones is a primary tool (fig. 14). The probe should be about 7 feet long with a handle at the top. Metal shops can make these tools easily. Something red or orange should be tied near the handle of these probes, as they are easily misplaced in the field. A rock struck by a probe feels hard and makes a "chink"; a piece of wood feels softer and makes a "thunk"; a bone feels and sounds somewhere in between. One learns rapidly how to recognize the sensations derived from the end of a probe.

A shovel is necessary to dig away muck and marl surrounding vertebrate bones, and different sized bags are needed to carry bones and sample peat and marl back home or to the laboratory for further study. One may need to have a pump available, as these low places may rapidly fill with water as the digging process goes on.

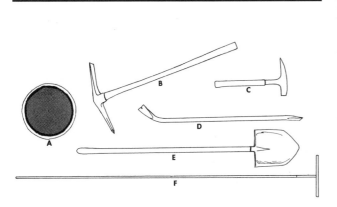

FIGURE 14. Some hardware used for collecting vertebrate fossils. A, sieve; B, marsh pick; C, rock hammer; D, crow bar; E, shovel; F, probe.

Bones from kettle bog or shallow depression sites are often very soft, so they must be removed with great care. As one digs up vertebrate bones from such sites, one should be constantly on the lookout for human artifacts, such as antler or flint spear points or even flint chips that might be in association with such finds. It is also important to save all the other fossil remains from the site such as wood, seeds, nuts, clams, snails, and beetle wings, as all of these yield paleoecological information.

CAVES

Caves often yield large numbers of Pleistocene vertebrate fossils, yet caves can be very dangerous places, and the exploration of these features may require extraordinary skills. Because of this, most paleontologists who study cave bones are also trained speleologists (cavers or spelunkers). There are so many techniques for negotiating difficult caves that I will not attempt to address them here but will only mention some minimal precautions and equipment.

One should never investigate a cave alone. Within a cave, one should also stay near a partner or group. A hard hat is a essential equipment, not only for protection against falling rocks but to avoid serious injury from the numerous head-bumps that seem to be a part and parcel of cave exploration.

The cave paleontologist needs a headlamp, extra headlamp batteries and bulbs, and at least one other source of light (such as a small flashlight) in case the headlamp malfunctions. There is no experience as black and disorienting as having the light go out deep in a cave.

The paleontological caver typically uses a trowel rather than a shovel because of low ceilings and other space limitations in caves, and geological hammers and crowbars are often needed to remove fallen rocks. Goggles should be worn to protect the eyes from sharp pieces of rocks, which are likely to fly up in one's face when larger pieces are struck with a rock hammer.

Small fossil vertebrates are usually recovered from caves by removing bulk samples of matrix (fossil-containing material) to be dried and screened at a later time; thus bags are essential. Many have learned that a cheap pocketknife is a handy tool for scraping around cave sediments, as one can soon recognize the feel of small bones in contact with a knife.

Screening

Small vertebrate fossils in caves usually occur in the form of single bones rather than complete skeletons. Pleistocene frogs, salamanders, lizards, snakes, birds, bats, and rodents are common vertebrate fossils in caves, and most of them are derived from the pellets of Pleistocene owls that were roosting on cave ledges. These small bones are normally collected by the removal of bulk samples of matrix, which are later washed and sieved.

Bulk matrix from caves usually contains many more small vertebrate bones than matrix from other types of fossil deposits. The kinds of cave sediments that are easiest to screen are sands, silts, and clays or combinations of these. It is almost impossible to wash and screen the matrix from glacially derived kettle bogs and shallow basins because of the high fiber contents of these sediments.

Cave sediments are usually sacked in burlap bags and then transported to a washing site. Most cave sediments break down easily when washed through a sieve, although certain clays must be dried before they will be broken down by the washing process. Occasionally, the washing step may be omitted in the case of very dry, sandy matrix, and the material may be sifted through screens in the field.

Screening boxes are easy to make if metal sieves are not available. The sides of the boxes should be about 1 inch thick, 5 inches high, and about 18 inches long. Window screen is tacked to the bottom. Finer mesh screen may be used, but this results in a reduction in the amount of material that can be processed in the field.

The ideal place to wash sediments through a screen box is a shallow, slow-moving stream, but washtubs or old bathtubs and a hose may be used as a substitute. The matrix should be shaken gently in a slow up-and-down motion. When no more material will pass through the screen, the concentrated material in the screen should be placed in the sun to dry. Drying can be done on tarps, pieces of toweling, or on specially built frame drying racks that get the concentrate off the ground. The concentrate should be rewashed if clay balls are present. After drying, the concentrate is ready to be picked through for small bones. Some prefer to sort through the concentrate in the field, whereas others transport the concentrate back home for the picking process.

LAKE AND STREAM DEPOSITS

Small Pleistocene vertebrates from lake and stream deposits are also mainly derived from bulk matrix that is washed and sieved. Because these deposits are usually easily accessed and occur in open spaces, large amounts of fossiliferous sediments can often be processed. A crew of four or five persons can process as much as 40 tons of sandy matrix in a five-week period if conditions are right.

Collecting Large Vertebrate Fossils

Large fossils often present different problems, especially if they occur in consolidated matrix. The cardinal rule for collecting such specimens is take your time and collect it right. Most large bones are laced through with a myriad of small cracks and fall into many pieces when pulled out of the ground by persons seeking instant gratification. Thousands of important vertebrate fossils have been needlessly destroyed in this way.

If a large fossil is found, the would-be collector must first attempt to determine the nature and extent of the fossil. A pocketknife or putty knife may be used to expose the bone further, and a small brush may be used to clean off the bone as it is being exposed. Since newly exposed bone is likely to break apart, some sort of cementing agent that can be brushed onto the fossil should be available. Some vertebrate fossil cementing agents and solvents are:

CEMENT	SOLVENT
Polyvinyl Acetate (PVAC)	Acetone
Duco	Acetone
Alvar	Acetone
Shellac	Alcohol

Shellac, diluted so that it penetrates quickly into the bone, is an inexpensive and readily available cement for the field, but polyvinyl acetate is the much preferred agent. All cements should be applied out of doors or under a fume hood indoors.

When the fossil has been exposed enough so that its dimensions have been established, it should be put in a plaster jacket according to the following instructions. Apply one or two coats of the cementing agent over both the matrix and the fossil. This step may take a couple of days, for it may take the cement a long time to penetrate deeply into the bone and its matrix. Then apply several thicknesses of moist tissue or toilet paper, as this ensures the proper separation of the plaster jacket from the fossil later. The next step is to dig a straight-sided trench completely around the block of hardened matrix that holds the fossil. This trench should be dug far enough away from the block that slumping of the matrix does not occur. The softer the sediments, the farther away the trench should be dug from the block.

Cheap materials that may be used to make the plaster jackets are toilet paper, strips of burlap bags about 3 inches wide, plaster of Paris, and water. First cover the exposed part of the fossil with moist toilet paper so that the fossil will separate from the jacket later. Then mix the plaster until it is about the consistency of thick cream. Always add plaster to the water, not the opposite. Put the strips of burlap into the plaster and squeeze the strips between the fingers so that the plaster works its way into the cloth. Remove the strips by pulling them between the fingers to remove excess plaster.

Next put the strips over the top of the block, allowing the strips to hang down to the bottom of the block. Begin at the middle and work rapidly to the sides of the block. Try to make the plaster strips fit the contour of the specimens. The strips should overlap each other by about half an inch. For specimens in unconsolidated matrix or for very large specimens, two layers of burlap strips at right angles to one another should be used. One must work quickly, as the plaster sets in a few minutes. After the plaster sets, the matrix under the level of the block should be undercut. This leaves the block standing mushroomlike on a pedestal.

The following step is most critical and must be done quickly and gently. Roll the specimen off its pedestal onto a flat surface that has been prepared for the block. The excess matrix now can be removed from the underside of the block to expose the bottom of the fossil. Then the bottom is prepared and bandaged the same way the top was prepared.

The removal of the fossil from the plaster jacket and preparation of the fossil for safe storage require different techniques. The plaster jacket should be marked in the field to show where the top is, the locality and date of the collection, and information about what is inside. Sometimes rough sketches of where the specimen lies within the jacket are made.

Before the specimen is removed, the plaster jacket is usually soaked in water for a few hours to soften it. A hacksaw is used to remove the top of the jacket. Acetone (or other solvent depending which cementing agent was used) may be used to remove the toilet paper or tissue coating. Next, remove the matrix from the fossil by diluting the cement with the proper solvent and then separate the matrix from the specimen, using a wide variety of straight or curved awls and dental tools. This usually is a long, tedious process.

MAINTAINING THE COLLECTION

The most important part of working with vertebrate fossils is the labeling, cataloging, and storage phase. These activities change a fossil collection from a whatnot shelf conversation piece to a valuable scientific collection. The label should be written in permanent ink and easily fit inside the box that contains the fossil. It should contain the following information as a minimum.

1. The number the collector assigns the fossil.
2. The fossil's scientific name (e.g., *Mammut americanum* if it is a mastodont bone).
3. The exact location, preferably to the nearest quarter section (e.g., the NW 1/4, NE 1/4, sec. 21, T 33 S, R 28 W, Cave County, Indiana).
4. The name of the geological formation (e.g., Halton Till, Ontario)
5. The age of the deposit (e.g., Pleistocene, Late Wisconsinan).
6. The name of the collector and the date of collection. (e.g., Collector: I. M. A. Digger, August 2, 2001).

The catalog should be made of high quality, acid-free paper. In the catalog, the fossils may be recorded by scientific name in the order of the numbers that they were given and with the inclusion of all of the data that was recorded on the label. One should keep duplicate copies of fossil catalogs in case of fire or water loss.

The vertebrate fossils should be stored in well-made boxes. A fossil collection has a much more professional look if the boxes are of a uniform type, and there are companies that sell a variety of sizes of sturdy boxes. The outside of the box should have the scientific name of the specimen and the specimen number on it. A label should be inside the box with the fossil; if large enough, the fossil should have its number written on it with permanent ink. The fossil collection should be stored in a moderately dry area on shelves or in study cabinets. The specimens should be arranged in their ordinal and family groups, and the genera and species may be arranged in alphabetical order. Many amateur collectors maintain vertebrate collections that outdo many museums in their proper application of data and in their storage facilities.

Dating the Fossils

The most scientifically important Pleistocene vertebrate fossils are those that have absolute dates to go with them. Today, there are several methods of absolute dating of Pleistocene events. Dates based on a time scale arrived at by counting cyclic events such as the formation of yearly varves (layers of dark sediment deposited in a lake or pond) or yearly tree rings have been successful in northern regions or in areas where trees have lived to be very old, such as in Arizona (bristlecone pine) and California (giant redwoods). Unfortunately, trees have not been found that have lived to be old enough to date Pleistocene events.

Radiocarbon dating (the carbon-14 method) is the most commonly used method of dating Great Lakes region vertebrate sites. This method is accurate back to about 50,000 B.P. Radiocarbon dates are based on the fact that plants incorporate small amounts of the unstable radioactive isotope carbon-14 in their tissues. Plants, of course, are producers at the base of the food chain, and plant carbon-14, therefore, is incorporated in all the higher trophic (feeding) levels. When any organism dies, carbon-14 no longer is incorporated into its tissues. On the other hand, the residual carbon-14 is lost through the process of radioactive decay.

Thus, if the amount of carbon-14 in a sample of previously living tissue is measured, the time that has elapsed since the death of the organism may be calculated. Radioactive decay decreases the carbon-14 by half at regular intervals, so there is usually not enough left to measure after about 50,000 years.

All sorts of organic material may be used to obtain radiocarbon dates, but the most common tissues used in Great Lakes region Pleistocene vertebrate assemblages are wood and bone. Historically, wood has been used more than bone, but in the last decades, bone dates have become rather common because of better radiocarbon dating techniques.

The use of wood for carbon-14 dating can be problematic in certain cases. Pleistocene wood usually has such a high organic content that it may attract other organisms such as tiny roots and mold, especially if the fossil wood lies near the top of the deposit. Abnormally recent dates may be obtained where carbon has been added to the wood by the invasion of these organisms. Moreover, in kettle bog and shallow basin sites, where fossil mastodonts and mammoths are often found, the probability is that the large mammals sank into older sediments where the fossil wood was already in place.

For bone radiocarbon dates, two components, collagen and bone apatite, have been used. Collagen, the complex protein that gives bone its tensile strength, is the most dependable component, although the tendency of collagen to weather out of bone in certain situations creates problems. In cases where no collagen remains, bone apatite has been used. This method, however, is of questionable accuracy because carbon in the groundwater may have replaced the original collagen in the apatite. Even with these problems, the recent dating of bone is generally much more accurate than the bone dates of a few years ago because methods of extracting collagen and removing organic contaminants have improved.

Methods using the decay of potassium to argon have been used to date Pleistocene sites in places with potassium-rich volcanic ash layers, and these dates go back much farther than 50,000 years. But the Great Lakes area lacks definable layers of Pleistocene volcanic ash, and the potassium to argon method has not been used here.

Obtaining carbon-14 dates is rather expensive. It costs well over $200 to get a standard radiocarbon date for a sample that contains about 1 gram (1 ounce = 28.4 grams) of carbon after the pretreatment process by the company (which, unfortunately, destroys the sample). The following table indicates the minimal amount of material in dry weight one needs to get a normal radiocarbon date. Of course, it is better to send more.

If it is important to obtain dates and the organic material from the site is not large, special accelerator mass spectrometer (AMS) techniques are available that use much smaller amounts of material. Obtaining these

charcoal - 3–5 grams (optimum 10–50 grams)

wood - 5 grams (optimum 30-100 grams)

shells - 15 grams (optimum 50-100 grams)

bone - 150 grams (optimum 200–500 grams)

peat - 150 grams (optimum 100–200 grams)

dates, however, is much more expensive and for bone may cost well in excess of $600.

If you are so excited about your bone or wood that you are willing to spend the money, your state or provincial geological survey or the geology department of the nearest university will be able to provide you with the addresses of companies that do radiocarbon dating.

We are now ready to proceed into the heart of the book, which is the bestiary section where the Pleistocene vertebrates of the Great Lakes region are individually discussed. It is very important to realize that all of the forthcoming Pleistocene fish, amphibian, reptile, and bird species (with the exception of an extinct giant tortoise) are still alive in the Great Lakes region today. This is in great contrast to the Pleistocene mammalian assemblage, which had many large members that are now extinct as well as extant northern invading species that do not presently occur in the region.

7

A Bestiary of Great Lakes Region Ice Age Vertebrates

This chapter presents a systematic discussion of genera and species of Ice Age vertebrates known from the province and states surrounding the Great Lakes. References to the sites referred to in this chapter are listed alphabetically by author and date under provincial and state headings in the bibliography of chapters 7 and 8. If they are known, both the Land Mammal Ages (LMA) and the classical Pleistocene ages, in that order, are referred to in the site references at the end of the taxonomic accounts.

FISHES
Class Osteichthyes

Fishes are moderately well represented in the Pleistocene of the Great Lakes region (figs. 15 and 16) as twenty-eight species of Pleistocene fishes occured here. This may seem surprising in the light of the area's abundant lakes, rivers, and streams. The lack of fish fossils may be due in part to the fact that it is very difficult to wash and screen the fibrous sediments in the kettle bog and shallow basin Pleistocene sites of Subregion I. It also may be due to the fact that most fish bones are small and fragile and may have been overlooked by collectors.

Nevertheless, there must be other unexplained reasons for this situation. I have washed and picked through sediments from many Michigan sites over the past decades and have found fish bones in only one of them. Garfish scales, often the most common fish remains in many fossil sites, appear to be lacking in the Great Lakes region.

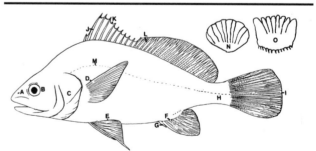

FIGURE 15.   Important external structures of a fish. A, nostril; B, eye; C, operculum; D, pectoral fin; E, pelvic fin; F, anal fin; G, anal fin spine; H, caudal peduncle area; I, caudal fin; J, spiny ray of first dorsal fin; K, membrane between spiny rays; L, soft ray of second dorsal fin; M, lateral line; N, cycloid scale; O, ctenoid scale.

MUDMINNOWS, PIKES SALMON, AND SMELTS
Order Salmoniformes

One character found in almost all salmoniform species is that the maxillary bone in the upper jaw is included

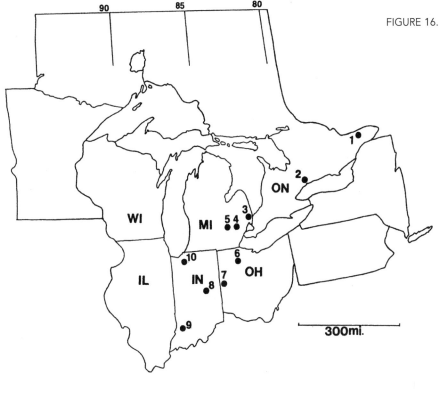

FIGURE 16. Pleistocene fish localities in the Great Lakes region. 1. Green Creek Site, Carlton County, Ontario, Late Wisconsinan. 2. Don Valley Brickyard Site, Toronto, Ontario, Sangamonian. 3. Mill Creek Site, St. Clair County, Michigan, pre-Late Wisconsinan. 4. Shelton Mastodont Site, Oakland County, Michigan, Late Wisconsinan. 5. Charles Adams Mastodont Site, Livingston County, Michigan, Late Wisconsinan. 6. Sheriden Pit Cave Site, Wyandot County, Ohio, Late Wisconsinan. 7. Carter Site, Darke County, Ohio, Late Wisconsinan. 8. Dollens Mastodont Site, Madison County, Indiana, Late Wisconsinan. 9. Prairie Creek D Site, Daviess County, Indiana, Late Wisconsinan. 10. Kolarik Mastodont Site, Starke County, Indiana, Late Wisconsinan (fish currently unidentified).

in the gape of the mouth (fig. 17). In many "higher" modern fish groups the maxilla is excluded from the jaw gape. The order Salmoniformes occurs worldwide and is so diverse that it has been suggested that it is not a natural group. Two salmonid families, mudminnows (Umbridae) and pikes (Esocidae), occur in the Pleistocene of the Great Lakes region.

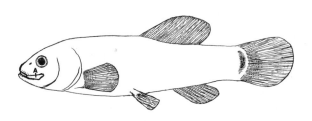

FIGURE 17. Central mudminnow (*Umbra limi*). A, maxilla.

MUDMINNOWS
Family Umbridae

Mudminnows are small fishes with short snouts and rounded tails. Presently, they occur in Europe and Asia and from subtropical to arctic North America. Mudminnows are hardy little animals that are able to breathe oxygen from the air and thereby survive in water too stagnant for other fishes. One genus, *Umbra*, is known from the Pleistocene of the Great Lakes region. Fossil Umbridae first occur in the Paleocene of Asia and North America.

Central Mudminnow
Umbra limi (fig. 17)

The central mudminnow is a small fish (2–4 inches) with a blunt snout, a round tail, and a body that is almost circular in the region behind the head. Central mudminnows prefer quiet ponds or pools of streams where the bottom has a thick organic layer. These robust little fish eat small invertebrate animals such as insect larvae, isopods, and small snails. While feeding, the mudminnow tends to move its body either by quick darts or by paddling slowly along with its pectoral fins. Like the pike, the mudminnow feeds actively under the ice in winter.

This species presently occurs mainly in eastern Canada and in the northeastern and north-central United States. Central mudminnow remains have been found in a Late Wisconsinan deposit in west-central Ohio. This species would have been a small predator on invertebrate animals during the Pleistocene of the Great Lakes region and would have been preyed upon itself by larger fishes such as pike, bass, and walleye.

Site. Carter, Darke County, Ohio, Rancholabrean LMA–Late Wisconsinan (Todd 1973; McDonald 1994).

PIKES
Family Esocidae

Pikes are elongate, long-snouted freshwater fishes that are restricted to the Northern Hemisphere in their distribution. Only one genus, *Esox*, is recognized. *Esox* is an ancient genus that is known from Paleocene to modern times.

Northern Pike
Esox lucius (fig. 18)

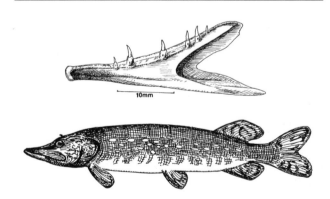

FIGURE 18. Upper, right dentary of a northern pike (*Esox lucius*); lower, northern pike.

The northern pike, sometimes called the great northern pike, is a large, predatory fish with an elongate body and a long-snouted head with large, sharply pointed teeth. Pike that are encountered usually are about 20 to 30 inches long and weigh 1 or 2 pounds, but specimens weighing 20 pounds or more are occasionally caught by anglers.

The northern pike occupies lakes, ponds, rivers, and streams but is restricted to situations where it has access to shallow water in which to spawn. The pike is sought by many Great Lakes region anglers, and many are taken through the ice in the winter with smelt, large minnows, or small suckers as bait. The northern pike often lies in wait for its prey, which it seizes with a quick rush.

The northern pike occurs throughout the Northern Hemisphere. In the Americas, it occurs from northern Canada and Alaska south through southern Illinois and Missouri. *Esox lucius* has been reported from the Late Wisconsinan of southeastern Michigan. Bones assigned to *Esox* sp. have been identified from the Sangamonian of the Toronto region of Ontario, from the pre-Late Wisconsinan of southeastern Michigan, and from the Late Wisconsinan of west-central Ohio. The northern pike would have been one of the top aquatic predators during the Pleistocene of the Great Lakes region.

Sites. Don Formation, near Toronto, Ontario, Rancholabrean LMA–Sangamonian (*Esox* sp. Harington 1990). Mill Creek, St. Clair County, Michigan, Rancholabrean LMA–pre-Late Wisconsinan (*Esox* sp., Karrow et al. 1997). Carter, Darke County, Ohio, Rancholabrean LMA–Late Wisconsinan (*Esox* sp., McDonald 1994). Shelton Mastodont, Oakland County, Michigan, Rancholabrean LMA–Late Wisconsinan (*Esox lucius,* Shoshani et al. 1989).

Muskellunge
Esox masquinongy

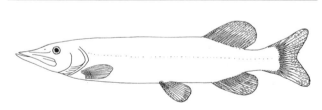

FIGURE 19. Muskellunge (*Esox masquinongy*).

The muskellunge, or musky, is the giant of the pike family. It has an elongate body and long snout with large, sharp teeth. It has been reported to reach a length of 6 feet and a weight of 100 pounds within historic times, and modern records of 30 or 40 pounds are not unusual. The muskellunge is a prize trophy fish in the Great Lakes region and is widely sought after by anglers. It sometimes lies near the surface of the water awaiting its prey and at times will strike surface lures savagely, much to the delight of the angler.

Presently, this species occurs mainly in southeastern Canada and the northern portion of the Great Lakes region but in historic times was more common than it is now in more southern areas. The muskellunge has been reported from the Late Wisconsinan of west-central Ohio. It was certainly an awesome top predator during the Pleistocene of the Great Lakes region.

Site. Carter, Darke County, Ohio, Rancholabrean LMA–Late Wisconsinan (McDonald 1994).

SALMON GROUP
Family Salmonidae

The family Salmonidae contains the familiar salmon, trout, whitefishes, and their relatives. They are medium- to large-sized fishes, ranging from species that weigh only a few ounces to those that reach a weight of 100 pounds. Salmonids mainly occur in freshwater situations, and those that venture into marine waters always return to freshwater to spawn. The Salmonidae is presently the dominant fish family in the cold northern waters of North America, Europe, and Asia. The family is known from the Late Cretaceous in the days of the dinosaurs to modern times. A single genus, *Salvelinus*, is known from the Pleistocene of the Great Lakes region. A "whitefish," which could represent any of a plethora of species (see Page and Burr 1991), has been reported from the Sangamonian Don Formation, near Toronto (Harington 1990).

Lake Trout
Salvelinus namaycush
The familiar lake trout is a large trout with a deeply forked tail. Typically, it has many light spots on a darker background that ranges from light green to almost black. The average lake trout weighs only a few pounds, but in some habitats individuals reach 50 pounds or more. In the northern part of its range, the lake trout may occur in rather shallow rivers and lakes, but in the southern Great Lakes region it is restricted to relatively deep lakes. The lake trout occasionally enters saltwater in areas of relatively low salinity. Lake trout eat a large variety of fishes including whitefish, smelt, shiners, suckers, perch, and sculpin.

Presently, lake trout occur naturally from Alaska and Canada south into the Great Lakes. In the Pleistocene of the Great Lakes region, they are known from the Late Wisconsinan Champlain Sea in Ontario. Certainly, lake trout would have been important predators of smaller fishes in the Late Wisconsinan Champlain Sea. A "trout" listed by Harington (1990) from the Sangamonian Don Formation near Toronto could be this species.

Sites. Green Creek, near Ottawa, Ontario, Rancholabrean LMA–Late Wisconsinan (Harington 1977, 1978). An old record (Hussakof 1918) from the "Late Pleistocene" of Wisconsin needs to be confirmed.

SMELTS
Family Osmeridae

Smelts are small, silvery fishes, some of which are adaptable to both freshwater and saltwater situations. They occur in the Northern Hemisphere in near-coastal situations. Two genera, *Mallotus* and *Osmerus*, occur in the Pleistocene of the Great Lakes region. The smelt family is known from Miocene to modern times.

Capelin
Mallotus villosus
The capelin is a very small, silvery smelt with very small scales, feeble teeth, and broad pectoral fins (the fins directly behind the gills). A large individual is 6 or 7 inches long. Capelins spawn in saltwater and rarely, if ever, enter freshwater. The eggs are put on shore to incubate. Capelins feed mainly on small crustaceans.

Presently, capelins occur in the arctic seas of the Northern Hemisphere, although there are reports of them reaching as far south as Cape Cod in Massachusetts. In the Pleistocene of the Great Lakes region, capelins have been reported from the Late Wisconsinan Champlain Sea in Ontario. These little fishes would have been predators of small crustaceans and would have been eaten themselves by many larger fishes in the Late Wisconsinan Champlain Sea.

Site. Green Creek, near Ottawa, Ontario, Rancholabrean LMA–Late Wisconsinan (Harington 1997, 1978).

Rainbow Smelt
Osmerus mordax (fig. 20)

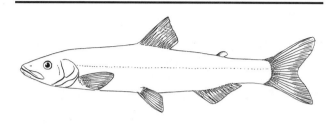

FIGURE 20. Rainbow smelt *(Osmerus mordax)*.

The rainbow smelt is widely caught in dip nets and seines in late winter and early spring in the Great Lakes region today. It is a small, silvery fish with large scales, and strong teeth and has smaller pectoral fins than in the related capelin. Unlike capelins, smelt enter freshwater and brackish water to spawn in late winter and early spring, and have often adjusted to a permanent life in large freshwater lakes. Rainbow smelt range in size from about 8 inches to 1 foot in length. A relatively small individual is able to produce as many as 50,000 eggs. Rainbow smelt are often seen in large schools and are voracious feeders that eat small fishes and actively swimming crustaceans.

Presently, rainbow smelt occur in coastal areas in the Northern Hemisphere, except for a rather large gap in the arctic seas of North America. Introductions into freshwater lakes have greatly extended the range of these small fishes in historic times. Rainbow smelt have been tentatively identified from the Late Wisconsinan Champlain Sea in Ontario. This smelt would have been a voracious little predator of the Late Wisconsinan Champlain Sea, and, in turn, their large schools would have provided a staple food for large, predatory fishes.

Site. Green Creek, near Ottawa, Ontario, Rancholabrean LMA–Late Wisconsinan (McAllister et al. 1981; Wagner 1984).

CODLIKE FISHES
Order Gadiformes

Codlike fishes are robust and elongate with relatively large heads. A barbel is often present on the chin. The pelvic fins (see fig. 15), when present, are positioned below or in front of the pectorals and have up to sev-

enteen rays. There are no spines in the fins. Long dorsal and anal fins are characteristic of the group. This order is widely distributed in cool and cold seas, mainly in the Northern Hemisphere.

CODS
Family Gadidae

Cods are elongate fishes with large heads and a slender barbel at the tip of the chin. The teeth are numerous and small and occur on bands on both the jaws and roof of the mouth. The scales are small and cycloid (see fig. 15). The swim bladder has anterior projections. These are bottom-dwelling fishes that mainly inhabit cool and cold seas in the Northern Hemisphere. A single species is known from the Pleistocene of the Great Lakes region.

Atlantic Tomcod
Microgadus tomcod (fig. 21)

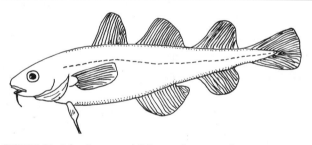

FIGURE 21. Atlantic tomcod *(Microgadus tomcod)*.

The tomcod, one of the smallest species in the family Gadidae, is an elongate fish about 1 foot long with three dorsal fins and a barbel on the tip of the chin. The scales are small and partially embedded in the body. This fish is generally brownish or olive-brown in color. The tomcod is mainly a marine species but at present seasonally invades freshwaters on the eastern seaboard and has become landlocked in a few Canadian lakes.

The tomcod is a voracious fish and feeds mainly on small crustaceans, although it readily eats marine worms, small mollusks, and small fishes. Presently, the tomcod is widely sought after as a food fish and may be caught in large numbers through the ice in December and January during the spawning season. In the

Late Wisconsinan of the Great Lakes region, two specimens were found in concretions in Ontario east of Ottawa. These small cods would have been important predators on crustaceans, small mollusks, and small fishes in the Late Wisconsinan Champlain Sea.

Site. Nodule, "East of Ottawa," Ontario, Rancholabrean LMA–Late Wisconsinan (McAllister et al. 1981; Wagner 1984).

Burbot
Lota lota
The burbot has a large, wide head and a long slender body that is strongly compressed posteriorly. A long barbel occurs on the tip of its chin. The burbot occurs in deep freshwater lakes and rivers down to about 300 feet. This fish spawns in mid-winter under the ice. The burbot is a voracious predator that eats a variety of invertebrates and fish and would have been an important deep-water predator in the Sangamonian of Ontario, where it has been reported from the Don Formation near Toronto.

Site. Don Formation, near Toronto, Ontario, Rancholabrean LMA–Sangamonian (Harington 1990).

MINNOWS, SUCKERS, AND RELATIVES
Order Cypriniformes

Cypriniform fishes usually have a protractile, toothless mouth. The Great Lakes region members of this group have pharyngeal (throat) teeth (fig. 22) on the gill skeleton that are very efficient in processing food. These toothed elements are the skeletal parts most often used to identify Pleistocene cypriniform fishes to the specific level. This order reaches its greatest diversity in Southeast Asia, and there are many species in China. The minnow family (Cyprinidae) and sucker family (Catostomidae) are known from the Pleistocene of the Great Lakes region.

MINNOWS
Family Cyprinidae

Minnows and their relatives the carps have their pharyngeal teeth (fig. 22) arranged in one to three rows

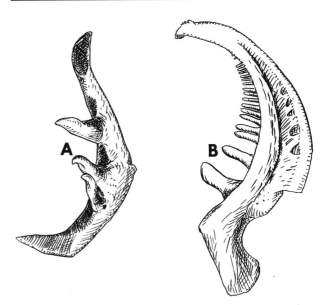

FIGURE 22. Pharyngeal teeth of Cyprinidae and Catostomidae compared. A, left pharyngeal arch with teeth of creek chub (*Semotilus atromaculatus*), Cyprinidae. B, left pharyngeal arch with teeth of white sucker (*Catostomus commersoni*), Catostomidae.

with never more than eight teeth in any single row. Their lips are usually thin. These fishes presently occur in Europe, Africa, and North America. The Cyprinidae forms a huge group. Five genera have been reported from the Pleistocene of the Great Lakes region. The family Cyprinidae is first known from the Eocene of Europe. Minnows unidentified to the generic level have been reported from the Rancholabrean LMA–pre-Late Wisconsinan Mill Creek Site in St. Clair County, Michigan (Karrow et al. 1997).

Central Stoneroller
Campostoma anomalum
Central stonerollers are minnows with a rather large, swollen head and an arch in the body that occurs just behind the head in the nape region. Prominent tubercles occur on the head and nape region of breeding males. This species normally grows to about 6 inches in length. This minnow prefers small to medium-sized streams where it often occurs in pools near riffles. Stonerollers have a ridge on the lower jaw, which they use to scrape algae and other organic bits from rocks on the stream bottom. They have a long intestine that aids them in digesting algae.

This species occurs today in the northeastern and

central United States, entering Mexico through south-central Texas. Central stonerollers have been identified from the Late Wisconsinan of northwestern Ohio. This species would have been an important algal and detritus gatherer during the Pleistocene of the Great Lakes region and would have been preyed upon by larger fishes such as pike, bass, and walleye.

Site. Sheriden Pit Cave, Wyandot County, Ohio, Rancholabrean LMA–Late Wisconsinan (Ford et al. 1996).

Silver Chub
Macrhybopsis storeriana

Silver chubs have very large eyes that occur near the top of the head; a short, rounded snout; and a barbel in the corner of its undercut mouth. This slender species may reach about 9 inches in length. This minnow occupies pools and oxbows of rivers and large streams as well as lakes. The silver chub is a carnivore that feeds on small invertebrates. Adults are reported to be particularly fond of mayflies.

The silver chub occurs from central southern Canada south into the central United States. It has been reported from the Late Wisconsinan of northwestern Ohio. This form would have been an important secondary consumer during the Pleistocene of the Great Lakes region and would have provided food for the larger fishes of the time.

Site. Sheriden Pit Cave, Wyandot County, Ohio, Rancholabrean LMA–Late Wisconsinan (Ford et al. 1996).

Hornyhead Chub
Nocomis biguttatus (fig. 23)

The hornyhead chub has a rather sharply rounded snout. There is a prominent red spot behind the eye of males and a brassy yellowish spot behind the eye of females. The male has many large tubercles on top of its head during the breeding season. This species may reach a length of 9 or 10 inches, but most individuals are smaller. This minnow is at home in the rocky pools of streams and small rivers where they eat a variety of plant and animal material. Plants and algae are important dietary items in small individuals, but as the fish grow larger they begin to feed upon snails, insect larvae, annelids, and even small crayfish and fishes.

Presently, hornyhead chubs occur from central southern Canada south into the central United States.

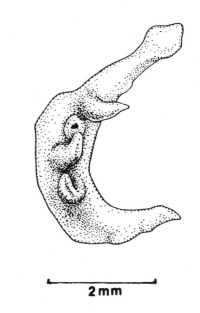

2 mm

FIGURE 23. Right pharyngeal arch with teeth of hornyhead chub (*Nocomis biguttatus*) from the Late Wisconsinan Sheriden Pit Cave Site, Wyandot County, Ohio.

The hornyhead chub has been identified from the Late Wisconsinan of northwestern Ohio. This species would have been an omnivore in its Pleistocene community and would have been eaten by large fishes such as pike, bass, and walleye.

Site. Sheriden Pit Cave, Wyandot County, Ohio, Rancholabrean LMA–Late Wisconsinan (Ford et al. 1996).

Golden Shiner
Notemigonus crysoleucas (fig. 24)

Golden shiners are rather large, silvery minnows with very compressed (flattened from side to side) bodies, large scales, and an upturned mouth. They average about 4 or 5 inches long but may be as much as 1 foot in length in some habitats.

These minnows prefer vegetated lakes and ponds but also occur in pools of creeks and small rivers. In acidic, coffee-colored water, their sides and fins take on a golden hue. This species feeds at the surface and in the middle part of the water column. Golden shiners are mixed feeders, eating small invertebrates as well as filamentous algae. They are probably the most popular bait fish in North America, as their shiny color and active swimming on the hook tend to attract such game fishes as pike, bass, and walleye.

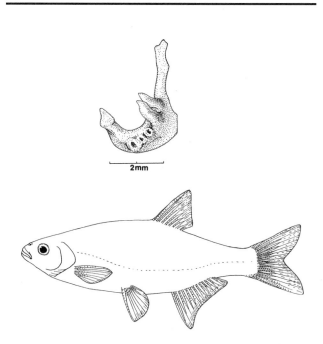

FIGURE 24. Upper, left pharyngeal arch with teeth of golden shiner (*Notemigonus crysoleucas*) from the Late Wisconsinan Sheriden Pit Cave Site, Wyandot County, Ohio. Lower, golden shiner.

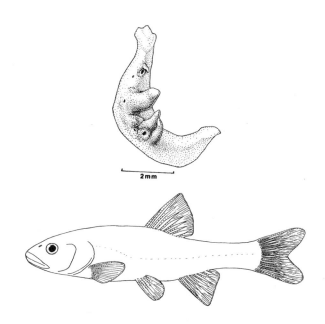

FIGURE 25. Upper, right pharyngeal arch with teeth of creek chub (*Semotilus atromaculatus*) from the Late Wisconsinan Sheriden Pit Cave Site, Wyandot County, Ohio. Lower, creek chub.

The species has a wide range in North America, occurring in about the eastern two-thirds of Canada south to southern Florida and southern Texas in the United States. The golden shiner occurs as a fossil in the Late Wisconsinan of northwestern Ohio. This species would have acted as both a primary and a secondary consumer in its Pleistocene habitat and would have been eaten by the large, predatory fishes of the Ice Age community.

Site. Sheriden Pit Cave, Wyandot County, Ohio, Rancholabrean LMA–Late Wisconsinan (Ford et al. 1996).

Shiner (undetermined species)
cf. *Notropis* sp.

Fish remains of another species of shiner were recovered from Sangamonian Don Formation near Toronto, Ontario. The fossil bones compare well with those of *Notropis*, a genus that consists of a plethora of species of shiners.

Site. Don Formation, near Toronto, Ontario, Rancholabrean LMA–Sangamonian (Harington 1990).

Creek Chub
Semotilus atromaculatus (fig. 25)

The creek chub, a chunky minnow with a down-turned mouth, has a large black spot on the front part of its dorsal fin. The breeding males are quite colorful, having pink and orange colors on the lower body as well as bluish cheeks. The heads of the breeding males are tuberculate. Average specimens are about 4 or 5 inches long, but occasional individuals may reach a size of 10 or 12 inches. This species prefers the rocky or sandy pools of streams and small rivers. It feeds on a wide variety of plant and animal material, including worms, insect larvae, algae, and even small crayfishes and fishes. Creek chubs are favorite bait fishes as they are hardy on the hook and attract a variety of game fishes, especially pike.

Creek chubs are widely distributed in North America, occurring in about the eastern two-thirds of southern Canada and in about the eastern three-fourths of the United States south to Florida and Texas. The species occurs in the Late Wisconsinan of northwestern Ohio. Creek chubs would have been important omnivores during the Pleistocene of the Great Lakes region and would have provided food for large, predatory fishes.

Site. Sheriden Pit Cave, Wyandot County, Ohio, Rancholabrean LMA–Late Wisconsinan (Ford et al. 1996).

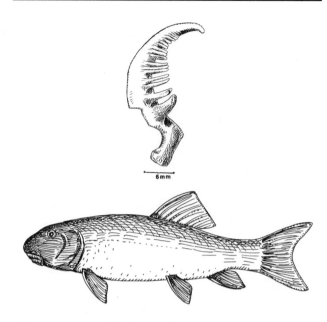

FIGURE 26. Upper, left pharyngeal arch with teeth of modern white sucker (*Catostomus commersoni*). Lower, white sucker.

SUCKERS
Family Catostomidae

Suckers have their pharyngeal teeth arranged in a single row of sixteen or more teeth (see fig. 22). Their lips are down-turned, thick, and fleshy. These fishes presently occur in freshwater habitats in China, Siberia, and North America. Two sucker genera, *Catostomus* and *Moxostoma*, occur in the Pleistocene of the Great Lakes region. The Catostomidae are first known from the Eocene of North America and China. *Catostomus* itself is known from the Eocene. Suckers unidentified to the generic level have been reported from the pre-Late Wisconsinan Mill Creek Site in St. Clair County, Michigan (Karrow et al. 1997).

White Sucker
Catostomus commersoni (fig. 26)

The familiar white sucker is a robust, whitish gray fish with a down-turned fleshy mouth and with its lower lip about twice as thick as its upper lip. Average specimens weigh about 1.5 to 2 pounds, but some may reach a length of over 2 feet and a weight of 6 or 7 pounds. They occupy a large variety of habitats, from small creeks to lakes as large as the Great Lakes. White suckers are bottom feeders that mainly eat invertebrates such as worms, snails, and small clams. They are bony and said to be best for human consumption when they are smoked. Small suckers are often used as bait for pike during winter ice fishing.

White suckers range throughout most of Canada and occur in about the eastern three-fourths of the United States south to northern Georgia and Texas. This species occurs in the Late Wisconsinan of the south-central part of the Lower Peninsula of Michigan. White suckers would have been important consumers of invertebrate animals in Ice Age communities of the Great Lakes region, and small suckers would have been a major food item for the large, predatory fishes of the time.

Site. Charles Adams Mastodont, Livingston County, Michigan, Rancholabrean LMA–Late Wisconsinan (Holman 1979).

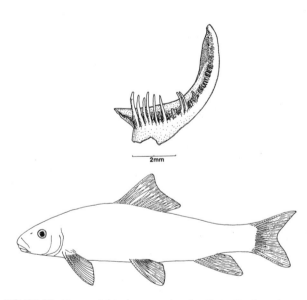

FIGURE 27. Upper, right pharyngeal arch with teeth of modern shorthead redhorse (*Moxostoma macrolepidotum*) from the Late Wisconsinan Sheriden Pit Cave Site, Wyandot County, Ohio. Lower, shorthead redhorse.

Shorthead Redhorse
Moxostoma macrolepidotum (fig. 27)

The shorthead redhorse has a moderately robust body that terminates in a bright red tail fin. Large papillae occur on its lower lips. These fish may reach a length of about 25 inches and a weight of 5 or 6 pounds, but most individuals are about 15 inches long and weigh 1 or 2 pounds. Shorthead redhorse occur in rivers and lakes and often prefer rocky bottoms. Like the white sucker, they are bottom feeders that eat a variety of invertebrate species, which they suck up and strain. These fish are eaten by humans, but the flesh is very bony. They are said to be quite good when they are smoked. The young of this species are eaten by several species of large game fishes, but the adults are probably immune from most predators except for very large pike and musky.

The species occurs throughout much of Canada, except the northern parts of some of the western provinces, and in the eastern two-thirds of the United States south to Texas. The shorthead redhorse occurs in the Late Wisconsinan of northwestern Ohio. *Moxostoma* sp. has been identified from the Late Wisconsinan of Darke County, Ohio, and may represent this species. The shorthead redhorse was undoubtedly important as a consumer of invertebrate animals during the Pleistocene of the Great Lakes region, while its young must have been preyed upon by large, predatory fishes.

Sites. Sheriden Pit Cave, Wyandot County, Ohio, Rancholabrean LMA–Late Wisconsinan (Ford et al. 1996). Carter, Darke County, Ohio, Rancholabrean LMA–Late Wisconsinan (*Moxostoma* sp., McDonald 1994).

CATFISHES
Order Siluriformes

Catfishes are known for the sensory barbels that occur on the snout and also for the spines that are associated with fins on the back (dorsal spines) and behind the head (pectoral spines) (fig. 28). These spines are capable of introducing toxins that produce irritating to fatal reactions in humans, depending upon what species of catfish is represented and how the particular human reacts to the toxin. Catfishes usually have naked skins, but some are armored with bony plates. The Wels catfish of large European rivers commonly reaches a length of 10 feet. Catfishes occur in both freshwater and saltwater habitats in almost all parts of the world. A single family, the Ictaluridae, is recognized from the Pleistocene of the Great Lakes region.

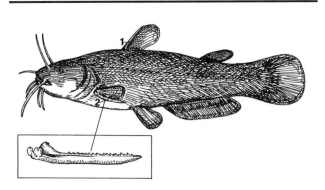

FIGURE 28. Yellow bullhead (*Ameiurus natalis*) showing sensory barbels on the snout and dorsal (1) and pectoral (2) spines.

NORTH AMERICAN FRESHWATER CATFISHES
Family Ictaluridae

The family Ictaluridae is recognized on the basis of the eight barbels on the head (see fig. 28), by a naked skin, and, except for a single Mexican species, for having its dorsal and pectoral spines on the front edges of the fins. Fossil ictalurids are often identified on the basis of the structure of their pectoral spines (see fig. 28). In the past, native North Americans fashioned spines of large catfishes into tools. Awls for leather work were made by rounding the base of the spines and removing the serrations. These spines could also be used as needles if the hole in its base remained intact. Tools such as these have been found in North American deposits that were at least 3,000 years old. The family is presently very widespread in North America, except for far northern regions. Fossil ictalurids are first known from the Paleocene.

Brown Bullhead
Ameiurus nebulosus (fig. 29)

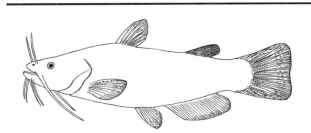

FIGURE 29. Brown bullhead (*Ameiurus nebulosus*).

The brown bullhead is a very robust fish with a big head and brown or black mottling or spots on the sides of the body. The rear end of the tail fin (caudal fin) is straight. The brown bullhead also may be identified on the basis of the five to eight large sawlike teeth on the back of its pectoral spines. Brown bullheads may reach a length of about 20 inches, but most individuals encountered are about 1 foot long and weigh 1 or 2 pounds. The brown bullhead feeds near the bottom, usually at night. The barbels around the mouth help it locate suitable food. This species eats a wide variety of animal food ranging from invertebrates to fishes and frogs. It also acts as a scavenger, feeding on offal and recently dead bodies of other aquatic organisms. The brown bullhead is probably the tastiest of the three bullhead species and is widely eaten by humans. Its spines do not always protect it from predators, and small individuals are eaten from time to time by most of the large, predatory species, including pike, musky, bass, and walleye.

Brown bullheads occur in northeastern Canada to Nova Scotia but barely enter southern Manitoba and Alberta in the west. They mainly occur in the eastern United States south to southern Florida and west to North Dakota. Fossil brown bullhead remains have been identified from Late Wisconsinan sites in northwestern and west-central Ohio. The species probably was an important predator and scavenger in its Ice Age habitat and was eaten from time to time by large, predatory species.

Sites. Sheriden Pit Cave, Wyandot County, Ohio, Rancholabrean LMA–Late Wisconsinan (Ford et al. 1996). Carter, Darke County, Ohio, Rancholabrean LMA–Late Wisconsinan (McDonald 1994).

Channel Catfish
Ictalurus punctatus

The channel catfish is streamlined compared to the bullhead species. The head is more pointed and not as pronounced as in the bullheads, and the tail is forked. It tends to have light sides with scattered black spots on them. Channel catfish may reach a length of 50 inches and a weight of over 30 pounds, but most individuals encountered weigh from about 2 to 4 pounds. The channel catfish is not as restricted to the bottom and is more active than the bullhead species.

Channel catfish are thought to depend more on sight in their feeding behavior than do most other catfishes. This species eats insects, mollusks, crayfishes, and various minnows and small fishes. Large individuals of this species are mainly free from predation, but the young are probably eaten by various predatory species. The channel catfish is an important food fish for humans and lends itself well to pond culture.

In Canada, channel catfishes are mainly restricted to the southeast but reach southwestern Ontario and southeastern Manitoba in the west. Channel catfish occur in about the eastern two-thirds of the United States, reaching south-central Florida, Texas, and northern Mexico in the western part of their range. Oddly, they do not occur along the Atlantic coastal region north of Georgia. Channel catfish have been found in the Sangamonian of the Toronto region of Ontario and from the Late Wisconsinan of northwestern Ohio. This species undoubtedly was an important predator during the Pleistocene of the Great Lakes region, and its young were probably eaten by various predator fishes.

Sites. Don Formation, near Toronto, Ontario, Rancholabrean LMA–Sangamonian (Harington 1990; Karrow et al. 1980). Sheriden Pit Cave, Wyandot County, Ohio, Rancholabrean LMA–Late Wisconsinan (Ford et al. 1996).

STICKLEBACKS AND TUBESNOUTS
Order Gasterosteiformes

The fishes of the order Gasterosteiformes are found in marine, brackish, and freshwater habitats in the Northern Hemisphere. Only the stickleback family is known from the Pleistocene of the Great Lakes region.

STICKLEBACKS
Family Gasterosteidae

The stickleback family has very well developed dorsal spines followed by a normal dorsal fin (fig. 30). This family is especially familiar to scientists because of the behavioral and physiological studies made on some of its member species that are easy to manage in the laboratory. Sticklebacks are first known from the Miocene.

Threespine Stickleback
Gasterosteus aculeatus

The threespine stickleback has three (occasionally four) isolated, serrated spines on its back, each of which has a posterior (toward the tail) fin membrane attached to it. These tiny fishes are about 2 inches long in the adult form. Threespine sticklebacks have complex breeding patterns in which the male coaxes the female into the nest with a courtship ritual, which includes touching the nest-opening with his snout. The male guards and fans the nest and then continues to guard the hatchlings until they are able to fend for themselves. The threespine stickleback feeds voraciously on worms, crustaceans, and even eggs and young fishes, including its own species.

Threespine sticklebacks occur in both freshwater and saltwater habitats and are present in coastal regions throughout the world, except in the arctic coasts of Siberia and most of North America. In the Pleistocene of the Great Lakes region, this species has been found in the Late Wisconsinan Champlain Sea in Ontario. The threespine stickleback would have been a voracious little predator in the Late Wisconsinan Champlain Sea.

Site. Green Creek, near Ottawa, Ontario, Rancholabrean LMA–Late Wisconsinan (Harington 1977; 1978).

Ninespine Stickleback
Pungitius sp. (fig. 30)

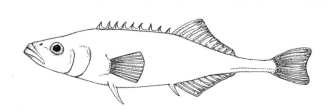

FIGURE 30. Ninespine stickleback (*Pungitius* sp.).

The ninespine stickleback usually has nine short spines that are alternately arranged from left to right. Its body is very slender and compressed. These pugnacious little fishes are usually about 2 to 2.5 inches long. This species prefers shallow, well-vegetated areas of lakes, ponds, and sluggish streams. Its food consists mainly of aquatic insects and small crustaceans. This is a north-ern fish that ranges from Alaska and boreal Canada to the Great Lakes region. In the Pleistocene, it was probably an important tiny predator. The ninespine stickleback has been reported from the pre-Late Wisconsinan of southeastern Michigan.

Site. Mill Creek, St. Clair County, Michigan, Rancholabrean LMA–pre-Late Wisconsinan (Karrow et al. 1997).

HIGHER FISHES
Order Perciformes

The Perciformes comprise the largest vertebrate order in the world. Perciform fishes dominate modern saltwater habitats and are the dominant freshwater species in many places. These fishes may be distinguished from other groups in having spines in the fins, two dorsal fins with supporting rays, and ctenoid-type scales (scales whose free portions are covered with small teeth; see fig. 15), or with no scales. Three perciform families—sunfishes and basses (Centrarchidae), perches (Percidae), and drums and croakers (Sciaenidae)—are represented by Ice Age fishes in the Great Lakes region.

SUNFISHES AND BASSES
Family Centrarchidae

Centrarchid species have three or more fin spines in their anal fin (see fig. 31). In most species, the male hollows out a nest with his tail and then guards the eggs from predators. Hybrid centrarchid species are not rare in the wild, and even hybrid genera are occasionally reported. Centrarchids, including such species as the large- and smallmouth bass, rock bass, bluegills, sunfishes, and crappies, are favorite food and game fishes in North America today.

This family occurs throughout most of southern Canada and natively throughout the United States except for a few of the western states. Both species of basses and a crappie occur as Pleistocene fossils in the Great Lakes region. The family is first known from the Eocene of North America.

Smallmouth Bass
Micropterus dolomieu (fig. 31)

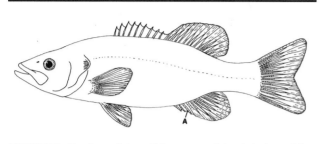

FIGURE 31. Smallmouth bass (*Micropterus dolomieu*). A, anal fin.

The smallmouth bass is a robust predatory fish with a bronze coloration and a series of broken or sometimes fused brownish vertical bars on its sides. The mouth ends at about the level of the front of the eye. Smallmouth bass may grow to a length of about 25 inches and attain a weight of about 8 or 9 pounds, but fishes of this size are rare. Smallmouth taken by anglers in the Great Lakes region usually weigh 1 to 2 pounds. These fishes prefer clear, gravel-bottom streams or rocky areas in lakes. They are active, predatory species that are especially fond of crayfishes, but they will also eat many kinds of minnows and small fishes as well as an occasional frog. The smallmouth bass has very firm, pleasant-tasting flesh and is the favorite game fish of many anglers in the Great Lakes region. Smallmouth are very strong fighters compared to most freshwater species.

Smallmouth bass occur in southeastern Canada and in the eastern half of the United States south to northern Alabama and Arkansas. The species occurs as an Ice Age fossil in Late Wisconsinan deposits in northwestern and west-central Ohio. It would have been an important predator in its Pleistocene habitat, and its young, in turn, would have been food for other predatory fishes.

Sites. Sheriden Pit Cave, Wyandot County, Ohio, Rancholabrean LMA–Late Wisconsinan (Ford et al. 1996). Carter, Darke County, Ohio, Rancholabrean LMA–Late Wisconsinan (McDonald 1994).

Largemouth Bass
Micropterus salmoides
The largemouth bass is a robust fish with a greenish coloration and a broad, black (sometimes broken) stripe along its sides. The mouth extends behind the eye. Largemouth bass may occasionally reach a weight in excess of 15 pounds in southern waters, but those caught by anglers in the Great Lakes region usually weigh 1 to 2 pounds. This species occurs in vegetated ponds and lakes and in the slower reaches of rivers and streams, usually over muddy or sandy bottoms. The largemouth is an aggressive, predatory species, often feeding near the surface where it causes a characteristic swirl as it engulfs its prey. Fishes and crayfishes are among its favorite food, but a hungry largemouth will probably take about any animal it can swallow. This species has a pleasant-tasting flesh and is the favorite game fish of many anglers because of its leaping, head-shaking style of resisting the hook.

The largemouth bass occurs in about the eastern half of Canada north to but not including Lake Superior and in most of the eastern two-thirds of the United States south to Florida and Texas, with the exception of many of the northeastern seaboard states. The largemouth bass has been identified as a fossil from a Late Wisconsinan site in west-central Ohio. This bass would have been a top predator in the ponds, lakes, and swamps of the Pleistocene of the Great Lakes region.

Site. Carter, Darke County, Ohio, Rancholabrean LMA–Late Wisconsinan (McDonald 1994).

Crappie
Pomoxis sp.
Crappies are very compressed, somewhat silvery fishes with from six to eight dorsal spines. Two species of crappie, the white crappie (*Pomoxis annularis*) and the black crappie (*P. nigromaculatus*; fig. 32) presently occur in North America, where their range is mainly in southeastern Canada and the eastern half of the United States. Both crappie species prefer quiet water situa-

FIGURE 32. Black crappie (*Pomoxis nigromaculatus*).

tions and prefer minnows as food. The white crappie is much more tolerant of turbid water than is the black crappie. Fossil crappie material from a Late Wisconsinan site in the south-central part of the Lower Peninsula of Michigan was identified only to the generic level. Either of these species would have been predators of minnows and small fishes during the Pleistocene.

Site. Charles Adams Mastodont, Livingston County, Michigan, Rancholabrean LMA–Late Wisconsinan (Holman 1979).

PERCHES
Family Percidae

The fishes of the family Percidae have only one or two spines in their anal fins, and if two are present, the second is usually weak. The group includes important food and game fishes such as the yellow perch, walleye, and sauger, as well as myriad species of darters. The family is widely distributed in the Northern Hemisphere. Only the yellow perch represents this family as a Pleistocene fossil in the Great Lakes region. The Percidae first appeared in the Eocene of Europe and North America.

Yellow Perch
Perca flavescens (fig. 33)

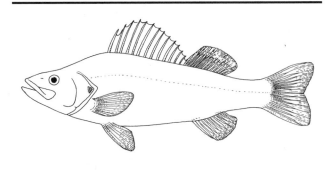

FIGURE 33. Yellow perch (*Perca flavescens*).

Yellow perch are robust fishes with about six to nine brownish bands extending down the sides of their yellowish or sometimes greenish-white bodies. They have a large mouth that extends to the middle of the eye. Yellow perch may grow to be 15 or 16 inches long and weigh several pounds, but most of those caught by anglers in the Great Lakes region are about 8 to 10 inches long and weigh .5 pound or less. Stunted populations of yellow perch, where individuals rarely exceed 5 or 6 inches in length, are rather common in ponds or small lakes in the region. Yellow perch feed on small invertebrates, minnows, small fishes, and small crayfishes. They are in turn eaten by many larger predatory fishes such as bass, pike, walleye, and musky. They have been a very important food fish in North America, as their flesh is flaky and delicious.

The species occurs in most of Canada except for British Columbia. In the United States, this species is found mainly in the northeastern and Great Lakes region, extending into South Carolina along the Atlantic seaboard states. This species occurs as a Pleistocene fossil in the Sangamonian of Ontario near Toronto, and in the Late Wisconsinan of southeastern Michigan and west-central Ohio. It was undoubtedly an important small carnivore in the Ice Age of the Great Lakes region and prominent in the diet of larger predatory species.

Sites. Don Formation near Toronto, Ontario, Rancholabrean LMA–Sangamonian (Harington 1990). Shelton Mastodont, Oakland County, Michigan, Rancholabrean LMA–Late Wisconsinan (Shoshani et al. 1989). Carter, Darke County, Ohio, Rancholabrean LMA–Late Wisconsinan (McDonald 1994).

DRUMS AND CROAKERS
Family Sciaenidae

The drum and croaker family has a very long dorsal fin that is divided by a notch that separates a front spiny part from a rear soft portion (fig. 34). In most species, the body is deep and arches highly at the level of the spiny part of the dorsal fin. Most members of this family live on the continental shelves of temperate and tropical oceans, but one species, the freshwater drum (see fig. 34), is restricted to freshwater habitats in North America. This species occurs as a Pleistocene fossil in the Great Lakes region. The family gets its name because its members are able to produce drumming or croaking sounds by using their gas bladders as sound-producing chambers. Many sciaenids are excellent food fishes. The family is first known from the Upper Cretaceous of West Africa.

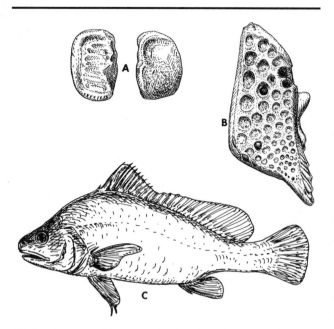

FIGURE 34. A, otolith (views of each side) of freshwater drum (*Aplodinotus grunniens*). B, pharyngeal tooth plate (in occlusial view) of freshwater drum. C, freshwater drum.

Freshwater Drum
Aplodinotus grunniens (fig. 34)

Freshwater drum are large, robust, whitish gray fishes with an arched body and a down-turned mouth. They may grow to over 2 feet in length and attain a weight of up 25 or 30 pounds, but most individuals in the Great Lakes region today range from about 2 to 5 pounds. Freshwater drum have large, circular, calcified ear plates called otoliths and large tooth plates that are occasionally found as Pleistocene fossils and often turn up in archaeological sites (fig. 34). This species prefers rivers and lakes and spends most of its time on the bottom. It feeds upon insects, mollusks, and crayfishes and sometimes eats small fishes. Larger individuals have few predators. Most people do not savor the flesh of the freshwater drum, but some are eaten locally.

The freshwater drum has an extensive range in North America, extending from northern Manitoba south through the central part of the United States and then along the Gulf coastal region into Guatemala. This species has been identified as a fossil from the Sangamonian near Toronto, Ontario, and from the pre-Late Wisconsinan of southeastern Michigan. The drum would have been a bottom-feeding carnivore in its Ice Age habitat in the Great Lakes region.

Sites. Don Formation, near Toronto, Ontario, Rancholabrean LMA–Sangamonian (listed as cf. *Aplodinotus grunniens* by Harington (1990). Mill Creek, St. Clair County, Michigan, Rancholabrean LMA–pre-Late Wisconsinan (listed as *Aplodinotus* sp. by Karrow et al. 1997).

SCULPIN
Family Cottidae

The sculpin have large, armored heads and stout bodies with very large, fanlike pectoral fins. Sculpin are bottom fishes that are primarily found in saltwater, with the exception of the genus *Cottus*, which is widely distributed in cold freshwater situations in the Northern Hemisphere where they are widely distributed. Sculpin are generally small fishes, but some species may reach a length of about 2 feet. The family is known from Oligocene to modern times. A member of the Cottidae, unidentified as to genus or species occurs in the Sangamonian Don Formation near Toronto, Ontario (Harington 1990).

Hook-eared Sculpin
Artediellus uncinatus

The hook-eared sculpin has a process on the back of the head that hooks strongly upward. Its back and sides are naked, and its tail fin is quite small. Hook-eared sculpin only reach a length of about 4 inches and have little commercial value. Presently, this species occurs from Labrador to Cape Cod on the East Coast, but in the Pleistocene it occurred in the Late Wisconsinan Champlain Sea in Ontario. This bottom fish would have provided a bony dinner for any predatory fish in the Late Wisconsinan Champlain Sea. American populations of this genus are sometimes referred to as a separate species, the Atlantic hook-eared sculpin (*A. atlanticus*) (Robbins et al. 1986).

Site. Green Creek, near Ottawa, Ontario, Rancholabrean LMA–Late Wisconsinan (Harington 1977, 1978; McAllister et al. 1981; Wagner 1984).

LUMPFISH AND SNAILFISHES
Family Cyclopteridae

The lumpfish and snailfishes are a strange group of marine fishes that have fused fins on the belly modified as a sucker. The lateral line is usually absent in the region behind the head, and the opening to the gills is reduced in size. Many of these fishes have a lumpy shape. This family occurs in Antarctic and Arctic seas in both the Atlantic and Pacific Oceans. Cyclopterids are known in the fossil record only from the Pleistocene, but they obviously originated much farther back in time. A single genus, *Cyclopterus*, has been identified from the Pleistocene of the Great Lakes region.

Lumpfish
Cyclopterus lumpus (fig. 35)

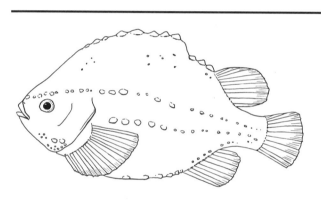

FIGURE 35. Lumpfish (*Cyclopterus lumpus*).

The short, thick lumpfish has a rough, lumpy body and a sucker on its belly made of fused fins. Lumpfish reach a length of about 2 feet and a weight of up to 13 pounds. These odd fishes are sluggish bottom-dwellers that normally live among the rocks, but sometimes they are found in mats of floating vegetation. The food of the lumpfish consists mainly of invertebrates, but jellyfish are included as food items. A large female of this species may produce as many as 130,000 eggs. Males guard the nest and circulate water around the eggs.

Lumpfish occur in northern waters on both coasts of the Atlantic, occasionally occurring as far south as Chesapeake Bay in North America. In the Pleistocene of the Great Lakes region, fossils as-signed to the lumpfish have been reported from the Late Wisconsinan Champlain Sea in Ontario. Lumpfishes would have been predators on invertebrate life of the rocky regions in the Late Wisconsinan Champlain Sea.

Site. Green Creek, near Ottawa, Ontario, Rancholabrean LMA–Late Wisconsinan (McAllister et al. 1981).

AMPHIBIANS
Class Amphibia

Records of Pleistocene amphibians are rare in the Great Lakes region, as only nine species from eleven localities are reported here (fig. 36). This is somewhat surprising in light of the fact that wetlands must have been very abundant in the region during unglaciated parts of the Pleistocene. On the other hand, amphibian fossils are small and fragile, and it is possible that they have been overlooked by collectors. Also, the lack of Pleistocene amphibians in the region may be due in part to the fact that it is very difficult to wash and screen the fibrous sediments of kettle bog and shallow basin sites that occur in Subregion I.

SALAMANDERS
Order Caudata

Salamanders have distinct heads; elongate, cylindrical bodies; and long tails (fig. 37). Their limbs are usually short compared to the length of their bodies, and their skulls contain a reduced number of bones. Fossil salamanders are generally identified on the basis of their vertebrae (fig. 38). Salamanders are mainly animals of the temperate regions of the Northern Hemisphere, except for one group that has reached South America. The eastern United States, especially the Appalachian region, has more species of salamanders than any comparable region in the world. Two families, the mole salamanders (Ambystomatidae) and the lungless salamanders (Plethodontidae), have been reported from the Pleistocene of the Great Lakes region.

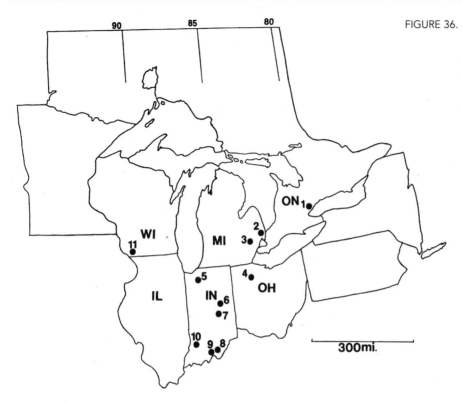

FIGURE 36. Pleistocene amphibian localities in the Great Lakes region. 1. Kelso Cave Site, Halton County, Ontario, Late Wisconsinan. 2. Meskill Road Water Well Site, St. Clair County, Michigan, Late Wisconsinan. 3. Shelton Mastodont Site , Oakland County, Michigan, Late Wisconsinan. 4. Sheriden Pit Cave Site, Wyandot County, Ohio, Late Wisconsinan. 5. Kolarik Mastodont Site, Starke County, Indiana, Late Wisconsinan. 6. Dollens Mastodont Site, Madison County, Indiana, Late Wisconsinan. 7. Christensen Bog Mastodont Site, Hancock County, Indiana Late Wisconsinan. 8. King Leo Pit Cave, Harrison County, Indiana, Late Wisconsinan. 9. Megenity Peccary Cave Site, Crawford County, Indiana, Sangamonian and Late Wisconsinan. 10. Prairie Creek D Site, Daviess County, Indiana, Late Wisconsinan. 11. Moscow Fissure Site, Iowa County, Wisconsin, Late Wisconsinan.

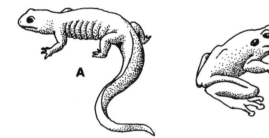

FIGURE 37. A, typical salamander, a tailed amphibian (note the lack of claws). B, typical treefrog, a tailless amphibian.

MOLE SALAMANDERS
Family Ambystomatidae

Mole salamanders have a stout body with a broad head and a compressed tail (see fig. 37). Lungs are present. Many of these salamanders burrow, and the rest are secretive. Mole salamanders are widely distributed in North America, ranging from southern Canada well into Mexico in the mountains. They are absent, however, from some of the far western states of both the

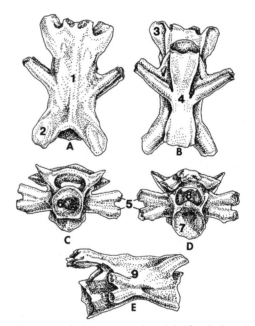

FIGURE 38. A typical Pleistocene salamander fossil, the vertebra of a mole salamander (*Ambystoma*). A, dorsal; B, ventral; C, anterior; D, posterior; E, lateral. 1, neural spine; 2, prezygapophyseal accessory facet; 3, postzygapophyseal accessory facet; 4, bottom of centrum; 5, rib-bearing processes; 6, anterior cotyle; 7, posterior cotyle; 8, neural canal; 9, lateral wall of neural arch.

United States and Mexico. Ambystomatids are first known from the Late Eocene of North America.

Blue-spotted Salamander complex
Ambystoma laterale complex

Both *Ambystoma laterale*, the blue-spotted salamander, and A. jeffersonianum, Jefferson salamander, are very similar to one another, and to make the situation more confusing, these animals hybridize throughout most of the Great Lakes region. Both of these species and the hybrids between them are black or bluish with light bluish spots or flecks on their sides and tail and are usually about 4 or 5 inches long. Animals of this complex feed on small invertebrates of the forest floor and during moist spring and early summer times can often be found living under rotting logs.

Vertebrae identified from the Late Wisconsinan of northwestern Ohio appear to be identical to those of the blue-spotted salamander, but of course the possibility exists that they might represent the Jefferson salamander or a hybrid between the two species. Other fossil salamander vertebrae from the Late Wisconsinan of southwestern Indiana are assigned to the genus *Ambystoma*, but a species is not suggested. These small salamanders would have been predators of small invertebrates of the woodland floor during the Pleistocene of the Great Lakes region and would have been eaten by snakes and other small predators.

Sites. Sheriden Pit Cave, Wyandot County, Ohio, Rancholabrean LMA–Late Wisconsinan (*Ambystoma laterale* complex, Holman 1997). Prairie Creek D, Daviess County, Indiana, Rancholabrean LMA–Late Wisconsinan (*Ambystoma* sp., Holman and Richards 1993).

LUNGLESS SALAMANDERS
Family Plethodontidae

Lungless salamanders have a long, slender body with a narrow head and a long, cylindrical tail. They are salamanders of the woodland floor, caves, and mountain streams. Lungless salamanders occur in southeastern Canada and about the eastern third of the United States. They also occur in the western Pacific coast area of North America and through Central America into northern South America. A single genus occurs in southeastern France, Italy, and Sardinia. But the center

of diversity of the plethodontids is in the Appalachian region of the United States. This is the most advanced group of salamanders, and the family only extends back as far as the Early Miocene.

Cave Salamander
Eurycea lucifuga

The cave salamander is a long-tailed, reddish salamander with small black spots or flecks on the head, body, and tail. It often inhabits the mouths of caves and on occasion may be found in the lightless regions. Cave salamanders may also be found in rock crevices or under logs, litter, or flat rocks on the woodland floor. These animals have a prehensile tail, which allows them to hold onto objects when they are climbing on rocky ledges inside or outside of cave habitats. Cave salamanders feed mainly on small invertebrates.

These salamanders occur in the northern parts of the southeastern United States from southern Indiana and Ohio to northern Alabama and Georgia. Vertebrae identified from a Late Wisconsinan cave site in southern Indiana represent the cave salamander. These small amphibians would have been troglodytic (cave-dwelling) or woodland predators of small invertebrate during the Pleistocene and would have in turn been preyed upon by snakes and other small predators.

Site. King Leo Pit Cave, Harrison County, Indiana, Rancholabrean LMA–Late Wisconsinan (Richards and McDonald 1991).

TOADS AND FROGS
Order Anura

Anurans have shortened bodies, and adults lack tails (see fig. 37). The back legs are highly developed for hopping or jumping, and, like the salamanders, their skulls contain a reduced number of bones. Unlike the salamanders, the anurans reach the height of their diversity in tropical regions. Fossil anurans are very often identified on the basis of the ilium (fig. 39), a part of the pelvic (hip) girdle. This family has almost a worldwide distribution, except for polar regions and extremely dry deserts. Three families—toads (Bufonidae), true frogs (Ranidae), and tree frogs (Hylidae)—have been identified from the Pleistocene of the Great Lakes region.

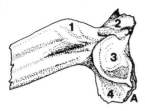
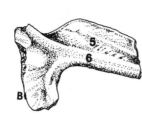
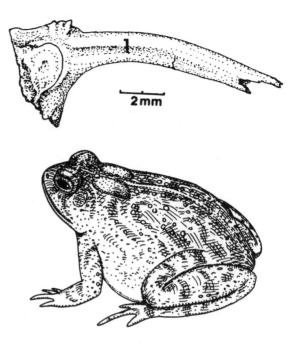

FIGURE 39. A typical Pleistocene anuran fossil, a partially broken left ilium of a true frog (*Rana*). A, lateral; B, medial. 1, tuber superior; 2, dorsal acetabular expansion; 3, acetabular fossa; 4, ventral acetabular expansion; 5, ilial crest or blade; 6, ilial shaft. The ilial crest, or blade, is lacking in North American toads (*Bufo*).

TOADS
Family Bufonidae

Toads are short, robust anurans with a rough, warty skin (fig. 40). They have a prominent pair of rounded, parotoid glands behind the head. Toads occur almost worldwide except for Madagascar, Australia, New Guinea, New Zealand, the polar regions, and some extremely dry deserts. Toads are known from as far back as the Paleocene of South America.

FIGURE 40. Upper, right ilium of an American toad (*Bufo americanus*) from the Late Wisconsinan Meskill Road Water Well Site, St. Clair County, Michigan. North American Toads (*Bufo*) lack the ilial crest, or blade, on the ilial shaft (1) found on the true frogs (*Rana*). Lower, American toad.

American Toad
Bufo americanus (fig. 40)

The American toad is a brownish, grayish, or reddish brown toad with dark spots on its back. It may be distinguished from the closely related Fowler's toad in that it has only one or two warts within each of the dark spots, whereas Fowler's toad has three or more warts in each spot. American toads occur in a remarkable variety of habitats, from city gardens and backyards to remote wild areas. To exist, American toads need shallow pools in which to breed, adequate insect and other small invertebrate food, and places to hide.

The American toad is widespread in eastern North America, ranging north to Hudson's Bay and Newfoundland and south to Louisiana, Mississippi, Alabama, and Georgia. Fossils of the American toad have been found in the Late Wisconsinan of southeastern Michigan, northwestern Ohio, and extreme southwestern Wisconsin. Bones assigned to *Bufo* sp. from the Late Wisconsinan of southwestern Indiana may represent either the American toad or Fowler's toad. The Wisconsin American toad probably represents a northern subspecies known as *Bufo americanus copei*. American toads obviously played the role of small insectivores during the Pleistocene and were preyed upon by small predators, such as garter snakes and hognose snakes, that are not bothered by the toxic substances in toad skins.

Sites. Meskill Road Water Well, St. Clair County, Michigan, Rancholabrean LMA–Late Wisconsinan (Holman 1988b). Sheriden Pit Cave, Wyandot County, Ohio, Rancholabrean LMA–Late Wisconsinan (Holman 1997). Moscow Fissure, Iowa County, Wisconsin, Rancholabrean LMA–Late Wisconsinan (*Bufo americanus* [?] *copei*, Foley 1984). Prairie Creek D, Daviess County, Indiana, Rancholabrean LMA–Late Wisconsinan (*Bufo* sp., Holman and Richards 1993).

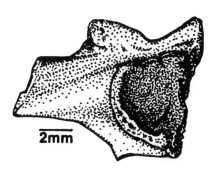

FIGURE 41. Left ilium of Fowler's toad (*Bufo fowleri*) from the Late Wisconsinan Sheriden Pit Cave Site, Wyandot County, Ohio.

Fowler's Toad
Bufo fowleri (fig. 41)

Fowler's toad is a brownish or grayish toad with dark spots on its back. It may be distinguished from the closely related American toad on the basis of having three or more warts within each of these dark spots, whereas the American toad only has one or two warts per spot. Fowler's toad generally occurs in more sandy areas than does the American toad. Like the American toad, Fowler's toad feeds mainly on insects and other small invertebrate species.

Fowler's toad occurs in Canada only in southern Ontario, north of Lake Erie. It occurs in about the eastern third of the United States south to northern Florida. Fossils of Fowler's toad have been found in the Late Wisconsinan of northwestern Ohio. Fowler's toad would have been a small insectivore during the Pleistocene of the Great Lakes region and would have been eaten by small predators, such as garter snakes and hognose snakes that are resistant to the toxic substances in toad skins.

Site. Sheriden Pit Cave, Wyandot County, Ohio, Rancholabrean LMA–Late Wisconsinan (Holman 1997).

TREEFROGS
Family Hylidae

Treefrogs are slender frogs with long legs and depressed (flattened) bodies (see fig. 37). Most have adhesive, circular disks on the fingers and toes. Treefrogs occur in most tropical and temperate regions of the world, with the exception of most of Africa, southern India, and a large part of southern Australia. The family is first known from the Late Eocene of western Canada.

Striped Chorus Frog
Pseudacris triseriata

The tiny striped chorus frog is about 1 inch long and normally has three dark stripes down the back and small finger and toe disks. Striped chorus frogs are seldom seen outside of the breeding season in the spring when they are vocalizing, usually in grassy areas near or in pools and marshes. The rest of the year this species tends to be very secretive. They are able to exist in some surprisingly dry habitats. These animals feed upon small insects and other invertebrates.

Striped chorus frogs range from southern Ontario and New York to the Gulf of Mexico in the eastern part of North America and from Arizona to near the Arctic Circle in the West. The striped chorus frog has been identified from the Late Wisconsinan of northwestern Ohio. This species was a predator on very small insects and other invertebrates during the Pleistocene and was eaten by a large variety of small predators.

Site. Sheriden Pit Cave, Wyandot County, Ohio, Rancholabrean LMA–Late Wisconsinan (Holman 1997).

TRUE FROGS
Family Ranidae

True frogs are usually green or brownish frogs with slim, streamlined bodies and a tapered head (fig. 42). The eardrum tends to be large, the legs long, and the toes webbed (at least in the North American species). True frogs occur in most regions of the world, with the exception of southern South America, most of Africa, most of Australia, and most truly boreal areas, although some species range north of the Arctic Circle. The family occurs as early as the Eocene of Europe, where the modern genus, *Rana*, is found. The family may possibly have occurred as early as the Late Cretaceous in North America.

Bullfrog
Rana catesbeiana

Bullfrogs are large anurans with a body length that may exceed 6 inches. They are greenish frogs that lack the ridges (dorsolateral folds) on the upper sides of the body that distinguish some other species of the genus. Bullfrogs generally prefer permanent bodies of water and spend much of their time sitting near the edge of the water waiting for their prey to come to them. Bullfrogs will eat almost any moving animal they can swallow, and birds and bats have been recorded in their diet.

Bullfrogs occur in southeastern Canada and about the eastern half of the United States, but they have been widely introduced into other areas both in North America and abroad. Bullfrogs have been recorded from the Late Wisconsinan of southeastern Michigan, northwestern Ohio, and southwestern Indiana. Bullfrogs were small predators during the Pleistocene, and smaller bullfrogs probably fell prey to snakes, herons, raccoons, and other predators.

Sites. Shelton Mastodont, Oakland County, Michigan, Rancholabrean LMA–Late Wisconsinan (DeFauw and Shoshani 1991). Sheriden Pit Cave, Wyandot County, Ohio, Rancholabrean LMA–Late Wisconsinan (Holman 1997). Prairie Creek D, Daviess County, Indiana, Rancholabrean LMA–Late Wisconsinan (Holman and Richards 1993).

Green Frog
Rana clamitans (fig. 42)

FIGURE 42. Green frog (*Rana clamitans*).

The green frog is a large greenish or greenish brown frog that is often mistaken for the bullfrog, but the green frog differs in having dorsolateral folds. Male green frogs have bright yellow chins. This species may grow to about 4 inches in body length, but most individuals are about 3 inches long. Green frogs occur along borders of streams, ponds, and lakes and, like the bullfrog, will eat about any other animal species they can swallow. Green frogs are attracted by the calls of other smaller frogs, which they hop up to and eat.

Green frogs occur in southeastern Canada north to Newfoundland in the East and occur in about the eastern third of the United States from Maine and Minnesota south to southern Florida and eastern Texas. Green frogs have been reported from the Late Wisconsinan of southeastern Michigan, northwestern Ohio, and southwestern Indiana. This species would have been an important small predator of invertebrates and small vertebrates during the Pleistocene and would have been fed upon by predators such as snakes, birds, and small mammals.

Sites. Shelton Mastodont, Oakland County, Michigan, Rancholabrean LMA–Late Wisconsinan (Defauw and Shoshani 1991). Sheriden Pit Cave, Wyandont County, Ohio, Rancholabreran LMA–Late Wisconsinan (Holman 1997). Prairie Creek D, Daviess County, Indiana, Rancholabrean LMA–Late Wisconsinan (Holman and Richards 1993).

Northern Leopard Frog
Rana pipiens (fig. 43)

Northern leopard frogs are greenish or brownish frogs with dark spots. They have prominent dorsolateral folds. Leopard frogs are less aquatic than many true frogs, ranging far into moist meadows in the summertime. Thus, they may be locally called grass or meadow frogs. Leopard frogs eat a wide variety of small invertebrate species.

Northern leopard frogs range widely through Canada and the northern United States, mainly south of boreal regions, and they occur as far south as northern New Mexico and Arizona in the West. Northern leopard frogs have been reported from the Late Wisconsinan of northwestern Ohio. Frogs that represent the leopard frog complex of species, but not necessarily the northern leopard frog, have been reported from the Late Wisconsinan of northwestern, east-central, and southwestern Indiana. Frog remains identified only as *Rana* sp. from the Late Wisconsinan of a site near

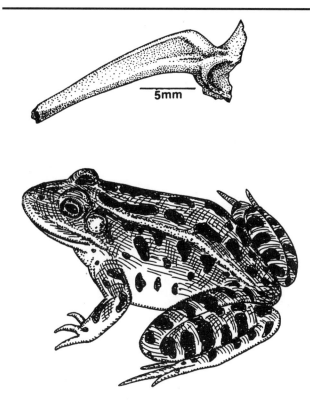

FIGURE 43. Upper, left ilium of leopard frog (*Rana pipiens*) from the Late Wisconsinan Sheriden Pit Cave Site, Wyandot County, Ohio. Lower, leopard frog.

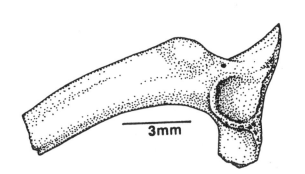

FIGURE 44. Left ilium of a wood frog (*Rana sylvatica*) from the Late Wisconsinan Sheriden Pit Cave Site, Wyandot County, Ohio.

Toronto, Ontario; from east-central Indiana; and from the Late Wisconsinan of extreme southwestern Wisconsin may represent the northern leopard frog. Leopard frogs would have played the role of a predator of invertebrate species during the Pleistocene, and leopard frogs would have been eaten by several kinds of snakes, birds, and small mammals.

Sites. Sheriden Pit Cave, Wyandot County, Ohio, Rancholabrean LMA–Late Wisconsinan (Holman 1997). Kolarik Mastodont, Starke County, Indiana, Rancholabrean LMA–Late Wisconsinan (*Rana pipiens* complex, Holman 1995b). Christensen Bog, Hancock County, Indiana, Rancholabrean LMA–Late Wisconsinan (*Rana pipiens* complex, Graham et al. 1983). Prairie Creek D, Daviess County, Indiana, Rancholabrean LMA–Late Wisconsinan (*Rana pipiens* complex, Holman and Richards 1993). Glacial Lake Iroquois near Toronto, Ontario, Rancholabrean LMA–Late Wisconsinan (*Rana* sp., Churcher and Peterson 1982). Dollens Mastodont, Madison County, Indiana, Rancholabrean LMA–Late Wisconsinan (*Rana* sp., Richards et al. 1987). Moscow Fissure, Iowa County, Wisconsin, Rancholabrean LMA–Late Wisconsinan (*Rana* sp., Foley 1984).

Wood Frog
Rana sylvatica (fig. 44)

The wood frog is a small anuran of the woodland floor with a light brown body and a black mask running through the eye and over the top of the mouth. Wood frogs are usually about 2 inches long. They are very cold-tolerant and in fact may freeze solid in the winter with no ill effects. These frogs are among the first amphibians to emerge in the spring, making the northern woods resound with their ducklike, quacking calls. Wood frogs feed on small invertebrates of the forest floor.

Wood frogs range farther north than any other North American amphibian or reptile, occurring across the northern part of the continent from Labrador to Alaska and penetrating southward in the United States to the southern Appalachian region. Wood frogs were probably among the first, if not the first, amphibian or reptile species to have reoccupied deglaciated areas in the postglacial Pleistocene.

This species has been identified from the Late Wisconsinan of northwestern Ohio and southwestern Indiana. Wood frogs would have been important predators of invertebrates of the woodland floor during the Pleistocene and would have been the prey for a large variety of predators from larger frogs to small mammals.

Sites. Sheriden Pit Cave, Wyandont County, Ohio, Rancholabrean LMA–Late Wisconsinan (Holman 1997). Prairie Creek D, Daviess County, Indiana, Rancholabrean LMA–Late Wisconsinan (Holman and Richards 1993).

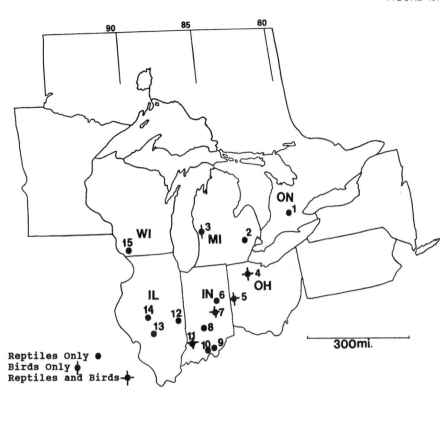

FIGURE 45. Pleistocene reptile and bird localities in the Great Lakes region. 1. Innerkip Site, Oxford County, Ontario, Late Sangamonian to Early Wisconsinan. 2. New Hudson Mastodont Site, Oakland County, Michigan, Late Wisconsinan. 3. Casnovia Duck Site, Muskegon County, Michigan, Late Wisconsinan. 4. Sheriden Pit Cave, Wyandot County, Ohio, Late Wisconsinan. 5. Carter Site, Darke County, Ohio, Late Wisconsinan. 6. Dollens Mastodont Site, Madison County, Indiana, Late Wisconsinan. 7. Christensen Bog Mastodont Site, Hancock County, Indiana, Late Wisconsinan. 8. Harrodsburg Crevice, Monroe County, Indiana, Sangamonian. 9. King Leo Pit Cave, Harrison County, Indiana, Late Wisconsinan. 10. Megenity Peccary Cave, Crawford County, Indiana, Sangamonian stratum. 11. Prairie Creek D Site, Davies County, Indiana, Late Wisconsinan. 12. Polecat Creek Site, Coles County, Illinois, Late Wisconsinan. 13. Hopwood Farm Site, Montogomery County, Illinois, Sangamonian. 14. Clear Lake Site, Sangamon County, Illinois, Late Wisconsinan. 15. Moscow Fissure, Iowa County, Wisconsin, Late Wisconsinan.

REPTILES
Class Reptilia

Reptiles are moderately well represented in the Pleistocene of the Great Lakes region, as at least twenty species have been identified. Probably more reptiles would have been recognized from the Pleistocene of the Great Lakes region if not for the fact that small forms such as lizards and snakes have tiny bones that may have been overlooked by paleontological collectors. In general, Great Lakes region Pleistocene reptiles represent rather cold-tolerant species.

TURTLES
Order Testudines

A turtle is tough and it's stout,
for its skeleton's turned inside out.

When the dinos appeared all the turtle clan jeered
"What's this oddball new group all about?"

Turtles are easy to recognize because of their bony (sometimes leathery) shells (fig. 46). Turtles are the most primitive living reptiles and indeed are very near the ancestral stock of all reptiles, fossil and living. Turtles evolved millions of years before the dinosaurs and happily are still with us 65 million years after the dinosaurs became extinct. Many people believe turtles enrich our lives more than any other reptile group. They have been symbols of longevity and sagacity in many human cultures and have sometimes been considered to be sacred. Five families of turtles—the mud and musk turtles (Kinosternidae), snapping turtles (Chelydridae), New World pond turtles (Emydidae), tortoises (Testudinidae), and softshell turtles (Trionychidae)—have been identified from the Pleistocene of the Great Lakes region. Pleistocene turtles are most often identified on the basis of shell bones.

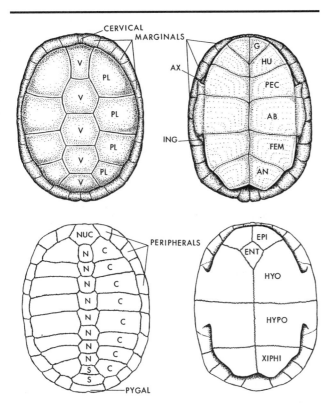

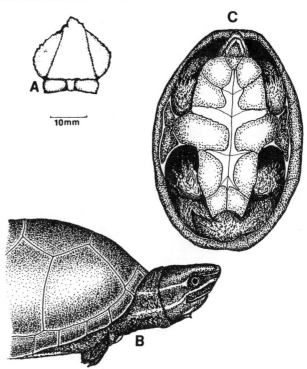

FIGURE 46. Terminology of turtle shell parts. The upper shell (two left views) of a turtle is called the carapace, the lower shell (two right views) is called the plastron. Upper left, dorsal view of epidermal scutes; upper right, ventral view of epidermal scutes. Lower left, dorsal view of bony plates; lower right, ventral view of bony plates. Abbreviations: V, vertebral scutes; PL, pleural scutes; AX, axillary scute; ING, inguinal scute; G, gular scute; HU, humeral scute; PEC, pectoral scute; AB, abdominal scute; FEM, femoral scute; AN, anal scute; NUC, nuchal bone; N, neural bones; S, suprapygal bones; EPI, epiplastral bone; ENT, entoplastral bone; HYO, hyoplastral bone; HYPO, hypoplastral bone; XIPHI, xiphiplastral bone.

FIGURE 47. A, nuchal bone in dorsal view of common musk turtle (*Sternotherus odoratus*) from the Late Wisconsinan Prairie Creek B Holocene Site, Daviess County, Indiana. B, lateral view of common musk turtle. C, ventral view of common musk turtle.

MUD AND MUSK TURTLES
Family Kinosternidae

The family Kinosternidae is composed of small, rather drab species with blackish, brownish, or olive-colored shells. They have large heads and small, fleshy barbels on their chins. Kinosternids are aquatic animals that walk along the bottom of quiet aquatic situations. Some species have two hinges in their plastron that allow them to close the shell around their entire body. The family occurs in eastern southern Canada through about the eastern half of the United States south to Ar-

gentina. Kinosternids are first known from the Eocene of North America. Only one genus, *Sternotherus*, is known from the Pleistocene of the Great Lakes region.

Common Musk Turtle
Sternotherus odoratus (fig. 47)

The little common musk turtle has a dull brownish or blackish shell, a large head with two stripes down the side, and barbels on the chin and throat. These animals have a strong odor, which is derived from musk secreted by a pair of glands in the skin near the junction of the upper and lower parts of the shell. The shell of this species averages about 3.5 to 4 inches long in adult specimens. Common musk turtles walk slowly along the bottom of ponds, lakes, and very slow-moving streams, looking for plants, small invertebrates, or even dead material. They will take small minnows or fishes if they are able to catch them. All baby turtles are tiny, have soft shells, and are eaten by predators ranging from giant water bugs and crayfishes to fishes and birds. Newborn musk turtles are among the smallest baby turtles in North America, having a shell a little more than .5 inch long.

The common musk turtle ranges from Maine and southern Ontario southward through Florida and southern Texas. This species has been identified from the Late Wisconsinan of southwestern Indiana. The common musk turtle would have been a bottom-dwelling omnivore and scavenger of quiet water situations during the Pleistocene.

Site. Prairie Creek D, Daviess County, Indiana, Rancholabrean LMA–Late Wisconsinan (Holman and Richards 1993).

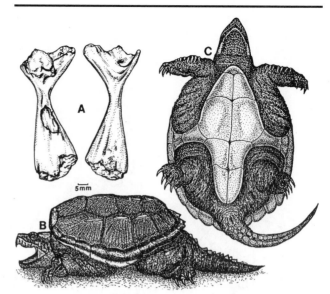

FIGURE 48. A, anterior (left) and posterior (right) views of the left humerus of the common snapping turtle (*Chelydra serpentina*) from the Late Wisconsinan Christensen Bog Mastodont Site, Hancock County, Indiana. B, lateral view of the common snapping turtle. C, ventral view of the common snapping turtle.

SNAPPING TURTLES
Family Chelydridae

Snapping turtles are big turtles with dull blackish or brownish shells. They have very large heads with hooked beaks, a very small lower shell, and a long, saw-toothed tail. On land they snap viciously at anything that approaches them, but they are very reluctant to bite under water. Snapping turtles eat about any animals they can overpower but evidently eat quite a bit of plant material as well. There are only two genera and two species of snapping turtles in the family Chelydridae, and their composite range is from about the eastern two-thirds of southern Canada south through the United States to northwestern South America. The family is first known from the early Paleocene of North America. Only one genus, *Chelydra*, is known from the Pleistocene of the Great Lakes region.

Snapping Turtle
Chelydra serpentina (fig. 48)

The snapping turtle, or snapper, is hard to mistake for any other turtle. It has a large head with a hooked beak; a dull-colored, flattened shell with three ridges on top; and a long, saw-toothed tail. Individuals that one usually sees have a shell that ranges from about 8 to 10 inches in length, but occasionally very large individuals that are 18 or 19 inches long and that weigh over 30 pounds are encountered. Snappers are able to live in about any kind of freshwater situation and sometimes enter brackish water. They are bottom crawlers that eat about any animal they can catch and subdue as well as some plant material.

This species occurs from about the eastern two-thirds of southern Canada through the United States to northwestern South America. Pleistocene fossils of the snapping turtle have been found in the Late Wisconsinan of northwestern, central, and west-central Ohio; east-central and southwestern Indiana; and east-central and central Illinois. Snapping turtles would have been a predator to be reckoned with in many aquatic situations during the Pleistocene. Adult snappers would have had few predators.

Sites. Sheriden Pit Cave, Wyandot County, Ohio, Rancholabrean LMA–Late Wisconsinan (Holman 1997). Johnston Mastodont, Licking County, Ohio, Rancholabrean LMA–Late Wisconsinan (Hansen 1992). Carter, Darke County, Ohio, Rancholabrean LMA–Late Wisconsinan (McDonald 1994). Dollens Mastodont, Madison County, Indiana, Rancholabrean LMA–Late Wisconsinan (Richards et al. 1987). Christensen Bog, Hancock County, Indiana, Rancholabrean LMA–Late Wisconsinan (Graham et al. 1983). Prairie Creek D, Daviess County, Indiana, Rancholabrean LMA–Late Wisconsinan (Holman and Richards 1993). Clear Lake, Sangamon County, Illinois, Rancholabrean

LMA–Late Wisconsinan (Holman 1966). Polecat Creek, Coles County, Illinois, Rancholabrean LMA–Late Wisconsinan (Preston 1979; Holman 1995b).

NEW WORLD POND TURTLES
Family Emydidae

The so-called New World pond turtles range from terrestrial species to very aquatic ones. The shells of the emydids are all well–developed, and some emydids have a hinge on the plastron that allows the shell to completely enclose the body. With the exception of one Old World genus, *Emys*, emydids occur from Canada to central South America. All of the emydid turtles have a single central bone in the front part of the lower shell called the entoplastron (see fig. 46), which they share with the Old World pond turtle family Bataguridae and with the land tortoise family Testudinidae. Both the Old World and New World pond turtles differ from tortoises in that they lack the stumpy, elephant-like hind legs of the Testudinidae. The family is first known from the Eocene.

Painted Turtle
Chrysemys picta (fig. 49)

The painted turtle is the most colorful turtle in North America, and many people think it is the most beautiful. It has a smooth, streamlined, blackish or olive-colored shell with attractive red marks around the edges. There are also bright red marks on the legs and bright yellow stripes and spots on the head. The shell ranges from about 5 to 6 inches long in average-sized specimens in the Great Lakes region, but occasionally specimens reach 6.5 to about 7 inches in length. Painted turtles prefer quiet, shallow water where there is plenty of aquatic vegetation. The species is omnivorous and feeds upon aquatic plants, insects, small mollusks, and on occasional worms or small fish. Painted turtles often bask on logs or other objects that project from the water, especially in the spring and late fall when the water is cold. These turtles are very cold-tolerant and probably were among the first reptile species to reinvade previously glaciated regions during the Pleistocene. The young are able to overwinter in the nest above the frost line and, like the wood frog, can often tolerate freezing solid.

Painted turtles occur from southern Canada south

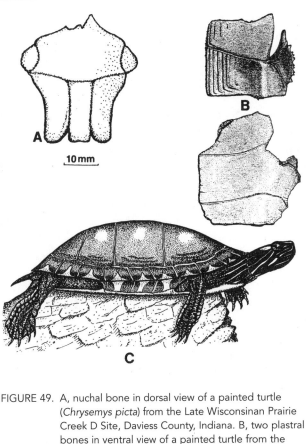

FIGURE 49. A, nuchal bone in dorsal view of a painted turtle (*Chrysemys picta*) from the Late Wisconsinan Prairie Creek D Site, Daviess County, Indiana. B, two plastral bones in ventral view of a painted turtle from the Late Wisconsinan New Hudson Mastodont Site, Oakland County, Michigan (upper, right hypoplastron with annual rings; lower, left hyoplastron). C, painted turtle in lateral view.

to the Gulf Coast in Louisiana. They are absent from the deep southeastern United States and are distributed discontinuously in the West. In the Pleistocene of the Great Lakes region, painted turtles occurred in the Late Wisconsinan of southeastern Michigan, northwest and west-central Ohio, east-central and southwestern Indiana, and east-central Illinois. Undoubtedly, painted turtles were important small omnivores of quiet, well-vegetated, shallow bodies of water during the Pleistocene.

Sites. New Hudson Mastodont, Oakland County, Michigan, Rancholabrean LMA–Late Wisconsinan (Holman and Fisher 1993). Sheriden Pit Cave, Wyandot County, Ohio, Rancholabrean LMA–Late Wisconsinan (Holman 1997). Carter, Darke County, Ohio, Rancholabrean LMA–Late Wisconsinan (Holman 1986). Christensen Bog, Hancock County, Indiana, Rancholabrean LMA–Late Wisconsinan (Graham et al.

1983). Prairie Creek D, Daviess County, Indiana, Rancholabrean LMA–Late Wisconsinan (Holman and Richards 1993). Polecat Creek, Coles County, Illinois, Rancholabrean LMA–Late Wisconsinan (Holman 1995b).

Blanding's Turtle
Emydoidea blandingii (fig. 50)

Blanding's turtles are unique North American turtles with oblong, dark shells with numerous light dots and flecks; a flattened head with bulging eyes on the end of a very long neck; and a bright yellow chin and throat. A somewhat movable hinge is found on the lower shell, but this turtle seldom retreats within the shell like the true box turtles. Blanding's turtles commonly reach a shell length of 7 or 8 inches. These turtles prefer quiet, shallow, well-vegetated water. Blanding's turtles' long necks dart out very quickly at the same time their jaws gape and suck in crayfishes, fishes, or other small animals.

Blanding's turtles occur discontinuously from Nova Scotia to central Nebraska but are most common in the Great Lakes region. These turtles have been reported from a Late Sangamonian/Early Wisconsinan site in southern Ontario and in the Late Wisconsinan of northwestern Ohio and southwestern Indiana. Blanding's turtles had a much wider distribution during the Pleistocene than they do today and occurred as far south as northern Mississippi. These turtles would have been an effective small predator of the marshes and shallow ponds of the Pleistocene of the Great Lakes region.

Sites. Innerkip, Oxford County, Ontario, Rancholabrean LMA–Late Sangamonian/Early Wisconsinan (Churcher et al. 1990). Sheriden Pit Cave, Wyandot County, Ohio, Rancholabrean LMA–Late Wisconsinan (Holman 1997). Prairie Creek D, Daviess County, Indiana, Rancholabrean LMA–Late Wisconsinan (Holman and Richards 1993).

Map Turtle
Graptemys sp. (fig. 51)

Map turtles are large, shy turtles of lakes and rivers. All have a dorsal ridge or keel at the top of the shell. Females may reach 8 or 10 inches in shell length in this genus, but males are often much smaller. Map turtles tend to feed upon crayfishes and mollusks. Three species of map turtles occur today in the Great Lakes

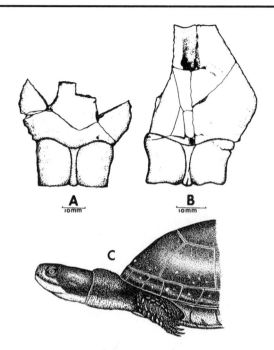

FIGURE 50. A and B, two nuchal bones in dorsal view of a Blanding's turtle (*Emydoidea blandingii*) from the Late Wisconsinan Prairie Creek D Site, Daviess County, Indiana. C, Blanding's turtle in lateral view.

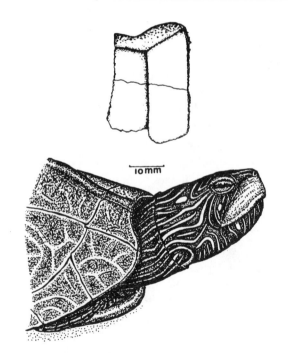

FIGURE 51. Upper, peripheral bone in dorsal view of a map turtle (*Graptemys* sp.) from the Late Wisconsinan Prairie Creek D Site, Daviess County, Indiana. Lower, lateral view of common map turtle (*Graptemys geographica*).

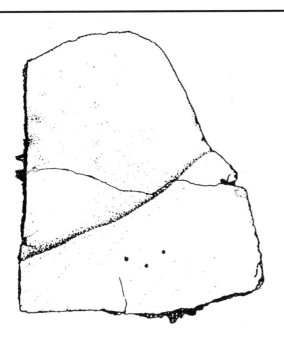

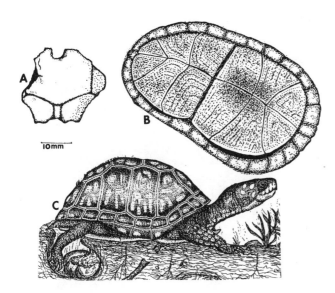

FIGURE 52. Right xiphiplastron in ventral view of a cooter (*Pseudemys* sp.) from the Late Wisconsinan Prairie Creek D Site, Daviess County, Indiana.

10mm

FIGURE 53. A, nuchal bone in dorsal view of an eastern box turtle (*Terrapene carolina*) from the Holocene Prairie Creek B Site, Daviess County, Indiana. B, upside-down closed shell of eastern box turtle. C, eastern box turtle digging nest.

region, but Pleistocene map turtle bones in the region have only been identified to the generic level. A single record of *Graptemys* sp. is known from a Late Wisconsinan site in southwestern Indiana.

Site. Prairie Creek D, Daviess County, Indiana, Rancholabrean LMA–Late Wisconsinan (Holman and Richards 1993).

Cooter
Pseudemys sp. (fig. 52)

Cooters are large, streamlined, river turtles that are fond of basking. Although bones from the Late Wisconsinan of southwestern Indiana were identified only as *Pseudemys* sp., it is likely that the bones represent the river cooter (*Pseudemys concinna*), as that is the only species of cooter that presently lives in the Great Lakes region. All forms of river cooters have circular marks on some parts of the upper shell (carapace). Adults often have a shell about 12 inches long. The staple diet of adult river cooters appears to be aquatic plants, but they occasionally eat invertebrates and small vertebrates and may be scavengers from time to time.

Site. Prairie Creek D, Daviess County, Indiana, Rancholabrean LMA–Late Wisconsinan (*Pseudemys* sp., Holman and Richards 1993).

Eastern Box Turtle
Terrapene carolina (fig. 53)

Box turtles are medium-sized turtles that may be easily distinguished from other North American turtles in that they may completely close the shell over all of the parts of the body by means of a single hinge in the middle of the plastron. The eastern box turtle (*Terrapene carolina*) may be distinguished from the more westerly distributed ornate box turtle (*Terrapene ornata*) on the basis of the keel that runs down the middle of the carapace in the eastern species and is lacking in the western one. Eastern box turtles spend most of their time on land, but they occasionally soak in shallow water for rather extended times in hot, dry weather. These turtles are omnivorous, feeding on a wide variety of plant and animal material on the woodland floor. Eastern box turtles favor rather open hardwood forest habitats.

At present, eastern box turtles occur from southern New Hampshire, northwestern Lower Michigan, and central Illinois south through Florida and west through eastern Kansas, Oklahoma, and Texas and then well south into Mexico as a separate subspecies. In the Pleistocene of the Great Lakes region, the eastern box turtle has been reported from the Late Wisconsinan of east-central Illinois. Moreover, turtle remains tentatively referred to the genus *Terrapene* from the Sangamonian of south-central Indiana probably represent the eastern box turtle. Eastern box turtles would have been important omnivores of open hardwood forests in the southern part of the Great Lakes region during the Pleistocene.

Sites. Harrodsburg Crevice, Monroe County, Indiana, Rancholabrean LMA–Sangamonian (tentatively referred to *Terrapene*, Parmalee et al. 1978). Polecat Creek, Coles County, Illinois, Rancholabrean LMA–Late Wisconsinan (Holman 1995b).

Slider
Trachemys scripta (fig. 54)
Sliders are medium-sized to large basking turtles with a prominent patch of red, orange, or yellow on each side

FIGURE 54. Upper, nuchal bone in dorsal view of a slider turtle (*Trachemys scripta*) from the Holocene Prairie Creek B Site, Daviess County, Indiana. Lower, juvenile slider turtle.

of the head. The only slider that presently occurs in the Great Lakes region has a red patch on the side of the head and is named the red-eared slider (*Trachemys scripta elegans*); thus the fossils reported herein probably represent this subspecies. Sliders also tend to have yellow stripes on the head, neck, and legs. Most sliders found in the Great Lakes region today are about 6 to 8 inches long. Sliders prefer quiet waters with lots of vegetation, soft bottoms, and floating logs or other emergent objects to bask upon. Baby sliders (fig. 54) are mainly carnivorous, but the adults are omnivores, feeding upon aquatic plants and small animals such as snails, insects, and minnows.

Sliders occur from Maryland, northern Indiana and Illinois, and Kansas south through northern Florida and southern Texas into northern Mexico. Sliders have been found in the Late Wisconsinan of central and east-central Illinois in the Great Lakes region. The slider would have been a significant omnivore of quiet ponds and lakes during the Pleistocene of the Great Lakes region.

Sites. Clear Lake, Sangamon County, Illinois, Rancholabrean LMA–Late Wisconsinan (Holman 1966). Polecat Creek, Coles County, Illinois, Rancholabrean LMA–Late Wisconsinan (Holman 1995b).

TORTOISES
Family Testudinidae

Tortoises are terrestrial reptiles that have a shell with the same plastral (lower shell) bones (including the entoplastron; fig. 46) as the Old World and New World pond turtles, but unlike the pond turtles, tortoises have stumpy, elephant-like hind legs. Modern tortoises range in size from species that weigh only a few ounces to the giant tortoise of the Galapagos and Aldabra Islands that weigh up to 500 pounds. Tortoises tend to live in rather arid habitats, and some, including the gopher tortoises of the United States, make extensive burrows. Tortoises occur mainly in subtropical and tropical areas and are found on all continents except Australia and Antarctica. The family is first known from the Eocene. The only tortoises that presently occur in North America live in the southeastern and southwestern United States and in Mexico. But in the Pleistocene, extinct giant tortoises ranged as far north as Pennsylvania, Indiana, and Illinois during interglacial times.

FIGURE 55. Upper, right xiphiplastron in dorsal view of a *giant land tortoise (*Hesperotestudo crassiscutata) from the Sangamonian Hopwood Farm Site, Montgomery County, Illinois. Lower, a giant land tortoise in a dry habitat.

*Giant Land Tortoise
*Hesperotestudo crassiscutata (fig. 55)

This is the first extinct animal I have documented in this chapter. Extinct vertebrates are prefixed with an * before both the common and the scientific name the first time they are referred to in the text, figures, and tables. All of the Great Lakes region fossil fishes, amphibians, and reptiles discussed previously in this chapter represent species presently living in the Great Lakes region. We shall not encounter extinct Ice Age vertebrates again until we reach the mammalian section of the chapter.

Giant tortoises (fig. 55) occurred in the United States during the Pleistocene but never reached Canada during the Ice Age. These reptiles occurred in both glacial and interglacial ages in southern states such as Florida and Texas but extended their ranges northward during interglacial times and then withdrew again in glacial times.

Giant tortoises are quite significant members of Pleistocene faunas in that they indicate that warm winters prevailed in the areas where the tortoises occurred. C.W. Hibbard pointed out that these large, cold-blooded (ectothermic) animals would not have been able to hibernate and thus would have not been able to exist in areas where temperatures dipped below freezing in the winter (Hibbard's Rule).

The giant tortoise Hesperotestudo crassiscutata has been referred to by the generic name Geochelone for many years, but vertebrate paleontologists have recently changed the name to Hesperotestudo. Hesperotestudo crassicuttata reached a very large size, and shells have been found in Florida and Texas that indicate that some individuals weighed several hundred pounds. The geographic range of this huge species was once considered to be from Florida to South Carolina and west to eastern Texas, but material that represents this species has been found recently in more northern states, including south-central Illinois and southern Indiana in the Great Lakes region.

In the Great Lakes region, Hesperotestudo crassiscutata was first unearthed from a Sangamonian site in south-central Illinois. These remains indicate that the area had much warmer winters than it has at present. Remains of the giant land tortoise were also discovered in a Pleistocene cave in southern Indiana of probable Sangamonian age. These huge reptiles undoubtedly were important Pleistocene herbivores. The implications of the presence of giant tortoises in the Great Lakes region will be discussed in more detail later in the book.

Sites. Hopwood Farm, Montgomery County, Illinois, Rancholabrean LMA–Sangamonian (King and Saunders 1986). Megenity Peccary Cave, Crawford County, Indiana, probable Sangamonian level (Richards and Whitaker 1997).

SOFTSHELL TURTLES
Family Trionychidae

Softshell turtles are highly aquatic reptiles with a soft, leathery shell; a long, retractable neck; and an elongate, snorkel-like nose (fig. 56). Softshells range in shell size from less than 1 foot long to over 3 feet in length in the African softshell. These reptiles occur in all kinds of

Spiny Softshell Turtle
Apalone spinifer (fig. 56)

The spiny softshell has large eyespots on its olive-colored shell whose carapace has a sandpapery feel. Most individuals of this species seen in the Great Lakes region have a shell about 8 to 12 inches long, but individuals 14 inches long may occasionally be found. The spiny softshell occurs in soft bottom or sandy bottom reaches of streams, rivers, ponds, and lakes. The favorite food of spiny softshells appears to be crayfish, but they are known to pursue and eat other animals, including both invertebrates and small vertebrates.

This species occurs from southeastern Canada and the midwestern United States south to northern Florida and through Texas into northern Mexico. This species occurs in the Late Wisconsinan of the south-central part of the Lower Peninsula of Michigan, southwestern Indiana, and central Illinois. *Apalone* sp. is reported from the Late Wisconsinan of east-central Indiana and may represent the spiny softshell, a species that would have been an important aquatic carnivore in the Pleistocene.

Sites. Heisler Mastodont, Calhoun County, Michigan, Rancholabrean LMA–Late Wisconsinan (Holman and Fisher 1993). Prairie Creek D, Daviess County, Indiana, Rancholabrean LMA–Late Wisconsinan (Holman and Richards 1993). Clear Lake, Sangamon County, Illinois, Rancholabrean LMA–Late Wisconsinan (Holman 1966). Christensen Bog, Hancock County, Indiana, Rancholabrean LMA–Late Wisconsinan (*Apalone* sp., Graham et al. 1983).

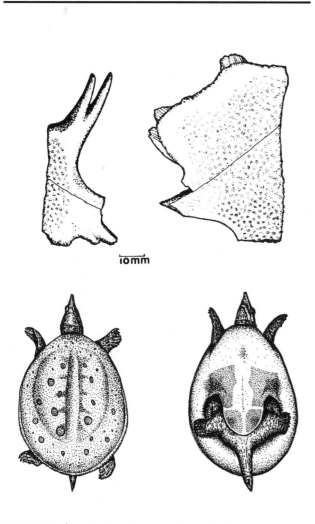

FIGURE 56. Upper left and right, portions of left hypoplastra of a spiny softshell turtle (*Apalone spinifer*) from the Late Wisconsinan Prairie Creek D Site, Daviess County, Indiana. Lower left, spiny softshell turtle in dorsal view; lower right, spiny softshell turtle in ventral view.

aquatic situations, from small ponds and streams to large lakes and rivers, but they prefer soft or sandy bottoms because their soft shells are easily bruised. The trionychids are mainly carnivores, and they are able to swim rapidly after very active prey. Softshells are widely distributed in temperate and tropical areas of North America, Africa, and Asia. Softshells are known from the Cretaceous, where they were contemporary with the dinosaurs.

LIZARDS
Order Squamata
Suborder Sauria

Lizards are scaly reptiles, most of which have legs and movable eyelids (fig. 57). Lizards occur on all continents except Antarctica but are much more abundant in tropical than in temperate regions. Fossil lizards are usually identified on the basis of dentary bones (lower jaw bones with teeth) or, in the case of fossil glass lizards (Family Anguidae), on the basis of individual trunk vertebrae (fig. 58).

ANGUID LIZARDS
Family Anguidae

Anguid lizards either have short limbs, very reduced limbs, or no limbs at all. This condition may occur in different species in a single group of closely related genera. They usually have a long, fragile tail. Some forms have a fold of skin along each side of the body, and most have bony plates called osteoderms under the scales. This family is well—represented in Eurasia, occurs in extreme northwest Africa, and has a rather fragmentary distribution in North and South America. The family is first known from the Late Cretaceous, when dinosaurs still existed.

Slender Glass Lizard
Ophisaurus attenuatus

Glass lizards are legless reptiles that most people think are snakes. But unlike snakes, they have eyelids, bony plates under the scales, and a prominent fold of skin along each side of the body. The tail of these lizards is longer than the body, and when a predator or a curious human grabs the animal, the tail often breaks off into several squirming sections while the body quietly crawls away. This makes it look like the animal has literally broken apart. The body then regenerates a new tail, but it is never as long as the original. Slender glass lizards range from about 20 to 30 inches long, although rarely individuals about 40 inches long are encountered. These animals are mainly insectivores, but large ones consume small reptiles, and a 40-inch individual can eat baby mice.

The slender glass lizard exists as two subspecies, one in the southeastern states from Kentucky and Virginia south through Florida and a grassland form that ranges from western Wisconsin and northern Kansas south through Texas. The slender glass lizard has been found in a Late Wisconsinan deposit in a cave in southwestern Indiana. It would have been an insectivore of grassy areas in the Great Lakes region during the Pleistocene.

Site. Mill Creek Cave, Greene County, Indiana, Rancholabrean LMA–Late Wisconsinan (Richards 1984b).

FIGURE 57. Typical lizard (*Anolis*, Iguanidae). Lizards have claws (1), unlike salamanders (see fig. 37).

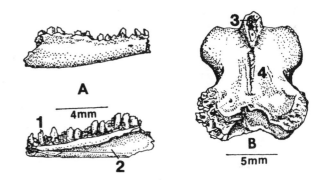

FIGURE 58. Typical Pleistocene lizard fossils (Pleistocene of Texas). A, iguanid lizard right dentary bone: upper, lateral view; lower, medial view. B, Glass lizard (*Ophisaurus*) trunk vertebra in dorsal view. 1, pleurodont tooth; 2, Meckelian groove; 3, neural spine; 4, left side of neural arch.

SNAKES
Suborder Serpentes

Snakes are long, scaly, legless reptiles (fig. 59). They lack eyelids and thus have a characteristic "stare." A transparent plate called a brille protects the eyeball.

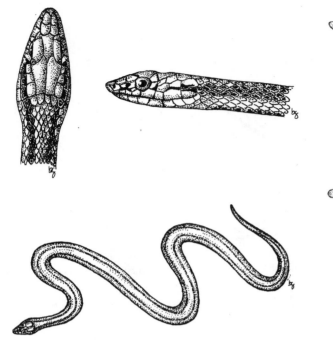

FIGURE 59. Typical advanced colubrid snake (coachwhip, *Masticophis taeniatus*). Upper left, head in dorsal view; upper right, head in lateral view. Lower, entire snake in dorsal view.

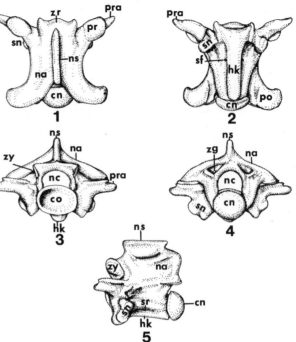

FIGURE 60. Vertebra of a typical advanced colubrid snake (Sonoran mountain kingsnake, *Lampropeltis pyromelana*) in 1, dorsal view; 2, ventral view; 3, anterior view; 4, posterior view; 5, lateral view. Abbreviations: cn, condyle; co, cotyle; hk, hemal keel; na, neural arch; nc, neural canal; ns, neural spine; po, postzygapophyseal articular facet; pr, prezygapophyseal articular facet; pra, prezygapophyseal accessory process; sf, subcentral foramen; sn, synapophysis; sr, subcentral ridge; zg, zygantral articular facet; zr, zygosphenal roof; zy, zygosphenal articular facet.

Snakes occur on all continents except Antarctica and, like lizards, are much more abundant in tropical than in temperate regions. Fossil snakes are usually identified on the basis of individual vertebrae (fig. 60).

COLUBRID SNAKES
Family Colubridae

The family Colubridae comprises a huge, very diverse group of snakes. In fact, most scientists think that the family is an unnatural assemblage that represents several families. Most colubrids have a distinct head with rather well developed eyes and a tapering body (see fig. 59). If venom is present, it is usually the kind that produces a relatively mild reaction in humans, although some humans have been killed by the venom of so-called rear-fang colubrid snakes. The Colubridae occur on all continents except Antarctica, although there are gaps in distribution in North Africa and most of Australia lacks colubrids. Fossil colubrids are first known from the Eocene of Thailand.

Racer
Coluber constrictor (fig. 61)

Racers are large, slender, shiny snakes that are bluish in the northern part of the Great Lakes region and blacker in southern Ohio, Indiana, and Illinois. Racers are nervous snakes that are usually seen in the field rapidly gliding away into low bushes. Most racers observed in the field are about 3.5 to 4 feet long, but 5-

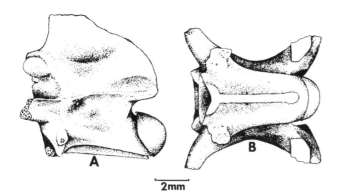

FIGURE 61. Vertebra of a racer (*Coluber constrictor*) from the mixed Pleistocene and Holocene Prairie Creek C Site, Daviess County, Indiana. A, lateral view; B, ventral view.

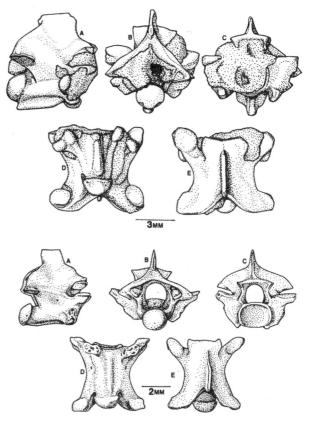

FIGURE 62. Two vertebrae (an upper and a lower group of five views of each) of a fox snake (*Elaphe vulpina*) from the Late Wisconsinan Sheriden Pit Cave Site, Wyandot County, Ohio. A, lateral; B, posterior; C, anterior; D, ventral; E, dorsal.

foot specimens are occasionally encountered. Racers are alert predators that eat both cold-blooded and warm-blooded prey, but frogs and mice seem to be a staple part of the diet. They do not kill their prey by constriction but hold it down with a loop of the body while it is being swallowed.

Racers occur in southern Ontario and throughout a large portion of the United State south into Mexico. Their distribution is spotty in many of the far southwestern states. The racer, presumably the blue racer subspecies, has been identified from the Late Wisconsinan of northwestern Ohio. The racer was undoubtedly an active predator of rather open situations during the Pleistocene of the Great Lakes region.

Sites. Sheriden Pit Cave, Wyandot County, Ohio, Rancholabrean LMA–Late Wisconsinan (Holman 1997). *Coluber constrictor* (fig. 61) is also known from the mixed Pleistocene and Holocene Prairie Creek C Site in Daviess County, Indiana (Holman and Richards 1993).

Fox Snake
Elaphe vulpina (fig. 62)

The fox snake is a large, robust, constricting species with a brownish or reddish head. Its upper body is covered with large, dark brown blotches on a yellowish background. The fox snake often vibrates its tail when disturbed and is sometimes mistaken for a rattlesnake based on this habit or as a copperhead based on its sometimes reddish head. This species often occupies the holes or burrows made by other animals. Fox snakes are predators that feed mainly on small rodents such as mice and voles, which they kill rapidly by constriction.

Fox snakes occur as two subspecies (sometimes considered to be full species). One, *Elaphe vulpina gloydi*, occupies marshy areas around Lake Huron and Lake Erie in Ontario, northern Ohio, and southeastern Michigan; another, *Elaphe vulpina vulpina*, ranges from the western Upper Peninsula of Michigan, to Wisconsin, Illinois, and northern Missouri west to eastern South Dakota and Nebraska. But fox snakes, like Blanding's turtles, had a much wider distribution during the Pleistocene than at present, occurring as far south as northwestern Georgia. Fox snakes have been reported from the Late Wisconsinan of northwestern Ohio and extreme southwestern Wisconsin. The fox snake would have been an important predator on mice and voles and other small rodents during the Pleistocene of the Great Lakes region.

Sites. Sheriden Pit Cave, Wyandot County, Ohio, Rancholabrean LMA–Late Wisconsinan (Holman 1997). Moscow Fissure, Iowa County, Wisconsin, Rancholabrean LMA–Late Wisconsinan (Foley 1984).

Milk Snake
Lampropeltis triangulum

The milk snake is a somewhat slender, medium-sized, constricting species with a light V or Y marking on the back of the top of its head. The back has large brown blotches that alternate with smaller ones on the sides. The background color is usually light grayish. Milk snakes in the Great Lakes region are usually about 30 inches long. The milk snake, like the fox snake, will vibrate its tail when confronted and is often mistaken for a rattlesnake. Milk snakes frequently hibernate in loose foundations around houses or barns and thus are often seen in the spring and the fall when they are entering or emerging from hibernation. Milk snakes feed upon small reptiles and small rodents, which they kill by constriction.

Milk snakes occur in southeastern Canada and are widespread throughout the United States into Mexico. Several United States and Mexican subspecies are recognized. Milk snakes are known from Late Wisconsinan sites in northwestern Ohio and extreme southwestern Wisconsin. Milk snakes would have been predators on small reptiles and small rodents during the Pleistocene of the Great Lakes region.

Sites. Sheriden Pit Cave, Wyandot County, Ohio, Rancholabrean LMA–Late Wisconsinan (Holman 1997). Moscow Fissure, Iowa County, Wisconsin, Rancholabrean LMA–Late Wisconsinan (Foley 1984).

Plain-bellied Water Snake
Nerodia erythrogaster

Plain-bellied water snakes are thick-bodied, rough-scaled snakes that tend to have uniformly colored bellies, or at least bellies that lack the complicated patterns found in other water snakes. *Nerodia erythrogaster* is usually about 3 feet long, but some individuals get to be over 4 feet in length. These aquatic snakes are sometimes mistakenly referred to as water moccasins in the Great Lakes region. Plain-bellied water snakes prefer river bottoms or swampy woodland habitats. They mainly feed on fishes, but amphibians are also an important part of their diet.

Plain-bellied water snakes occur in isolated colonies in northern Ohio, Indiana, and Illinois and then have a broad distribution from more southern parts of the Midwest south to Florida and through southern Texas into Mexico. They are absent from the Appalachian region of the United States and do not occur in Canada. Plain-bellied water snake vertebrae were identified from the Late Wisconsinan of southwestern Indiana. This species would have been an important fish-eating predator of woodland swamps and river bottoms during the Pleistocene of the Great Lakes region.

Site. Prairie Creek D, Daviess County, Indiana, Rancholabrean LMA–Late Wisconsinan (Holman and Richards 1993).

Northern Water Snake
Nerodia sipedon

Northern water snakes are thick-bodied, dull-colored, rough-scaled snakes that have a complicated pattern of marks on their bellies. They are usually about 3 feet long, and 4-foot individuals are now considered rather rare in the Great Lakes region. These aquatic snakes are also mistakenly referred to as water moccasins in the region. When they are not persecuted by humans, northern water snakes occupy a very wide variety of aquatic habitats, including streams, rivers, marshes, swamps, ponds, and lakes. This species mainly feeds upon fishes and amphibians.

Northern water snakes occur in southeastern Canada and from Michigan west through the plains states and south to northern Florida and northern Texas. The species has been identified from the Late Wisconsinan of northwestern Ohio. Northern water snakes undoubtedly occupied a variety of aquatic habitats during the Pleistocene of the Great Lakes region and were a predator upon fishes and amphibians.

Site. Sheriden Pit Cave, Wyandot County, Ohio, Rancholabrean LMA–Late Wisconsinan (Holman 1997).

Smooth Green Snake
Opheodrys vernalis

Smooth green snakes (sometimes given the scientific name *Liochlorophis vernalis*) are beautiful, slender little creatures with a bright green upper body and smooth scales. They are usually about 1 foot long, and 2-foot-long individuals are considered large. These snakes prefer grassy areas where they feed upon small invertebrates such as spiders and insects.

Smooth green snakes have the center of their distribution in southeastern Canada and the Great Lakes region, but there are many disjunct populations in the plains and Rocky Mountain states, and one remote population occurs in south Texas. Smooth green snakes have been found in the Late Wisconsinan of northwestern Ohio and south-central Indiana. Presently, in Indiana, the smooth green snake is restricted to the extreme northwestern part of the state. This little snake would have been a predator upon small spiders and insects in grassy areas during the Pleistocene of the Great Lakes region.

Sites. Sheriden Pit Cave, Wyandot County, Ohio, Rancholabrean LMA–Late Wisconsinan (Holman 1997). Anderson Pit Cave, Monroe County, Indiana, Rancholabrean LMA–Late Wisconsinan (Holman and Richards 1981).

Queen Snake
Regina septemvittata

The queen snake is a small water snake with a slim brown body that has a yellowish stripe along each side. Queen snakes are usually about 20 inches to about 2 feet in length. These reptiles favor small streams with overhanging willow trees. Crayfish are their principal food. Queen snakes occur from southern Ontario and northern Michigan south to northwest Florida and Mississippi. An isolated population occurs in southwest Missouri and northern Arkansas. Queen snake remains occur in the Late Wisconsinan of northwestern Ohio. The queen snake would have been a predator of crayfish in small streams in the Great Lakes region during the Pleistocene.

Site. Sheriden Pit Cave, Wyandot County, Ohio, Rancholabrean LMA–Late Wisconsinan (Holman 1997).

> Mr. Garter, a smartly dressed fellow,
> has pin-stripes of light white to bright yellow.
> But his forked tongue is red, and that gets him
>   killed dead,
> Even though he's disposed to be mellow.

Common Garter Snake
Thamnophis sirtalis (fig. 63)

Common garter snakes are medium-sized, moderately stout-bodied snakes with a light stripe on the lower part of each side of the body and another light stripe

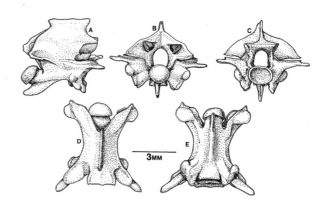

FIGURE 63. Common garter snake *(Thamnophis sirtalis)* vertebra from the Late Wisconsinan Sheriden Pit Cave Site, Wyandot County, Ohio. A, lateral; B, posterior; C, anterior; D, dorsal; E, ventral.

running down the middle of the back. This species has a bright red tongue with a black tip. Sometimes when people see this tongue extruded they think this snake is poisonous and kill it. It is actually harmless. The species is usually about 2 to 2.5 feet long, but individuals 3 feet in length or somewhat longer are sometimes encountered. Garter snakes are by far the most abundant snakes in the Great Lakes region today, and they are found in a wide variety of habitats, even backyards and vacant lots in some urban areas. Garter snakes eat a variety of small invertebrate and vertebrate food, but their ability to survive on an earthworm diet alone probably accounts for their ability to live in some intensely agricultural or urban areas.

This species occurs across provincial Canada and in most of the United States, except from some areas in the Southwest and Far West. Common garter snakes occur in the Late Wisconsinan of northwestern Ohio. Remains designated as *Thamnophis* sp. from the Late Wisconsinan of southwestern Indiana and extreme southwestern Wisconsin could represent this species or the ribbon snake. The common garter snake would have been an important predator of invertebrate and small vertebrate life in a wide variety of habitats during the Pleistocene of the Great Lakes region.

Sites. Sheriden Pit Cave, Wyandot County, Ohio, Rancholabrean LMA–Late Wisconsinan (Holman 1997). Prairie Creek D, Daviess County, Indiana, Rancholabrean LMA–Late Wisconsinan (*Thamnophis* sp., Holman and Richards 1993). Moscow Fissure, Iowa

County, Wisconsin, Rancholabrean LMA–Late Wisconsinan (*Thamnophis* sp., Foley 1984).

VIPERS
Family Viperidae

The family Viperidae includes those poisonous snakes that have a single pair of hollow fangs on a very short upper jaw bone (maxilla) that is able to flip up to bring the fangs forward into a striking position. The vipers are considered to be the most highly evolved group of snakes. The venom of the Viperidae produces serious or fatal reactions in humans. Viperids occur in all continents except Australia and Antarctica. This family is first known from the Miocene of Europe and North America.

Timber Rattlesnake
Crotalus horridus

The timber rattlesnake is a large but relatively slender rattlesnake that, in the northern United States, has dark crossbands on a yellowish, brownish, or blackish background. These snakes are usually about 4 feet long, but huge individuals about 6 feet in length have been recorded. Timber rattlers prefer woodland areas where there are rocky ledges for hibernation. This beautiful species has been eliminated from much of northeastern North America by the deliberate destruction of their hibernating sites by humans. Timber rattlers eat warm-blooded prey such as mice, voles, and the young of larger rodents or rabbits. They kill their prey by injection of venom.

This species occurs in Canada as a small population north of Lake Erie and then through about the eastern third of the United States, except for the Great Lakes Basin proper, and southern Florida. Timber rattler remains have been identified from the Middle Wisconsinan level of a cave in southern Indiana. The timber rattlesnake would have been an efficient predator of small mammals in rocky, wooded areas during the Pleistocene of the Great Lakes region.

Site. Megenity Peccary Cave, Crawford County, Indiana, Rancholabrean LMA–Middle Wisconsinan level (Richards 1990).

BIRDS
Class Aves

Birds are anatomically very similar to the small carnivorous dinosaurs of Mesozoic times. In fact, a few paleontologists have included birds in a class called Dinosauria. But let's get this straight: birds are not dinosaurs. Only eight species of birds are recognized from the Pleistocene of the region. Bird bones are fragile and hollow and unlikely to fossilize (figs. 64 and 65). Moreover, many birds live in trees, and tree-dwelling vertebrates usually occur in uplands away from the lower areas where most fossil deposition occurs. Since fossil sites are often former aquatic situations, it is not surprising that six of the eight Great Lakes region Pleistocene birds are aquatic species.

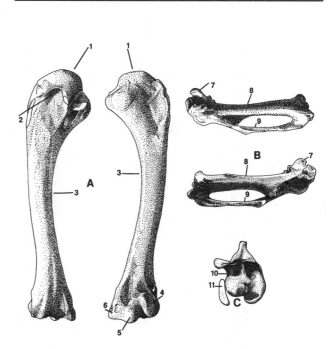

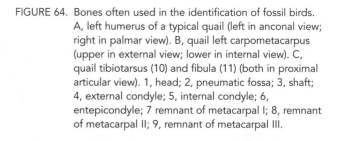

FIGURE 64. Bones often used in the identification of fossil birds. A, left humerus of a typical quail (left in anconal view; right in palmar view). B, quail left carpometacarpus (upper in external view; lower in internal view). C, quail tibiotarsus (10) and fibula (11) (both in proximal articular view). 1, head; 2, pneumatic fossa; 3, shaft; 4, external condyle; 5, internal condyle; 6, entepicondyle; 7 remnant of metacarpal I; 8, remnant of metacarpal II; 9, remnant of metacarpal III.

WATERFOWL
Order Anseriformes

The anseriforms are aquatic or semiaquatic birds, all of which are capable of swimming. Their nests are usually lined with down or feathers. This order has a worldwide distribution. The young are downy at birth and able to get about at once. There are only two families of anseriforms, a small South American family with only two genera and the very large family Anatidae, which is found in the Pleistocene of the Great Lakes region.

DUCKS, GEESE, AND SWANS
Family Anatidae

The family Anatidae comprises the familiar ducks, geese, and swans. They are all excellent fliers and good swimmers and are easily recognized by their ducklike beak. The family has a cosmopolitan distribution. This is an ancient group that is known from the Late Cretaceous.

Two old records of anatids from the "Pleistocene" of Illinois need to be confirmed on the basis of conditions consistent with the modern definition of the Pleistocene. These records are of the trumpeter swan (*Cygnus buccinator*) from Aurora (Wetmore 1935) and the common merganser (*Mergus merganser*) from the North Shore Channel of Chicago (Wetmore 1948).

Northern Shoveler
Anas clypeata

The northern shoveler is a small, surface-feeding duck that is easily recognized by its large, spoon-shaped bill and its habit of skimming for food. Shovelers prefer shallow, open-water ponds, lakes, or marshes, where they eat small invertebrates, plants, and other organic matter. Shovelers breed from western Canada south to Nebraska and Iowa and east to Pennsylvania and Delaware. They spend the winters in open ponds and marshes of more southern areas. Shoveler fossils are known from the Late Wisconsinan of west-central Ohio. It is probable that shovelers bred in west-central Ohio during postglacial Pleistocene times.

Site. Carter, Darke County, Ohio, Rancholabrean LMA–Late Wisconsinan (McDonald 1994).

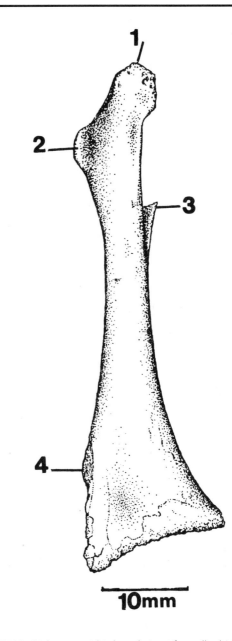

10mm

FIGURE 65. Right coracoid in lateral view of a mallard (*Anas* cf. *platyrhynchos*) from the Harper Site, Shiawassee County, Michigan. 1, furcular facet; 2, glenoid facet; 3, procoracoid process; 4, broken hyposternal process.

Mallard
Anas platyrhynchos (fig. 65)

The mallard, another surface-feeding duck, is the most abundant duck in the Great Lakes region today. The grayish male has a characteristic green head and a narrow white ring around the neck. The female is a mottled, brownish duck that has characteristic dark blue wing patches with white borders. Mallards are medium-sized ducks, a little smaller than the barnyard mallard, a derivative of the wild species. Mallards prefer shallow ponds and marshes but may be found in lakes, streams, and rivers. Mallards tip up in the water to get small invertebrates, plants, and other organic matter.

Mallards occur in most of Eurasia and North America as well as in North Africa and India. Mallard bones have been found in the Late Wisconsinan of west-central Ohio, and fossils that represent either the mallard or the black duck have been found in the Late Wisconsinan of east-central Indiana. Undoubtedly the mallard was part of the breeding bird population of the postglacial Pleistocene in the Great Lakes region.

Sites. Carter, Darke County, Ohio, Rancholabrean LMA–Late Wisconsinan (McDonald 1994). Christensen Bog, Hancock County, Indiana, Rancholabrean LMA–Late Wisconsinan (*Anas* sp., Graham et al. 1983).

Lesser Scaup
Aythya affinis (figs. 66 and 67)

The lesser scaup is a rather small, diving duck with a black head and white wing stripes (fig. 66). This scaup is more common in inland waters than its close relative, the greater scaup, and is mainly seen in ponds, lakes, and marshes as it migrates to its northern breeding grounds from Hudson's Bay to southeastern Ontario. Scaup dive deep in the water to get their food, which consists of small invertebrates, plants, and organic debris.

The lesser scaup migrates south to the Gulf of Mexico to winter. This species was identified on the basis of a left ulna (fig. 67) from a water well site in the west-central part of the Lower Peninsula of that was radiocarbon dated at about 25,000 B.P. Pollen associated with this interstadial site indicated a conifer-dominated plant community; thus it is probable that the lesser scaup were breeding in the area at the time.

Site. Casnovia Duck Site, Muskegon County, Michigan, Rancholabrean LMA–Late Wisconsinan interstadial (Holman 1976; Kapp 1978).

Ring-necked Duck
Aythya collaris

The ring-necked duck is a rather small, diving duck. It has a black back, a pointed head, and two white rings around the bill. The ring around the neck that gives the

duck its name is rather obscure. This species prefers woodland ponds and lakes and feeds on small invertebrates, plants, and organic debris.

The ring-necked duck breeds from Saskatchewan and western Ontario south to northern Nebraska and Iowa and then is found locally eastward to Pennsylvania. Like the lesser scaup, this species migrates south in the winter to the Gulf of Mexico. The ring-necked duck has been identified from the Late Wisconsinan of west-central Ohio. This duck probably bred in small woodland ponds and lakes of the Great Lakes region during the warmer parts of the Pleistocene.

Site. Carter, Darke County, Ohio, Rancholabrean LMA–Late Wisconsinan (McDonald 1994).

Canada Goose
Branta canadensis

The familiar Canada goose, a very abundant species in the Great Lakes region today because of frequent reintroductions by humans, is a very large bird with a black head and neck, a light breast, and a white patch on the cheek. Canada geese nest in a large variety of wetland habitats, and although some populations migrate northward, others stay to nest in parts of the Great Lakes region. The Canada goose often feeds in flocks in fallow fields on a wide variety of plant and animal material.

The Canada goose is the most abundant goose in the world and occurs over a large part of northern North America. It has been successfully reintroduced by humans in much of its original range, as well as south of its range, in the United States and into England, Scandinavia, and New Zealand. The Canada goose was identified from the Late Wisconsinan of east-central Indiana and probably bred in the Great Lakes region during the Pleistocene.

Site. Reeker Mastodont, Madison County, Indiana, Rancholabrean LMA–Late Wisconsinan (Richards 1984b).

FOWL
Order Galliformes

The order Galliformes consists of many familiar birds such as domestic chickens, guineas, and turkeys as well as grouse, quails, and pheasants. Galliformes have short, downturned bills and large feet. Both ground-

FIGURE 66. Lesser scaup (*Aythya affinis*) group from the Pleistocene of Michigan.

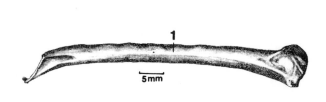

FIGURE 67. Left ulna of lesser scaup (*Aythya affinis*) from the Late Wisconsinan Casnovia Water Well Site, Muskegon County, Michigan. The wavy bumps on the top of the ulnar shaft (1) are cubital tubercles for the origin of flight feathers.

dwelling and tree-dwelling species are included in the order, but none are aquatic. Galliformes are cosmopolitan in their distribution except for Polynesia and Antarctica. Two families, the turkeys (Meleagridae) and grouse (Tetraonidae), are known from the Pleistocene of the Great Lakes region.

TURKEYS
Family Meleagridae

The family Melegridae consists of a single species, the familiar wild turkey *Meleagris gallopavo* and its domestic derivative. Turkeys occur in the eastern United States westward to South Dakota, Nebraska, Colorado, and Arizona south to the Mexican plateau. Bones believed to represent turkeys are known from as far back as the Oligocene.

Wild Turkey
Meleagris gallopavo

It is believed that wild turkeys originally occurred mainly in oak woodlands in southeastern Ontario and most of the eastern United States west to Arizona and south to the Mexican plateau. Turkeys eat a wide variety of food on the woodland floor, from acorns to insects. Wild turkeys were nearly driven to extinction by human hunters in historic times but have recently been reintroduced successfully into portions of their original habitat as well as north of their original habitat in Michigan. Wild turkey remains have been found in Late Wisconsinan sites in southeastern Michigan, northwestern and west-central Ohio, and east-central Indiana. Turkeys were probably important large omnivores of the oak woodlands during the Pleistocene of the Great Lakes region.

Sites. Shelton, Oakland County, Michigan, Rancholabrean LMA–Late Wisconsinan (Shoshani et al. 1989). Sheriden Pit Cave, Wyandot County, Ohio, Rancholabrean LMA–Late Wisconsinan (Ford 1994). Carter, Darke County, Ohio, Rancholabrean LMA–Late Wisconsinan (McDonald 1994). Christensen Bog, Hancock County, Indiana, Rancholabrean LMA–Late Wisconsinan (Graham et al. 1983).

GROUSE
Family Tetraonidae

Grouse are a northward-ranging group of birds whose nostrils are concealed by feathers and who sometimes also have feathered toes. This family is widespread in the Northern Hemisphere, especially in temperate and northern areas. The family occurs in the fossil record from the Miocene to Holocene times. Grouse remains assigned (tentatively) here to cf. ruffed grouse *Bonasa umbellus* are known from a site of probable Late Wisconsinan age in southeastern Ontario.

Site. Kelso Cave, Halton County, Ontario, Rancholabrean LMA–probably Late Wisconsinan ("grouse sp.," Churcher and Dods 1979).

Prairie Chicken
Tympanuchus cupido

The prairie chicken is a large, brown bird of the North American grasslands with a short, rounded tail. It was formerly known over a vast amount of prairie habitat from central and western Canada south to southwestern Michigan, Illinois, Missouri, and Arkansas, with coastal occurrences in Louisiana and Texas. But it has been extirpated from many areas within its former range because of human predation on adults and on the eggs that are laid in a nest on the ground. *Tympanuchus* sp., undoubtedly a prairie chicken, was recorded as a fossil from the Late Wisconsinan of west-central Ohio. This species would have occurred in grassy areas of the region during the Pleistocene.

Site. Carter, Darke County, Ohio, Rancholabrean LMA–Late Wisconsinan (*Tympanuchus* sp., McDonald 1994).

PERCHING BIRDS
Order Passeriformes

The perching birds comprise more than 60 percent of the bird species in the world. They are a very highly evolved group of birds that underwent a large diversification in the Miocene. Many perching birds dwell in small trees or bushes. One family, the Corvidae, has been identified from the Pleistocene of the Great Lakes region.

CROWS AND JAYS
Family Corvidae

The family Corvidae are intelligent, noisy, stout-beaked birds. They are cosmopolitan in their distribution except for Polynesia and Madagascar. Corvids are first known from the Miocene. Only one corvid species, the raven, is known from the Pleistocene of the Great Lakes region.

Raven
Corvus corax

The raven is a very large, shiny black bird, half again larger than a crow. Ravens can also be distinguished from crows on the basis of their ruffled throat feathers and the fact that their voices are croaks rather than caws. Ravens presently live in relatively wild areas from northern Canada south to Maine, Michigan, and Minnesota, with some local populations occurring in Appalachia and along the eastern seaboard states. Ravens, like crows, feed on a wide variety of plant and animal food and often are scavengers of dead mammals and other vertebrates. Raven fossils have been identified from the Late Wisconsinan of west-central Ohio, mainly south of their normal range today.

Site. Carter, Darke County, Ohio, Rancholabrean LMA–Late Wisconsinan (McDonald 1994).

MAMMALS
Class Mammalia

About a third more mammal species are recorded from the Pleistocene of the Great Lakes region than all of the Pleistocene nonmammalian vertebrate species put together. This is an artifact of fossil preservation and pa-

leontological bias, including the fact that mammalian teeth and bones are sturdier and more likely to fossilize than the delicate elements of fishes, amphibians, birds, and small reptiles; mammalian teeth are much more easily identified than most individual fossil elements of the other vertebrates; and there are many more Pleistocene mammalogists than Pleistocene experts on the nonmammalian groups.

Recorded Pleistocene Great Lakes Region Vertebrate Species
Fishes—29
Amphibians—9
Reptiles—20
Birds—9
Mammals—96

In many cases, records of individual Pleistocene mammalian sites in the Great Lakes region are much more numerous than those of nonmammalian sites. For instance, there are more than 200 mastodont sites recorded in Michigan alone, whereas there are often only one or a few site records of nonmammalian species in the state. A few mamalian species (e.g. mastodonts and mammoths) are represented by such a large number of sites that only the more important sites are listed in the individual species accounts that follow.

Only one of sixty-seven of the Pleistocene nonmammalian species discussed in this book is extinct (a giant tortoise), and relatively few Pleistocene nonmammalian species were found outside of their present ranges. We will find a different situation in the Pleistocene mammalian fauna, however, as many large species are extinct, and a number of extant ones were recorded from regions where they do not presently occur (extralimital species).

XENARTHRAN MAMMALS
Order Xenarthra

Xenarthrans (armadillos, anteaters, sloths, and their fossil relatives) are strange, primitive beasts that probably branched off the mammalian family tree very soon after mammals had evolved. Xenarthrans have small, tubular brains; teeth that are simple, single-rooted pegs with no enamel covering, or no teeth at all; posterior body vertebrae with extra connecting surfaces; an odd

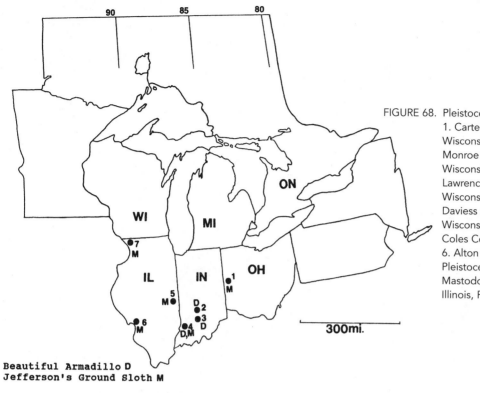

FIGURE 68. Pleistocene extinct xenarthran localities. 1. Carter Site, Darke County, Ohio, Late Wisconsinan. 2. Anderson Pit Cave, Monroe County, Indiana, Late Wisconsinan. 3. Sullivan's Cave, Lawrence County, Indiana, Late Wisconsinan. 4. Prairie Creek D Site, Daviess County, Indiana, Late Wisconsinan. 5. Polecat Creek Site, Coles County, Illinois, Late Wisconsinan. 6. Alton Site, Madison County, Illinois, Pleistocene indeterminate. 7. Galena Mastodont Site, Jo Daviess County, Illinois, Pleistocene indeterminate.

Beautiful Armadillo D
Jefferson's Ground Sloth M

number of neck vertebrae; short, heavy bones; and a low, poorly regulated body temperature. Xenarthrans presently occur widely in Central and South America, and one species, the nine-banded armadillo, extends northward as far as southern Nebraska.

Two families, the armadillos (Dasypodidae) and *megalonychid ground sloths (*Megalonychidae) occur in the Pleistocene of the Great Lakes region. Figure 68 is a map of extinct xenarthran Pleistocene sites in the region.

ARMADILLOS
Family Dasypodidae

The armadillos are easily recognized by the covering of jointed armor plates that protects the head, body, legs, and tail and by teeth that are simple, single-rooted pegs that lack an enamel covering. Presently, the family occurs from southern Nebraska south to Chile and Patagonia in South America. The family is first known from the Eocene of South America. A single extinct species,

the *beautiful armadillo (*Dasypus bellus), is known from the Pleistocene of the Great Lakes region.

Beautiful Armadillo
Dasypus bellus (figs. 69 and 70)
The extinct beautiful armadillo was very similar to the modern nine-banded armadillo (*Dasypus novemcinctus*), a species that has recently invaded the United States, except that the beautiful armadillo (fig. 69) was more than twice as large. The armadillo body can roll up in a shell composed of dermal plates of two major kinds. Six-sided buckler plates form the immovable section of the shell; elongate band plates form the movable parts of the shell (fig. 70).

The large beautiful armadillo dug in the ground with its strong claws for succulent plants and insects and probably ate an occasional mouse or small reptile. In turn, it has been shown in a Late Wisconsinan site in northwestern Georgia that the jaguar (with its massive jaws specialized for crunching shelled vertebrates) was feeding upon beautiful armadillos and large turtles.

The largest number of Pleistocene records for the

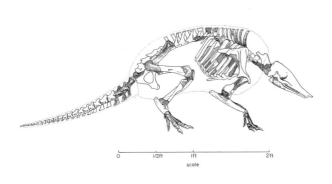

FIGURE 69. Reconstructed skeleton of a *beautiful armadillo (*Dasypus bellus*) from the Pleistocene in Florida.

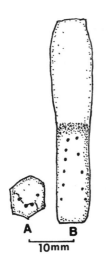

FIGURE 70. Shell (carapace) plates of the beautiful armadillo (*Dasypus bellus*). A, buckler plate; B, band plate.

beautiful armadillo are from Florida, but this species ranged as far north as Missouri and southern Indiana in the epoch. It has been suggested that the limiting factor in the distribution of *Dasypus bellus* was the availability of insects throughout the year. If this is true, the occurrence of the species in Missouri and southern Indiana raises some interesting questions about the Pleistocene winters in these states.

The beautiful armadillo occurred from the Illinoian to the end of the Pleistocene. In the Great Lakes region, the beautiful armadillo is known from Late Wisconsinan sites in south-central, southwestern, and southern Indiana. The species probably dug in the ground for

tender roots, insects, and small vertebrates and was consumed by the jaguars and other large Pleistocene predators that occurred in southern Indiana.

Sites, Anderson Pit Cave, Monroe County, Indiana, Rancholabrean LMA–Late Wisconsinan (Richards, in Graham and Lundelius 1994). Prairie Creek D, Daviess County, Indiana, Rancholabrean LMA–Late Wisconsinan (Tomak 1982; Richards 1984b). Sullivan's Cave, Lawrence County, Indiana, Rancholabrean LMA–Late Wisconsinan (Richards 1986).

*GROUND SLOTHS
*Family Megalonychidae

The extinct ground sloth ranged from cat-sized animals to those larger than a cow. Their teeth were simple and lacked an enamel layer, but a pair of anterior, ever-growing, self-sharpening teeth took on the form of canines, a curious character in mammals. Megalonychid sloths first appear in the Oligocene of South America, and these early forms may have been partly arboreal. These sloths reached North America in the Miocene. It is believed that this introduction occurred by accidental rafting, as North and South America were not connected at this time. The North American megalonychids greatly increased in size through time, reaching very large proportions by Pleistocene times. Only one species of megalonychid, *Jefferson's ground sloth (*Megalonyx jeffersonii*), has been reported from the Pleistocene of the Great Lakes region.

Jefferson's Ground Sloth
Megalonyx jeffersonii (figs. 71 and 72)
The giant, lumbering, Jefferson's ground sloth (fig. 71), with claws three times the size of a lion's (fig. 72), is the largest species in the genus, and large adults were bigger than a cow. The rather shapeless skull of adults is about 2 feet long and has a very blunt front end. This species was named in honor of Thomas Jefferson, who at first believed the claws and fragmentary remains of this animal represented a giant carnivore that might have been three times as big as a lion. Later, Jefferson realized that *Megalonyx* was related to South American armadillos and sloths. The paper he presented about the beast to the American Philosophical Society in 1797 was essentially one of the starting points of systematic vertebrate paleontology in North America.

FIGURE 71. *Jefferson's, ground sloth (*Megalonyx jeffersonii*) feeding in the trees.

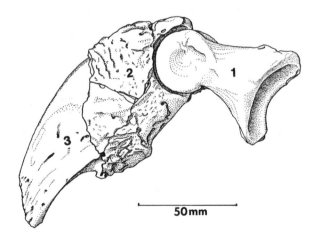

50mm

FIGURE 72. Terminal phalynx (1), claw sheath (2), and claw (3) of a Jefferson's ground sloth (*Megalonyx jeffersonii*) from the Late Pleistocene of Florida.

Jefferson's ground sloth lived in Pleistocene woodlands where it browsed on leaves and twigs and probably ate acorns and nuts as well. It is hard to imagine what Pleistocene predator might have taken on this bulky, massive-boned animal with its huge claws. The species is known from the Illinoian to the end of the Pleistocene. A radiocarbon date of about 9,400 B.P. on a specimen from Evansville in Vanderburg County, Indiana, is probably several hundred years too young. In the Great Lakes region, Jefferson's ground sloth is known from the Late Wisconsinan of north-central, west-central, and southern Ohio; several sites in southern Indiana; and extreme northwestern, east-central, and southwestern Illinois.

Selected Sites. Carter, Darke County, Ohio, Rancholabrean LMA–Late Wisconsinan (McDonald 1994). Prairie Creek D, Daviess County, Indiana, Rancholabrean

LMA–Late Wisconsinan (Richards 1984b). Galena Mastodont, Daviess County, Illinois, Rancholabrean LMA–Late Wisconsinan (Graham and Lundelius 1994). Polecat Creek Site, Coles County, Illinois, Rancholabrean LMA–Late Wisconsinan (Graham and Lundelius 1994). Alton, Madison County, Illinois, Rancholabrean LMA–Late Wisconsinan (Bagg 1909).

INSECTIVORES
Order Insectivora

Insectivores (shrews, moles, hedgehogs, and relatives) are small, primitive mammals with a flattened braincase; small brains; a long, narrow snout; minute eyes; and small ears (fig. 73). Insectivores occur widely throughout North America, northern South America, much of Eurasia, Africa, and Madagascar. Two families, shrews (Soricidae) and moles (Talpidae), occur in the Pleistocene of the Great Lakes region.

SHREWS
Family Soricidae

The shrew, a diminutive beast,
considers a large worm a feast.
The mammalogists game is to give him a name,
like the smoky, the masked, or the least.

Shrews are tiny, fidgety, hyperactive animals with minute eyes and very long, twitching snouts (fig. 73). The late mammalian paleontologist John Guilday aptly stated that a shrew looks like a mouse that got its nose caught in a pencil sharpener. Shrews are mainly insectivorous, have a very high metabolic rate, and eat vast quantities of food. Shrews feed on many kinds of small invertebrate animals as well as on some small vertebrates and are themselves eaten by many vertebrate predators, ranging from owls and snakes to frogs and fishes. Shrew teeth are capped with dark red or chestnut brown pigment produced from an iron compound that is harder than enamel (fig. 73). This pigment probably protects the teeth from the grit in the gut of earthworms, which are important items in the diet of most shrews. The pigment often survives the fossilization process and may be seen in the teeth of the Pleistocene shrews of the Great Lakes region. The Soricidae is a

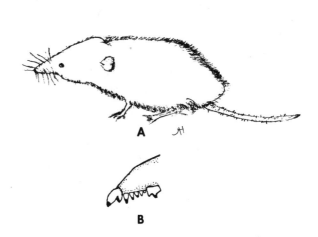

FIGURE 73. A, smoky shrew (*Sorex fumeus*). B, anterior upper teeth of Arctic shrew (*Sorex arcticus*) showing pigmented areas (black).

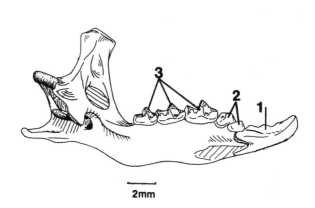

2mm

FIGURE 74. Left lower jaw in medial view of short-tailed shrew (*Blarina brevicauda*). 1, incisor; 2, premolars; 3, molars.

cosmopolitan family except for polar regions, the Australian region, and all of South America except for the extreme northwest. The fossil record of the group extends from the Late Eocene to the Holocene. Fossil shrews are most often recovered from fossil sites when washing and screening methods are employed.

Northern Short-tailed Shrew
Blarina brevicauda (fig. 74)
The northern short-tailed shrew is large as shrews go, with a total length of about 5 inches and a weight of almost 1 ounce. The very short tail is about a quarter of the total length of the animal. This fascinating animal

has many strange characters. They see no images, only light and dark; thus they locate their prey by touch and echolocation. They also produce a poison from their salivary glands. This poison flows along their lower front teeth (incisors; fig. 74–1) into their squirming prey, which is quieted by the process. Short-tailed shrews tunnel about under the leaf litter of the woodland floor, engaging in periodic feeding frenzies on insects and other invertebrates, as well as on smaller shrews, mice, small snakes, and small amphibians.

The northern short-tailed shrew ranges across the eastern two-thirds of southern Canada and southward in the United States to northern Georgia and Alabama. Short-tailed shrews have been identified from the Late Wisconsinan of northwestern Ohio; south-central, southwestern, and southern Indiana; and extreme northwestern Wisconsin. These little animals would have been fierce tiny predators of the leaf litter of the woodlands during the Pleistocene of the Great Lakes region.

Sites. Sheriden Pit Cave, Wyandot County, Ohio, Rancholabrean LMA–Late Wisconsinan (McDonald 1994). Anderson Pit Cave, Monroe County, Indiana, Rancholabrean LMA–Late Wisconsinan (Richards, in Graham and Lundelius 1994). Prairie Creek D, Daviess County, Indiana, Rancholabrean LMA–Late Wisconsinan (Richards 1992c). King Leo Pit Cave, Harrison County, Indiana, Rancholabrean LMA–Late Wisconsinan (Richards and McDonald 1991). Megenity Peccary Cave Unit C, Crawford County, Indiana, Rancholabrean LMA–Late Wisconsinan (Richards 1988b). Moscow Fissure, Iowa County, Late Wisconsin, Rancholabrean LMA–Late Wisconsinan (Foley 1984).

Least Shrew
Cryptotis parva (fig. 75)
The tiny least shrew is a grayish brown animal with a very short tail (less than a quarter of the total length) that distinguishes it from other tiny shrews. The total length of the least shrew is about 3 inches, and it weighs from about .1 to .2 of an ounce. This species favors grassy meadows, where it utilizes runways of voles. Least shrews feed upon invertebrate animals such as slugs, sow bugs, crickets, and grasshoppers.

The least shrew occurs in one colony in Ontario near Lake Erie's Long Point and then through much of the eastern United States, eastern Mexico, and most of Central America. It is presently rare in the Great Lakes

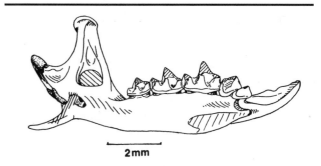

2mm

FIGURE 75. Left lower jaw in medial view of least shrew (*Cryptotus parva*).

region. This species has been identified from the Late Wisconsinan of southwestern and southern Indiana. These tiny animals would have been important insectivores in grassy meadows during the Pleistocene of the Great Lakes region.

Sites. Prairie Creek D, Daviess County, Indiana, Rancholabrean LMA–Late Wisconsinan (Richards 1992c). King Leo Pit Cave, Harrison County, Indiana, Rancholabrean LMA–Late Wisconsinan (Richards and McDonald 1991).

Arctic Shrew
Sorex arcticus (See fig. 73B)
The arctic shrew is larger than the least shrew and has a longer tail. It is about 4.5 inches long and weighs about .4 of an ounce. Its head and back are black or dark brown, and it is light brown along the sides. This shrew prefers moist areas near bogs, swamps, ponds, or lakes. It is mainly insectivorous, eating both the larvae and the adults.

The arctic shrew is an animal of the Far North, ranging from northern Canada to south-central Wisconsin, but it is presently absent from the Lower Peninsula of Michigan. It has been recorded from the Late Wisconsinan of southern Indiana, very much south of its present range, and from the Late Wisconsinan of extreme southwestern Wisconsin, where it is somewhat south of its present range in the south-central part of the state. The arctic shrew would have been an insectivore of moist regions near aquatic situations during the Pleistocene of the Great Lakes region.

Sites. Megenity Peccary Cave Unit C, Crawford County, Indiana, Rancholabrean LMA–Late Wisconsinan (Richards 1988b). Moscow Fissure, Iowa County,

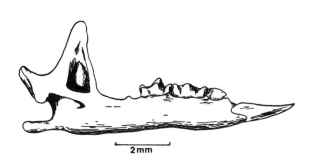

FIGURE 76. Fossil left lower jaw in medial view of masked shrew (*Sorex cinereus*) from the Pleistocene of Kansas. The first premolar and last molar are missing. From Hibbard (1940b).

Wisconsin, Rancholabrean LMA–Late Wisconsinan (Foley 1984).

Masked Shrew

Sorex cinereus (fig. 76)

The masked shrew is another tiny creature. It is brownish and has a tail that is much longer than a quarter of the total length of the body. It is about 3 inches long and weighs from about .1 to .2 of an ounce. The masked shrew is presently one of the most common woodland animals in the Great Lakes region, where it prefers the moister areas in its habitat. The food of this species consists mainly of insects and other small invertebrates, and it is said that it seldom attacks vertebrates.

The masked shrew occurs from Alaska and northern Canada south to the southern Appalachians in the East and to the mountains of New Mexico and Arizona in the West. It is known from the Late Wisconsinan of northwestern Ohio, southwestern and southern Indiana, and extreme southwestern Wisconsin. This shrew was no doubt an important insectivore of moist woodlands during the Pleistocene of the Great Lakes region.

Sites. Sheriden Pit Cave, Wyandot County, Ohio, Rancholabrean LMA–Late Wisconsinan (Ford 1994). Prairie Creek D, Daviess County, Indiana, Rancholabrean LMA–Late Wisconsinan (Richards 1992c). King Leo Pit Cave, Harrison County, Indiana, Rancholabrean LMA–Late Wisconsinan (Richards and McDonald 1991). Megenity Peccary Cave Unit C, Crawford County, Indiana, Rancholabrean LMA–Late Wisconsinan (Richards 1988b). Moscow Fissure, Iowa County, Wisconsin, Rancholabrean LMA–Late Wisconsinan (Foley 1984).

Smoky Shrew

Sorex fumeus (see fig. 73A)

The smoky shrew is a medium-sized member of the Soricidae with a relatively long, bicolored tail that is dark on top and tan on bottom. The animal is about 4.5 inches long and weighs from about .2 to .4 of an ounce. This is a woodland species that presently lives in the eastern part of the Great Lakes region. It subsists mainly on small invertebrates of the woodland floor, but some small salamanders and small reptiles such as newborn snakes are occasionally eaten.

Presently, the smoky shrew occurs through about the eastern third of southern Canada, ranging south through Appalachia into eastern Ohio and extreme southeastern Indiana. In the Pleistocene, the species occurred in the Late Wisconsinan of the south-central and southern tip of Indiana, somewhat northwest and southwest of its present range. The smoky shrew would have been a predator of the invertebrate life of the woodland floor during the Pleistocene of the Great Lakes region.

Sites. Anderson Pit Cave, Monroe County, Indiana, Rancholabrean LMA–Late Wisconsinan (Graham and Lundelius 1994). King Leo Pit Cave, Harrison County, Indiana, Rancholabrean LMA–Late Wisconsinan (Richards and McDonald 1991). Megenity Peccary Cave, Crawford County, Indiana, Rancholabrean LMA–Late Wisconsinan (Richards 1988b).

Pygmy Shrew

Sorex hoyi (fig. 77)

Pygmy shrews are the smallest mammals in North America, the tiny creatures weighing only about .1 of an ounce. Large pygmy shrews may resemble small masked shrews to the extent that proper identification

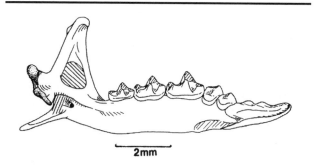

FIGURE 77. Left lower jaw in medial view of pygmy shrew (*Sorex hoyi*).

may require examination of the tiny teeth (fig. 77). The pygmy shrew occupies a wide variety of habitats, from deciduous and coniferous forests to fields, bogs, and swamps. It builds its own tunnels or uses those of voles or large beetles and eats a wide variety of small invertebrate life.

The pygmy shrew is mainly a northern animal, occurring in Alaska and across Canada south into the southern Appalachians in the East and into Montana, Idaho, and eastern Washington in the West. In the Great Lakes region, it occurs in Ontario, most of eastern Ohio, northern Michigan upwards from the northern part of the Lower Peninsula, and most of Wisconsin, except for the southwestern portion. Pygmy shrews occurred south of their present range in the pre-Late Wisconsinan of southeastern Michigan, and the Late Wisconsinan of northwestern Ohio and extreme southwestern Wisconsin, and considerably south of their present range in the Late Wisconsinan of the southern tip of Indiana. The pygmy shrew would have been an insectivore with a wide range of habitats during the Pleistocene of the Great Lakes region.

Sites. Mill Creek, St. Clair County, Michigan, Rancholabrean LMA–pre-Late Wisconsinan (Karrow et al. 1997). Sheriden Pit Cave, Wyandot County, Ohio, Rancholabrean LMA–Late Wisconsinan (McDonald 1994). King Leo Pit Cave, Harrison County, Indiana, Rancholabrean LMA–Late Wisconsinan (Richards and McDonald 1991). Moscow Fissure, Iowa County, Wisconsin, Rancholabrean LMA–Late Wisconsinan (Foley 1984).

Water Shrew
Sorex palustris
The water shrew is a relatively large shrew that may be recognized by its feet, which have hairs between the toes and are partially webbed. These animals are about 6 inches long and weigh about .5 of an ounce. Water shrews hunt both on land and in the water, where they are continuously poking their noses into the hiding places of insects, leaches, and snails.

The water shrew occurs from southern Alaska and northern Canada into the southern Appalachians in the East and northern New Mexico and southern California in the West. In the Great Lakes region, it occurs in most of Ontario, northern Michigan through the northern part of the Lower Peninsula, and the northern two-thirds of Wisconsin. It does not occur in Ohio. The water shrew has been identified from the Late Wisconsinan of extreme southwestern Wisconsin, south of

its present range in the state. The water shrew would have been a moist woodland and aquatic predator of small invertebrate life during the Pleistocene of the Great Lakes region.

Site. Moscow Fissure, Iowa County, Wisconsin, Rancholabrean LMA–Late Wisconsinan (Foley 1984).

MOLES
Family Talpidae

The mole is as "blind as a bat,"
but nose-hairs show him where he's at.
His front legs are big, form a burrowing rig,
and his claws are both sharpened and flat.

Moles (figs. 78 and 79) are silky-furred creatures that are marvelously adapted for underground life. They have large, flattened claws on the ends of their toes for digging and massive forefeet that act as scoops as the animals excavate their burrows. The upper humerus (upper arm bone) of the foreleg is so uniquely structured for the reception of muscles and tendons that it hardly resembles the humeri of other mammals. The eyes are tiny, and in some species they are covered by skin. External ears are absent. Moles occur only in the Northern Hemisphere, mainly in the temperate regions of North America and Eurasia. Moles are first known from the Eocene of North America. Three genera, each represented by a single species, are known both from the Holocene and the Pleistocene of the Great Lakes region.

Star-nosed Mole
Condylura cristata (fig. 78)
The star-nosed mole is unique among New World mammals in having its snout encircled by prominent tentacles (fig. 78), giving the animal a rather grotesque appearance. Star-nosed moles are 7 or 8 inches long and usually weigh about 2 ounces. Their eyes are larger and their tails are much longer than in other moles of the region. This species hunts for its food above ground, in tunnels, or in the water; often its tunnels open into a pond, lake, or stream. Star-nosed moles eat earthworms and slugs on land and a variety of invertebrate species in the water. The sensory tentacles enhance the effectiveness of the nose hairs found in other moles and aid the animal in getting about and procuring food in the dark.

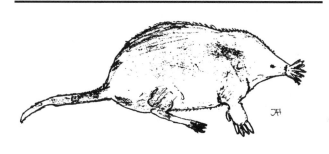

FIGURE 78. Star-nosed mole (*Condylura cristata*).

Star-nosed moles occur from northeastern Canada southward down the eastern seaboard to southern Georgia, into the southern Appalachians, and through the northern part of the Great Lakes region to central Ohio, northeastern Indiana, and all of Wisconsin, except for the southwestern portion of the state. Star-nosed moles are known from the Late Wisconsinan of southwestern Indiana, where they are well south of their modern range. These moles would have been important small predators upon both terrestrial and aquatic invertebrate life during the Pleistocene of the Great Lakes region.

Site. Prairie Creek D, Daviess County, Indiana. Rancholabrean LMA–Late Wisconsinan (Richards 1992c).

Hairy-tailed Mole
Parascalops breweri (fig. 79)

FIGURE 79. Hairy-tailed mole (*Parascalops breweri*).

Small size and a tail covered with stiff black hairs distinguish the hairy-tailed mole from others in the region (fig. 79). Hairy-tailed moles are about 6 inches long and usually weigh less than 2 ounces. This species spends most of its life in tunnels in a wide range of habitats, from woodlands to meadows. It feeds mainly

on earthworms, slugs, beetles, beetle larvae, and a variety of other subterranean invertebrate species.

This species ranges from southeastern Canada south into southern Appalachia and eastern Ohio. In the Great Lakes region, it has been identified from the Sangamonian and Late Wisconsinan of south-central Indiana. Both localities are west of the species' present range in eastern Ohio. This mole would have been a predator of the invertebrate life of the woodlands and meadows during the Pleistocene of the Great Lakes region.

Sites. Harrodsburg Crevice, Monroe County, Indiana, Rancholabrean LMA–Sangamonian (Richards 1982a). Anderson Pit Cave, Monroe County, Indiana, Rancholabrean LMA–Late Wisconsinan (Richards 1982a).

Eastern Mole
Scalopus aquaticus
The short, plump eastern mole lacks tentacles and also lacks a black, hairy tail. It is 6 or 7 inches long and weighs 3 or 4 ounces. The eastern mole has webbed feet, but it shuns the water, living in tunnels in the woodland or meadow floor. It also worries people by tunneling in their lawns or gardens. This species eats worms, insects, roots, and seeds as well as other small animal or plant items.

The eastern mole is mainly a southern species, occurring in Canada only in Ontario just north of Lake Erie, the Lower Peninsula of Michigan, southwestern Wisconsin, west to southern South Dakota, and south to southern Florida and Texas. This species has been recorded from the Late Wisconsinan of south-central and southwestern Indiana, well within its modern range. This mole would have been an important herbivore and underground predator of invertebrate life during the Pleistocene of the Great Lakes region.

Sites. Anderson Pit Cave, Monroe County, Indiana, Rancholabrean LMA–Late Wisconsinan (Graham and Lundelius 1994). Prairie Creek D, Daviess County, Indiana, Rancholabrean LMA–Late Wisconsinan (Richards 1992c).

BATS
Order Chiroptera

Bats are small, flying mammals with their wing membranes supported by the digits of the elongated fingers of the forelimbs (fig. 80). Most bats use echolocation

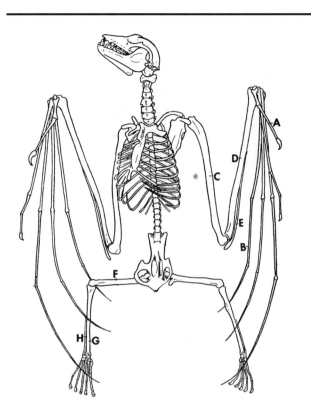

FIGURE 80. Skeleton of fruit bat (*Pteropus*) modified from Shipley and MacBride by Reynolds (1913). A, thumb; B, little finger; C, humerus; D, radius; E, ulna; F, femur; G, tibia; H, fibula.

to fly in the darkness and to locate their prey. Bats resemble insectivores in several ways and are thought to have evolved from them. Bats are present (except from some very small, remote islands) in all of the temperate and tropical regions of the world.

COMMON BATS
Family Vespertilionidae

Common bats are small, with the largest forms weighing about 2 ounces. All of them have a fleshy, pointed projection (tragus) at the bottom of the ear. The tragus is used for echolocation. Common bats are mainly insectivorous, but some eat fishes. Most common bats roost in caves, but they will often roost in other places such as rock ledges, hollow trees, or buildings. Vespertilionid bats are found in temperate and tropical regions throughout the world, with the exception of some islands in the South Pacific. Common bats are

first known from the Eocene. Fossil bat limbs are relatively easy to identify, as they are very elongate (fig. 80). Pleistocene bat fossils are often found in the remnants of the Pleistocene caves in which they were roosting. The first three bats in this section are the so-called mouse-eared bats of the genus *Myotis*.

Small-footed Bat
Myotis leibii
The tiny small-footed bat weighs about .2 of an ounce, has small feet, and has a black mask across the muzzle and eyes. This animal hibernates in the winter in cool caves, singly rather than in clusters like many species. Little is known of its summer habits, except that it feeds on a wide variety of small insects. The small-footed bat occurs in two separate populations, an eastern one and a western one, and many mammalogists think that each population should be regarded as a separate species.

The eastern population occurs from southeastern Canada south through the Appalachians and west to Arkansas. It does not occur in Michigan or in northern Ohio, Indiana, and Illinois. Fossils tentatively identified as *Myotis leibii* have been found in the Late Wisconsinan of southern Indiana. This bat would have existed as a tiny insectivore during the Pleistocene of the Great Lakes region.

Site. King Leo Pit Cave, Harrison County, Indiana, Rancholabrean LMA–Late Wisconsinan (*Myotis* cf. *Myotis leibii*, Richards and McDonald 1991).

Little Brown Bat
Myotis lucifugus
The little brown bat is a small species that weighs from .2 to about .4 of an ounce. This species has a uniformly colored, glossy coat and black ears. In the winter, the little brown bat hibernates in caves or abandoned mines. Some hibernating colonies number up to 300,000 individuals. In the summer, these bats leave their hibernating quarters and migrate to habitats as far as 200 miles away. The little brown bat does most of its feeding during the first few hours after sunset and favors insects with an aquatic larval stage, such as chironomid flies and mayflies.

At present, this bat has a huge range in North America, occurring from Alaska and northern Canada into northern Mexico. It is absent from most of Texas and Louisiana but reaches northern Florida. In the Pleistocene of the Great Lakes region, it is known from a probable Late Wisconsinan site in southeastern Ontario. This bat would have been an important small insectivore over a wide range of habitats during the Pleistocene of the Great Lakes region.

Site. Kelso Cave, Halton County, Ontario, Rancholabrean LMA, probably Late Wisconsinan (Churcher and Dods 1979; Mead and Grady 1996).

Northern Bat
Myotis septentrionalis

The tiny northern bat is distinguished from other small bats on the basis of its much larger ears. Northern bats weigh from about .2 to .3 of an ounce. After the northern bat emerges from hibernation, it migrates to its summer quarters, where it often feeds in woodlands between the canopy and the shrub layer. Northern bats eat a wide variety of small insects. This bat has been observed to eat insects directly off of leaves, and it has been suggested that it feeds off of the woodland floor as well.

The northern bat ranges from northern Canada to southern Georgia, eastern Oklahoma, and Kansas. It presently occurs throughout the Great Lakes region. This species has been identified from the Late Wisconsinan of northwestern Ohio. It would have been a tiny insectivore of the woodlands of the Great Lakes region during the Pleistocene.

Site. Sheriden Pit Cave, Wyandot County, Ohio, Rancholabrean LMA–Late Wisconsinan (Ford 1994).

Mouse-eared Bat Species
Myotis sp. indet.

Bat remains from the Rancholabrean LMA–Late Wisconsinan King Leo Pit Cave Site in Harrison County in extreme southern Indiana were quite indefinitely identified by Richards and McDonald (1991) as (1). *Myotis* cf. *lucifugus* (little brown bat), *sodalis*, (Indiana bat), or *austroriparius* (southern *myotis*) and (2) *Myotis* cf. *grisescens* (gray bat) or *septentrionalis* (northern bat). These kinds of indefinite identifications are not at all unusual in vertebrate paleontology, especially when the suite of species identified contains taxa that are very closely related to one another. However, such identifications are troublesome, especially when one is trying to tabulate the total number of species in a particular area or region, such as I shall be doing in a later chapter. Thus, from a rather conservative standpoint, I shall treat these indefinite King Leo Pit Cave *Myotis* identifications as one additional species of *Myotis* for the purpose of the forthcoming tabulations.

Eastern Pipistrelle
Pipistrellus subflavus

The eastern pipistrelle is a tiny brown bat with yellowish to reddish overtones in its dominant color. It weighs from about .1 to .2 of an ounce. Pipistrelles hibernate in caves where they hang singly from the walls. This bat flies over ponds and streams that occur in woodland breaks. For their size, they consume an incredible amount of insect food. These bats have a slower, less erratic flight than many small bats.

Eastern pipistrelles occur from southeastern Ontario south to southern Florida and west to Kansas, Oklahoma, and eastern Texas, and south into Mexico. In the Great Lakes region, they are absent from eastern Wisconsin, northwestern Ohio, northern Indiana, and Michigan, except for the western part of the Upper Peninsula. The eastern pipistrelle has been identified from the Late Wisconsinan of south-central and southern Indiana. This tiny bat would have been an efficient predator of small insects over ponds and streams in the Great Lakes region during the Pleistocene.

Sites. Anderson Pit Cave, Monroe County, Indiana, Rancholabrean LMA–Late Wisconsinan (Graham and Lundelius 1994). King Leo Pit Cave, Harrison County, Indiana, Rancholabrean LMA–Late Wisconsinan (Richards and McDonald 1991).

Big-eared Bat
Plecotus sp.

The big-eared bat may be recognized by its spectacular ears that are three-quarters as long as the body. This genus was reported from the Late Wisconsinan of southern Indiana, within its modern range, and probably represents Rafinesque's big-eared bat, *Plecotus rafinesquii*. This species occurs in the southeastern United States, reaching the Great Lakes region only in southern Ohio and Indiana and in extreme southeastern Illinois.

Site. King Leo Pit Cave, Harrison County, Indiana, Rancholabrean LMA–Late Wisconsinan (Richards and McDonald 1991).

PRIMATES
Order Primates
Family Hominidae

Evidence of primates in the Pleistocene of the Great Lakes region is mainly based on the Paleo-Indian

(*Homo sapiens*) artifacts that are widely scattered throughout the region. These artifacts consist of such objects as fluted spear points (see fig. 11), scrapers, and drills. The general lack of human skeletal elements in "natural" fossil deposits has been attributed to the fact that humans usually bury their dead.

Humans entered the Great Lakes region in the latter part of the Late Wisconsinan. These people were hunters and gatherers and did not form large settlements. They did not occupy the northern part of the Great Lakes region for extended periods of time in the Pleistocene. The role of humans in the Pleistocene ecosystem and how they affected the vertebrate populations of the area will be discussed later in the book.

CARNIVORES
Order Carnivora

Carnivores are medium- to large-sized mammals that have large brains but a generally primitive skeleton. All of them have the fourth upper premolar tooth and the first lower molar tooth modified as a shearing pair (carnassials; figs. 81 and 82), although the shearing capacity has been modified or lost in some evolutionary lines. The order is composed of both carnivores (those that eat animals such as insects, crustaceans, mollusks, and vertebrates) and omnivores (those that eat both plant and animal material). Many carnivores will supplement their carnivorous or omnivorous diet by eating carrion. None, however, are completely herbivorous. Both terrestrial and aquatic species exist. The order Carnivora occurs naturally on all continents but Australia, and aquatic species occur in coastal situations throughout the world. Living as well as extinct carnivores inhabited the Great Lakes region during the Pleistocene. Figure 83 is a map of large, extinct Pleistocene carnivore sites in the Great Lakes region.

DOGS AND RELATIVES
Family Canidae

Canids have small, unspecialized incisors but long canines, sharp premolars, and a well-developed pair of carnassials, with the remaining molars specialized for crushing (see fig. 81). They are mainly carnivorous, but some eat quite a bit of plant material from time to

time and others eat carrion. Canids are alert and intelligent and are known for their communicative vocalizations. They hunt singly, in pairs, or in packs and have an extraordinary sense of smell. Canids occur worldwide except for Antarctica, Australia (except for the dingo dog that was introduced by humans), and most oceanic islands. Canids occur from Early Eocene to Holocene times.

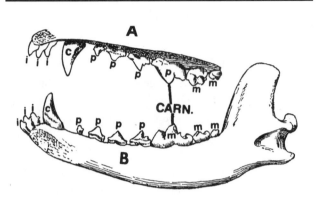

FIGURE 81. Upper (A) and lower (B) jaws of dog (*Canis*) modified from Reynolds (1913). i, incisors; c, canines; p, premolars; m, molars; CARN, the carnassial pair.

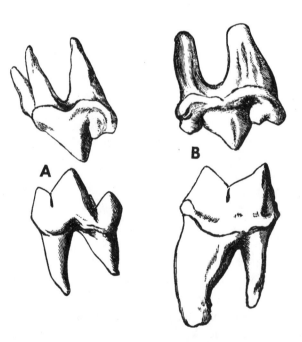

FIGURE 82. Carnassials of dog (*Canis*) (A) and cat (*Felis*) (B). Upper carnassials (top) and lower carnassials (bottom). From Reynolds (1913).

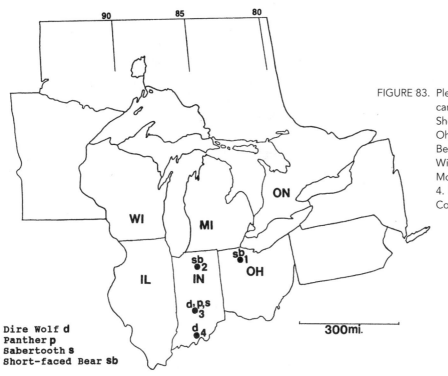

FIGURE 83. Pleistocene localities of large, extinct carnivores in the Great Lakes region. 1. Sheriden Pit Cave Site, Wyandot County, Ohio, Late Wisconsinan. 2. Short-faced Bear Site, Fulton County, Indiana, Late Wisconsinan. 3. Harrodsburg Crevice, Monroe County, Indiana, Sangamonian. 4. Megenity Peccary Cave, Crawford County, Indiana, Late Wisconsinan.

Dire Wolf **d**
Panther **p**
Sabertooth **s**
Short-faced Bear **sb**

*Dire Wolf
*Canis dirus

The extinct dire wolf was as large as a modern gray wolf but was much more robust as it had a much broader head and sturdier limbs. Dire wolf teeth were larger than in any other species of *Canis*, and the carnassials were much larger than those of the gray wolf. On the other hand, the braincase of the dire wolf was relatively smaller than that of the gray wolf, and it is possible that the extinct species might not have been as intelligent and cunning.

In the North American Pleistocene, the mammalian fauna resembled that of Africa today in that there were many large herbivores and carnivores that preyed upon them. There was even a lion (not in the Great Lakes region as far as we know) that was much bigger than the African lion. In these types of mammalian assemblages, animals such as hyenas, with a hunting-scavenging way of life, prosper. There were no hyenas in the North American late Pleistocene; thus it has been suggested that the dire wolf, with its huge carnassials, might have had hyena-like habits. These habits are also suggested based on the amazing concentration of Pleistocene dire wolf fossils at the Rancho La Brea asphalt pits in Los

Angeles, where more than 1,600 individuals were recovered. Hundreds of large herbivores were also trapped in the sticky Rancho La Brea mixture of ancient asphalt and oil.

Dire wolves suddenly appeared in the Late Illinoian glacial age of the Great Plains and did not become extinct until the Late Wisconsinan. During the interim, the range of the dire wolf extended through most of the United States and Mexico to Peru in South America. There was much geographic size variation in Pleistocene populations of this species. The largest dire wolves occurred in the Great Plains, the Appalachian region, and Florida. The dire wolves in California and Mexico were smaller.

Dire wolves are known from the Sangamonian of south-central Indiana, from a Late Pleistocene undesignated site in southern Indiana, and from the Late Wisconsinan of southern Indiana. The dire wolf was first named on the basis of a fossil from the undesignated Indiana site. Such a fossil is designated as a type specimen. The dire wolf was undoubtedly a powerful hunter-scavenger during the Pleistocene of the Great Lakes region.

Sites. Harrodsburg Crevice, Monroe County,

Indiana, Rancholabrean LMA–Sangamonian (Paramalee et al. 1978). Ohio River Site, Vanderburgh County, Indiana (type specimen), Late Pleistocene undesignated (Richards 1984b). Megenity Peccary Cave, Crawford County, Indiana, Rancholabrean LMA–Late Wisconsinan (Richards and Whitaker 1997).

Coyote
Canis latrans

Coyotes are slender, wolflike canids with very large, erect ears and a thick, bushy tail that is narrow at its base. The coat is grayish and the throat and underparts are whitish. A dark band extends from the back onto the tail. They are larger than foxes and smaller than dire wolves or gray wolves, weighing from about 25 to about 45 pounds. These animals prefer prairies and woodland edges. Coyotes have a lifetime mate, with which they often hunt, rather than hunting in organized packs like gray wolves. Unlike wolves, coyotes seldom attack animals larger than themselves but prefer to scavenge the carcasses of the larger mammals. They actively hunt for such small mammals as rabbits, squirrels, and even mice.

Presently, the range of the coyote comprises all of North America, except for the northernmost parts of Canada. But before the European settlers came, coyotes were mainly absent from the Great Lakes drainage basin proper. Coyote remains have been identified from the Late Wisconsinan of southern Indiana. The coyote would have been an important predator of small mammals in open areas in the southern part of the Great Lakes region during the Late Wisconsinan Pleistocene.

Site. Megenity Peccary Cave, Crawford County, Indiana, Rancholabrean LMA–Late Wisconsinan (Richards 1988a).

Gray Fox
Urocyon cinereoargenteus

The gray fox, a tree climber, is distinguished from the red fox by its grizzly, grayish coat. Individuals weigh from about 7 to 14 pounds. The gray fox is a resident of deciduous woodland habitats. These foxes feed mainly on small mammals, and the cottontail rabbit appears to be a favorite food. In the late summer and early fall, the gray fox consumes a large amount of vegetable material including domestic corn, apples, and grapes as well as natural fruits such as papaws and persimmons.

The gray fox is likely to be encountered in deciduous woodlands over most of North America, but its range ends rather abruptly north of Lake Superior where these habitats dwindle. The gray fox is presently much less common in the Great Lakes region than is the adaptable red fox. Gray fox remains have been identified from the Late Wisconsinan of south-central Indiana. The lower jaw of the gray fox may be separated from that of the red fox on the basis of a notch that occurs on the posterior part of the lower part of the bone (fig. 84). The gray fox would have been an important predator on small mammals in the deciduous woodlands during the Pleistocene in at least the southern part of the Great Lakes region.

Site. Anderson Pit Cave, Monroe County, Indiana, Rancholabrean LMA–Late Wisconsinan (Graham and Lundelius 1994).

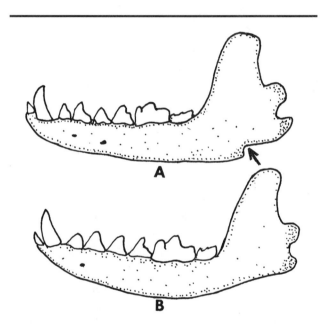

FIGURE 84. A, lower jaw of gray fox (*Urocyon cinereoargenteus*). B, lower jaw of red fox (*Vulpes vulpes*). The arrow indicates the notch in the lower jaw of the gray fox that distinguishes the two species.

Red Fox
Vulpes vulpes

The red fox has an attractive reddish coat, white underparts, and a bushy, reddish, white-tipped tail. The bottom part of the slim legs and the feet are black. Red foxes weigh from about 8 to 15 pounds. The red fox prefers open country and woodland edges and shuns closed canopy situations. It tends to hunt singly and feeds on a large variety of small mammals and ground-nesting birds, adding the occasional amphibian, reptile, or even insect to its diet.

This species occurs from Alaska through northern Canada south to central Texas and northern Florida. It is absent or scarce in some of the southwestern and far western states. Red fox remains are known from the Late Wisconsinan of northwestern Ohio. The lower jaw of the red fox lacks the notch on the lower part of the bone that occurs in the gray fox (fig. 84). The red fox would have been a predator mainly of small mammals and ground-nesting birds in the meadows and woodland edges during the Pleistocene of the Great Lakes region.

Site. Sheriden Pit Cave, Wyandot County, Ohio, Rancholabrean LMA–Late Wisconsinan (McDonald 1994).

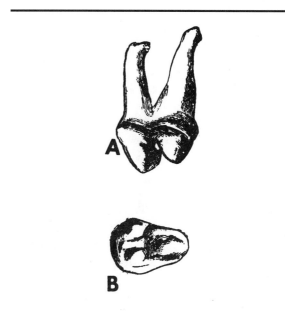

FIGURE 85. Upper carnassial of the Ursidae showing side view (A) and surface view (B). From Reynolds (1913). In the surface view, note the blunt biting surface.

BEARS
Family Ursidae

Bears are large to very large mammals that are usually omnivorous. This is reflected by their dentition, which typically has the molar teeth enlarged with bumpy surfaces, the premolars reduced, and the carnassials poorly developed (fig. 85). Bears tend to have a coat of a single color and to walk with a rather clumsy-looking, flat-footed gait. The ears are small and round and the tail extremely short. Bears occur today in North America and Eurasia as well as in the Atlas Mountains of North Africa and the South American Andes. There are many Oligocene records of the bear family and one questionable Late Eocene record from Asia. One extinct and two modern species of bears are known from the Pleistocene of the Great Lakes region.

FIGURE 86. The *short-faced bear (*Arctodus simus*).

*Short-faced Bear
Arctodus simus

The extinct short-faced bear is thought to have been the most powerful predator in the North American Pleistocene. It had a short face and very long legs and was taller than the Alaskan Kodiak bear. The skeleton of the short-faced bear was lighter than that of modern bears; thus it is thought that this extinct bear was faster and less clumsy than the modern species. The teeth of the short-faced bear resemble those of typical carnivores more than do those of the modern species of bears.

Short-faced bear remains are commonly found in Pleistocene cave deposits. It has been suggested that the piles of broken bones of large Pleistocene mammals often found in cave sediments represent the crushed remnants of animals drug into the cave by short-faced bears. It must have been a frightening experience for early humans seeking shelter in Pleistocene caves to be suddenly confronted by an upright, angry short-faced bear defending its territory.

The short-faced bear was very widely distributed in the Pleistocene. It occurred in Alaska, the Yukon, central and southern Alberta, and Saskatchewan, was widely scattered through the continental United States from east to west, and is known from northern Mexico. It occurred from the Early Pleistocene until the Late Wisconsinan. In the Great Lakes region, the short-faced bear has been found in the Late Wisconsinan of northern Indiana and in a Late Wisconsinan cave site in northwestern Ohio. The Indiana short-faced bear fossil has been referred to the northern subspecies, *Arctodus simus yukonensis*. Obviously, the short-faced bear was the top mammalian predator in the Great Lakes region during the Pleistocene; from the variety of mammals represented by broken bones (thought to have been refuse from short-faced bear kills), it must have eaten a large number of other mammalian species.

Sites. Rochester, Fulton County, Indiana, Rancholabrean LMA–Late Wisconsinan (*Arctodus simus yukonensis*, Richards and Turnbull 1995). Sheriden Pit Cave, Wyandot County, Ohio, Rancholabrean LMA–Late Wisconsinan (McDonald 1994).

Black Bear
Ursus americanus (fig. 87)
The familiar black bear usually has a shiny black or very dark brown coat, a much lighter colored muzzle, and a light patch on the throat. The eyes are small, and the short ears are erect. Black bears weigh from about 250 to 500 pounds or more. Black bears prefer deeply wooded areas (in either deciduous or coniferous forests) where there is a thick understory. They eat more plant than animal food and are fond of fruits of all kinds as well as nuts and acorns. They consume insects and insect larvae and raid the nests of bees for honey. Small mammals and an occasional fawn are taken, and carrion is also a staple part of the diet.

Before the arrival of European settlers, black bears occurred throughout the woodlands of the North American continent, but now they are restricted to the wilder woodland areas or protected situations in the North and the South. Black bears occur from Early Pleistocene to Holocene times and were the most common bears of the Late Pleistocene in North America. In the Great Lakes region, the black bear is known from the Late Wisconsinan of southeastern Michigan, northwestern Ohio, and south-central and southern Indiana. An extinct subspecies, *Ursus americanus amplidens*, was reported from the Parker Pit, an early Late Wis-

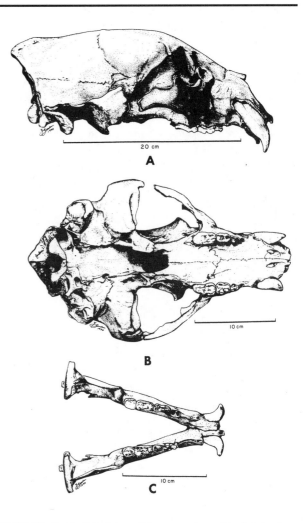

FIGURE 87. Skull of black bear (*Ursus americanus*) in A, lateral view, and B, palatal view. C, lower jaws in dorsal view of black bear.

consinan locality in southern Indiana. Presently, however, it is believed that this Pleistocene subspecies overlaps in osteological characters with the modern black bear. The black bear was undoubtedly a widespread large omnivore during the Pleistocene of the Great Lakes region.

Selected Sites. Parkers Pit, Harrison County, Indiana, Rancholabrean LMA–early Late Wisconsinan ("*Ursus americanus amplidens*," Richards 1984b). Green, Oakland County, Michigan, Rancholabrean LMA–Late Wisconsinan (Eshelman 1974). Sheriden Pit Cave, Wyandot County, Ohio, Rancholabrean LMA–Late Wisconsinan (McDonald 1994).

Brown and Grizzly Bears
Ursus arctos

Both the brown and grizzly bears are phases of a single species, *Ursus arctos*. This is a huge species with a brown coat (brown bear) or a grayish brown grizzly coat (grizzly bear). Both phases of *Ursus arctos* may be distinguished from the black bear by the dark muzzle, the longer claws, and a distinct hump with erect hairs that occurs on the back behind the shoulders. This species contains the largest living carnivores, with individuals that weigh more than 1,500 pounds. The brown bear mainly inhabits southwestern coastal Alaska and its islands. The grizzly bear is more widespread than the brown bear and ventures into higher terrain. Both phases are omnivorous, eating fruits, mosses, sedges, and other plant material, as well as mammals, salmon, and carrion.

Within historic times, *Ursus arctos* was common in Europe, Asia except for the tundra and southern peninsula, and western North America south into the mountains of Mexico. Presently, its last important strongholds are in Alaska, northwestern Canada, and the national parks in northwestern North America. The species is known in the fossil record from the Middle Pleistocene of Europe to Holocene times. At the end of the last glaciation, *Ursus arctos* spread its range to such far-flung places as southern California and southeastern Ohio. In the Great Lakes region, *Ursus arctos* is known from the Early, Middle, and Late Wisconsinan in Ontario and from the Late Wisconsinan of southeastern Ohio. This animal would have rivaled the short-faced bear as the most powerful predator of the Great Lakes region during the Pleistocene.

Sites. Don Brickyard, Toronto, Ontario, Rancholabrean LMA–Early Wisconsinan (Coleman 1933). Woodbridge, York County, Ontario, Rancholabrean LMA–Middle Wisconsinan (Churcher and Morgan 1976). Near Lake Simcoe, Ontario, Rancholabrean LMA–Late Wisconsinan (Peterson 1965). Overpeck, Butler County, Ohio, Rancholabrean LMA–Late Wisconsinan (Hansen 1992).

RACCOONS AND RELATIVES
Family Procyonidae

The family Procyonidae contains the raccoons and their relatives, the ringtails, coatis, and other less familiar forms. Procyonids are small to medium-sized mammals with long, fluffy tails that usually have alternating light and dark bands. All procyonids are good climbers that use their tails as a balance. These animals are omnivorous, with the rear teeth structured for crushing and the carnassials poorly developed. Procyonids are strictly New World animals, occurring from Canada to Argentina. The family is known from the early Oligocene to modern times. Only one species, the raccoon (*Procyon lotor*), is known from the Pleistocene of the Great Lakes region.

Raccoon
Procyon lotor

The raccoon, with its black eye mask separating its whitish muzzle and forehead and its ringed tail, is an animal that most people recognize. Raccoons range from about 15 pounds in trim individuals to about 45 pounds in fat adults. These animals may be found in wooded areas where there are hollow den trees, from wilderness areas to suburban parks. Raccoons are omnivores that eat a bewildering variety of plant and animal foods, as well as carrion and refuse from garbage cans. Raccoons often hunt along the edges of ponds and streams, and they are particularly fond of crayfish.

This species occurs from southern Canada south through most of the United States to Panama. Raccoons are known from the Middle Pleistocene to modern times and became widespread by Wisconsinan times. In the Great Lakes region, raccoons are known from the Late Wisconsinan of northwestern Ohio and south-central Indiana. These animals would have been important omnivores of woodland areas in the Great Lakes region by Late Wisconsinan times.

Sites. Sheriden Pit Cave, Wyandot County, Ohio, Rancholabrean LMA–Late Wisconsinan (McDonald 1994). Anderson Pit Cave, Monroe County, Indiana, Rancholabrean LMA–Late Wisconsinan (Graham and Lundelius 1994).

MUSTELIDS
Family Mustelidae

The mustelids are a diverse family of carnivores ranging from tiny to medium-sized animals. They are divided into four subfamilies: weasels and relatives (Mustelinae), badgers (Melinae), skunks (Mephitinae), and otters (Lutrinae). Although there are some notable exceptions, most mustelids have long bodies and short legs. Mustelid skulls are very characteristic as they have a short rostrum (snout) and a long, flattened braincase. Anal scent glands are well developed, especially in the skunks. Mustelids are mainly carnivorous but sometimes eat berries and other fruits in the summer. They usually have an enlarged pair of carnassials. Mustelids occur today on all continents except Antarctica and Australia and are known in the fossil record from Early Oligocene to modern times. All of the mustelid subfamilies, with the exception of the badgers (Melinae), are known from the Pleistocene of the Great Lakes region.

WEASELS AND RELATIVES
Subfamily Mustelinae

Marten

Martes americana (fig. 88)

The marten is a long-bodied, brown animal with dense, glossy fur and is about the size of a small cat. The face is lighter than the body, and the ears usually have white rims. The marten weighs from about 2 to about 3

FIGURE 88. Left lower jaw of a marten (*Martes americana*) from Holocene peat deposits of northwestern Illinois. A, lateral view; B, dorsal view. C, a marten.

pounds. It is a northern and mountain species of coniferous woodlands with a closed canopy and a dense understory. The martin hunts mammals such as shrews, voles, red squirrels, and hares and occasionally will eat an amphibian, reptile, or bird. Martens also eat fruits, such as blueberries, during the summer.

Martens presently occur in Alaska across northern Canada, south to the northern part of the Lower Peninsula of Michigan (reintroduced by humans), in the East, and into the Rocky Mountains and Sierra Nevada in the West. In the Pleistocene of the Great Lakes region, martins are known from the Late Wisconsinan near Ottawa, Ontario, as well as the Late Wisconsinan of northwestern and west-central Ohio, far south of their modern range. The marten would have been an efficient predator of the small mammals of the dense coniferous forests during the Pleistocene of the Great Lakes region.

Sites. Green Creek, near Ottawa, Ontario, Rancholabrean LMA–Late Wisconsinan (Harington 1972). Sheriden Pit Cave, Wyandot County, Ohio, Rancholabrean LMA–Late Wisconsinan (McDonald 1994). Carter, Darke County, Ohio, Rancholabrean LMA–Late Wisconsinan (McDonald 1994).

Fisher

Martes pennanti

The fisher is presently the largest weasel-like (Mustelinae) animal of the Great Lakes region. It is larger than the marten, weighing from about 5 to about 12 pounds. The dark brown coat of the fisher is coarser than that of the marten, and the ears are rounder. The fisher, like the marten, prefers dense, coniferous woodlands. Fishers eat shrews, voles, squirrels, and hares. It is the top predator on porcupines, which it weakens and eventually kills by biting the unquilled face.

Today, the fisher occurs across Canada and has been successfully reintroduced by humans into northern Michigan and Wisconsin. This species also occurs in an isolated population in the southern Appalachians and in the Rocky Mountains and Sierra Nevada in the West. In the Pleistocene of the Great Lakes region, the fisher is known from Late Wisconsinan sites in northwestern and west-central Ohio and southwestern and southern Indiana, all very far south of its present northern range in eastern North America. The fisher would have been the largest weasel-like predator of small mammals during the Pleistocene of the

Great Lakes region, where it probably hunted in dense coniferous woodlands.

Sites. Sheriden Pit Cave, Wyandot County, Ohio, Rancholabrean LMA–Late Wisconsinan (McDonald 1994). Carter, Darke County, Ohio, Rancholabrean LMA–Late Wisconsinan (McDonald 1994). Prairie Creek D, Daviess County, Indiana, Rancholabrean LMA–Late Wisconsinan (Richards 1992c). King Leo Pit Cave, Harrison County, Indiana, Rancholabrean–Late Wisconsinan (Richards and McDonald 1991). Megenity Peccary Cave, Crawford County, Indiana, Rancholabrean–Late Wisconsinan (Richards 1988c).

Ermine
Mustela erminea

The ermine is a very small weasel that has a summer coat of chocolate brown above and white below, except for the end of the tail which is ringed with black. In the winter, the ermine turns white, except for the black end of the tail. Ermine weigh from about 2 to about 4 ounces. Unlike the martin and the fisher, the ermine lives in a variety of habitats, from woodlands to open, shrubby areas. The ermine preys mainly on shrews, mice, and voles, which it quickly kills by a bite to the neck or the back of the head.

The ermine has a vast range in northern North America, occurring today in Canada and Alaska;, south to the northeastern states, Michigan, and Wisconsin; and through the mountains in the West to northern New Mexico. In the Pleistocene of the Great Lakes region, the ermine is known from the pre-Late Wisconsinan of southeastern Michigan and from the Late Wisconsinan of northwestern Ohio. The Ohio site is somewhat south of the present range of the ermine in south-central Michigan and southern Ontario.

Sites. Mill Creek, St. Clair County, Michigan, Rancholabrean LMA–pre-Late Wisconsinan (Karrow et al. 1997). Sheriden Pit Cave, Wyandot County, Ohio, Rancholabrean LMA–Late Wisconsinan (McDonald 1994).

Long-tailed Weasel
Mustela frenata

The long-tailed weasel is similar to the ermine but is larger, has a summer coat of reddish brown above and yellowish below, and has a longer tail. In the northern part of its range, like the ermine, the long-tailed weasel turns white in the winter except for the black end of its tail. But from the southern part of the Great Lakes re-gion southward, the long-tailed weasel is brown year-round. Long-tailed weasels weigh from about 3 to about 10 ounces. This weasel is ubiquitous in its choice of habitats, ranging from woodlands and woodland edges to farm buildings near moderately wooded areas. It is often found near marshes and bogs. The long-tailed weasel eats shrews, mice, and voles but is capable of killing cottontail rabbits.

This weasel has a very large distribution in the Western Hemisphere and ranges much farther into southern lands than the ermine. It occurs across southern Canada and through the United States into South America. In the Pleistocene of the Great Lakes region, the long-tailed weasel has been identified from the Late Wisconsinan of west-central Ohio and southwestern Indiana. This species would have been an important small predator of the region in the Pleistocene.

Sites. Carter, Darke County, Ohio, Rancholabrean LMA–Late Wisconsinan (McDonald 1994). Prairie Creek D, Daviess County, Indiana, Rancholabrean LMA–Late Wisconsinan (Richards 1992c).

Mink
Mustela vison

The mink, by far the largest weasel of the genus *Mustela*, is a rich brown color throughout the year. Its partially webbed toes also help to distinguish it from its relatives. Mink weigh from about 1.5 to about 3 pounds. They are found near ponds, lakes, and streams that have brushy cover nearby. In the summer, they eat crayfish, fishes, frogs, and small mammals. In the winter, small mammals and muskrats are important items in the mink's diet.

Mink occur throughout most of Alaska and Canada and nearly all of the continental United States except for the dry Southwest. In the fossil record, mink are known from Middle Pleistocene to modern times. In the Pleistocene of the Great Lakes region, mink are known from the Late Wisconsinan of northwest and west-central Ohio and south-central and southwestern Indiana. Mink would have been important predators of crayfish and aquatic vertebrates as well as small mammals during the Pleistocene of the Great Lakes region.

Sites. Sheriden Pit Cave, Wyandot County, Ohio, Rancholabrean LMA–Late Wisconsinan (McDonald 1994). Carter, Darke County, Ohio, Rancholabrean LMA–Late Wisconsinan (McDonald 1994). Anderson Pit Cave, Monroe County, Indiana, Rancholabrean LMA–Late Wisconsinan (Graham and Lundelius

1994). Prairie Creek D, Daviess County, Indiana, Rancholabrean LMA–Late Wisconsinan (Richards 1992c).

SKUNKS
Subfamily Mephitinae

Striped Skunk
Mephitis mephitis

The striped skunk is black with a white patch on the head and usually two broad stripes on the back. It has short legs and a head that appears too small for the body. Skunks weigh from about 4 to about 10 pounds. Both the striped skunk and the spotted skunk have a very strong musk that they use to defend themselves against predators. The striped skunk prefers habitats that include both woods and open areas, but it is able to exist in agricultural and suburban habitats. The striped skunk's diet changes with the seasons. In the spring and summer, it is mainly insectivorous, but as the season moves on it takes a wide variety of small vertebrates as well as ripening fruits. In the winter, it feeds mainly upon small mammals, especially voles and mice.

Striped skunks range through most of Canada and the United States south into northern Mexico. In the fossil record, the striped skunk is known from Late Pliocene to modern times. In the Pleistocene of the Great Lakes region, it is known from the Late Wisconsinan of northwestern Ohio. The striped skunk would have been an important insectivore and also a predator of small mammals in woodlands and open areas during the Pleistocene of the Great Lakes region.

Site. Sheriden Pit Cave, Wyandot County, Ohio, Rancholabrean LMA–Late Wisconsinan (McDonald 1994).

Spotted Skunk
Spilogale putorius

The spotted skunk is smaller than its relative, the striped skunk, and has a variable pattern of white stripes, bars, and spots on its back and sides, whereas the striped skunk never has more than two white stripes. Spotted skunks weigh between about .5 and about 1 pound. Spotted skunks prefer open woodland or shrubby woodland edge habitats. As in the striped skunk, the diet is seasonal. In the summer and fall, in-

sects, grains, and fruits are important, whereas in winter and spring, small mammals are the main diet.

At present, spotted skunks barely enter Canada in western Ontario and British Columbia. They range widely through the western two-thirds of the United States and through Mexico south to Panama. But they are mainly absent from the eastern seaboard states, except for Georgia and Florida, and are absent from the Great Lakes region, except for extreme southern Indiana and Illinois. In the fossil record, spotted skunks are known from the Late Pliocene to modern times. In the Pleistocene of the Great Lakes region, spotted skunks are known from the Sangamonian and Late Wisconsinan of south-central Indiana, somewhat north of their present range in the southern tip of that state. The little spotted skunk would have been an important predator of insects and small mammals in open woodland and brushy situations during the Pleistocene of the Great Lakes region.

Sites. Harrodsburg Crevice, Monroe County, Indiana, Rancholabrean LMA–Sangamonian (Richards 1985). Indun Rock Shelter, Monroe County, Indiana, Rancholabrean LMA–Late Wisconsinan (Richards and Munson 1988).

OTTERS
Subfamily Lutrinae

River Otter
Lutra canadensis

The river otter has the long body of the mustelid family but is quite modified for aquatic life. It has a round, very thick tail; fully webbed toes on all of its feet; reduced ears; and a shiny, waterproof coat. It is a medium-sized animal (large for a mustelid) that weighs from about 10 to about 30 pounds. The otter prefers rather deep ponds, lakes, and streams with steep banks. The main diet of otters is fishes, and they are said to mainly pursue the slower, nongame species. Otters are well known for their intelligence and playful nature.

The river otter occurred throughout most of Canada and the continental United States at one time, but it became extirpated in many areas after 1900. Otters are not uncommon today in some of the wilder areas of the Great Lakes region. In the fossil record, river otters are known from the Middle Pleistocene to

FIGURE 89. A jaguar feeding on an armadillo in the Pleistocene (Sangamonian) of south-central Indiana.

the Holocene. In the Pleistocene of the Great Lakes region, river otters are known from the Late Wisconsinan of northwestern Ohio and southern Indiana. The river otter would have been an important fish-eating animal of permanent ponds, lakes, and streams during the Pleistocene of the Great Lakes region.

Sites. Sheriden Pit Cave, Wyandot County, Ohio, Rancholabrean LMA–Late Wisconsinan (McDonald 1994). Megenity Peccary Cave, Crawford County, Indiana, Rancholabrean LMA–Late Wisconsinan (Richards and Whitaker 1997).

CATS
Felidae

The cats are carnivores in the narrowest sense, with long canines, an enlarged set of carnassials, and a re-

duced number of crushing teeth. Cat skulls are short compared to those of other carnivores. The claws are long and sharp and usually retract into a protective, fleshy sheath. Most felids have climbing ability. Felids range from small cats that weigh about 3 pounds to very large predators that weigh over 800 pounds. Presently, the family occurs naturally in most parts of the world accept Antarctica, Australia, and several large islands. Cats are known from Early Eocene to modern times. Two felids, the jaguar (*Panthera onca*) and the *sabertooth (*Smilodon fatalis*), are known from the Pleistocene of the Great Lakes region.

Jaguar
Panthera onca (figs. 89 and 90)
The jaguar is a large cat that weighs up to 300 pounds (fig. 89). It can be distinguished by markings on the sides and back that are formed by a ring of black with

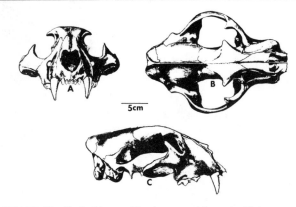

FIGURE 90. Skull of jaguar (*Panthera onca*) from the Pleistocene of Nebraska. The jaguar occurred in the Sangamonian of south-central Indiana. A, anterior view; B, dorsal (top) view; C, lateral view.

FIGURE 91. A Pleistocene *sabertooth (*Smilodon fatalis*). From Barbour and Cook (1914).

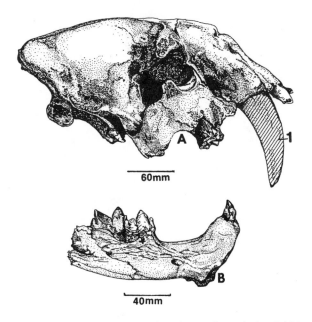

FIGURE 92. Cranial elements of a sabertooth (*Smilodon fatalis*) from the Pleistocene of Texas. A, skull in lateral view bearing saber (1) modified from canine. B, right lower jaw in lateral view. In the Great Lakes region, the sabertooth is known only from the Pleistocene (Sangamonian) of south-central Indiana.

a small spot in the center. The jaguar inhabits forests, grasslands, scrubby country, and even deserts in the northern part of its range. Jaguars stalk or ambush their prey, and it is said that capybaras (very large rodents) and peccaries (piglike animals) are their most important food. Nevertheless, studies of modern jaguars have shown that turtles, tortoises, and armadillos are important items in the diet of these large predators; the jaguar, with its wide head and extra strong jaws (fig. 90), is especially adapted for eating these shelled animals. I once studied a Late Pleistocene fissure site in northwest Georgia that was inhabited by jaguars. This site yielded a large mass of turtle-armadillo "tossed salad" and several large turtle shell bones with jaguar tooth marks on them.

Presently, jaguars occur from southern New Mexico and Arizona south through Mexico and Central America, far into South America. In the fossil record, jaguars are known from Early Pleistocene to modern times. In the Pleistocene, the jaguar is found far north of its present range, especially in pre-Wisconsinan times. In the Great Lakes region, the jaguar is known from a single Sangamonian site in south-central Indiana. These jaguar bones were identified as the large Pleistocene extinct subspecies *Panthera onca augusta*. The jaguar could have feasted upon peccaries, turtles, and armadillos in the Sangamonian Pleistocene of this part of the Great Lakes region.

Site. Harrodsburg Crevice, Monroe County, Indiana, Rancholabrean LMA–Sangamonian (Parmalee et al. 1978; Munson et al. 1980; Volz 1977).

*Sabertooth

Smilodon fatalis (figs. 91 and 92)

The extinct sabertooth is remarkably distinct from living cats in having the upper canines modified into long sabers that protruded downward from the chin when the mouth was closed (fig. 91). The sabertooth group of cats first appeared in the Eocene and is placed in its own subfamily (Machairodontinae). The Pleistocene sabertooth was as large as an African lion and had such robust limbs that it has been suggested that it stalked or ambushed its prey rather than trying to run it down. The manner in which the sabertooth used its canine sabers has been debated by the paleontological community for years. Some have suggested that the sabers (fig. 92) were used to stab and slice large prey from the top, causing the animals to bleed to death. More recent workers argue that the sabers were used to rip apart the soft bellies of the big cats' prey.

Sabertooths are known from the Middle Pleistocene to the Late Wisconsinan. They have been recorded from the Pleistocene of Saskatchewan, Canada, Indiana, Tennessee, Arkansas, Florida, Louisiana, Nebraska, Texas, New Mexico, Idaho, Oregon, Utah, and California. In the Pleistocene of the Great Lakes region, the sabertooth is known from a Sangamonian site in south-central, Indiana, the same site from which the jaguar was recovered. The big sabertooth, with its formidable canines, would have been a predator to have been reckoned with by most medium- and large-sized mammals of the area during the Pleistocene of the Great Lakes region.

Site. Harrodsburg Crevice, Monroe County, Indiana, Rancholabrean LMA–Sangamonian (Parmalee et al., 1978; Richards 1984b).

TRUE SEALS
Family Phocidae

The true seals have their hind limbs modified as flippers that extend straight backward and cannot be brought forward under the body. They also lack external ears. Phocids are large marine mammals that weigh from about 150 to 1,000 pounds. When on land, they can only move by muscular contractions of the body. At sea, they swim mainly by using hind flippers. True seals eat fishes, squid, octopuses, shellfish, and other marine creatures. Phocids occur along coastlines north of about 30 degrees N latitude and south of about 50 degrees S latitude, with some fragmentary populations in between. They also occur in Lake Baikal and the Caspian Sea. True seals are known from the Oligocene to the present, and they are thought to be descended from the otters. Three species of seals are known from the Pleistocene Champlain Sea.

Bearded Seal
Erignathus barbatus

The most recognizable character of the bearded seal is the bushy, mustachelike patch of bristles that occurs on the muzzle. The bearded seal may weigh up to 650 pounds by the time it gains its winter fat. Bearded seals prefer shallow waters near coastal areas, where they feed on bottom-living sea creatures such as crabs, sea cucumbers, mollusks, and fishes such as cod and flounders.

Bearded seals occur along the coasts and ice margins of the Arctic Ocean, south to northern Japan in the Eastern Hemisphere, and to Hudson's Bay and the Gulf of St. Lawrence in the Western Hemisphere. In the fossil record, bearded seals occur from Middle Pleistocene to modern times. They inhabited the Champlain Sea in the vicinity of Ottawa, Ontario, in the Late Wisconsinan. These large seals would have been important predators on bottom living invertebrates and fishes during Champlain Sea times in the Pleistocene of the Great Lakes region.

Site. McWilliams Gravel Pit near Finch, Stormont County, Ontario, Rancholabrean LMA–Late Wisconsinan (Harington 1988).

Harp Seal
Phoca groenlandica

Harp seals are light gray and have a dark face mask that extends to behind the eye. The mask is black in males and lighter in females. Large harp seals may reach a weight of over 350 pounds. This species lives in deep-sea coastal waters and is usually associated with drifting pack ice. The harp seal feeds on a wide variety of sea life, from large species of zooplankton (small sea animals) and crustaceans to both open-water and bottom-dwelling fishes. These seals are good swimmers and can dive more than 700 feet to get their food.

Harp seals occur from Hudson's Bay and the Gulf of St. Lawrence to northwestern Siberia. In the fossil record, this species is known from Late Wisconsinan to modern times. Harp seals inhabited the Champlain Sea in Late Wisconsinan times, where their remains have been identified near Ottawa, Ontario. This seal would have been an important predator of crustaceans and fishes in the Late Wisconsinan Champlain Sea.

Site. Green Creek Site, Carleton County, Ontario, Rancholabrean LMA–Late Wisconsinan (Harington 1988).

Ringed Seal
Phoca hispida

The ringed seal may be recognized by its black spots, many of which are surrounded by irregular lighter rings. The irregular rings give the animal a mottled appearance. This seal inhabits both the permanent ice of the arctic region and seasonally shifting ice packs. Ringed seals maintain breathing holes by scratching the ice with the claws on the front set of flippers. Like the harp seal, the ringed seal feeds on a wide variety of marine animals.

The ringed seal is the most common seal of the Arctic Ocean and adjoining seas. It occurs in a freshwater lake in both Finland and Russia. This species is known from the Middle Pleistocene to the Holocene. Ringed seals inhabited the Champlain Sea during the Late Wisconsinan, and their remains have been identified near Ottawa, Ontario. The presence of the ringed seal has been interpreted as suggesting that land-fast ice existed near the western margin of the Champlain Sea during the Late Wisconsinan. This seal would have been an important predator upon crustaceans and fishes in the cold waters of the Champlain Sea.

Site. Two Sand Pits near Ottawa International Airport, Carleton County, Ontario, Rancholabrean LMA–Late Wisconsinan (Harington 1988).

WHALES
Order Cetacea

Whales are large to giant aquatic mammals with cylindrical, essentially hairless bodies; a shortened neck; a telescoped skull; the forelimbs modified as clawless paddles; the hind limbs missing; and a tail that is modified as a horizontal fluke (fig. 93). As one can see, whales are a mammalian group that has very much departed from the basic mammalian body plan. But aside from their unusual appearance, many whales, such as the familiar porpoises and dolphins, are among the most intelligent of the nonprimate mammals. Some think that the whales should be considered to belong to two separate orders, but here I will recognize two suborders of the order Cetacea, one composed of toothed carnivores (Odontoceti) and the other composed of toothless forms that strain their food through a system of baleen plates (Mysticeti). Whales occur in all oceans and seas of the world. It is believed that whales evolved from some very odd- looking, primitive, four-legged

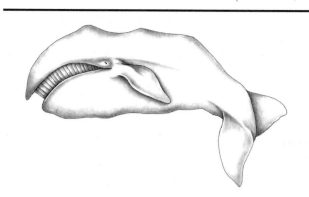

FIGURE 93. Bowhead whale (*Balaena mysticetus*).

mammals called mesonychids. Three families of whales—the white whales and narwhals (Monodontidae), the rorquals (Balaenopteridae), and the bowhead and right whales (Balaenidae)—occur in the Pleistocene of the Great Lakes region.

TOOTHED WHALES
Suborder Odontoceti

WHITE WHALES AND NARWHALS
Family Monodontidae

The family Monodontidae consists of toothed whales with high, rounded foreheads; a short, broad snout; and no dorsal fin. These animals attain a weight of over 3,000 pounds. Monodontid whales are characteristic of the arctic ice pack, occurring in the Arctic Ocean and adjacent seas. They often swim up large rivers in both the Eastern and Western Hemispheres. This family is composed of two genera, each with a single species. Monodontids are known from the Late Miocene to the Holocene. A single species, the white whale (*Delphinapterus leucas*), occurs in the Pleistocene of the Great Lakes region.

White Whale
Delphinapterus leucas

The white whale is easily recognized by its high, rounded forehead; short, rounded snout; and white color. White whales (also called beluga whales) regularly attain weights of over 3,000 pounds. This species inhabits deep, offshore waters as well as shallow coastal situations, and sometimes it swims far up into large rivers. White whales are particularly adept at sculling backward with their tails. The young of the species feed mainly on bottom-living crustaceans, whereas adults primarily prey upon fishes. White whales are very vocal, making trills, whistles, and clicks above water.

White whales occur mainly in the Arctic Ocean and adjoining seas. In the Great Lakes region, white whales are known from Champlain Sea Late Wisconsin sites near Ottawa, Ontario. These beautiful animals would have been important predators of crustaceans and fishes in the Late Wisconsinan Champlain Sea.

Sites. Cornwall, Ontario (nearly complete skeleton), Rancholabrean LMA–Late Wisconsinan (Harington 1988). Foster Sand Pit, Carleton County, Ontario, Rancholabrean LMA–Late Wisconsinan (Harington 1988). Parkenham, Lanark County, Ontario, Rancholabrean LMA–Late Wisconsinan (Harington 1988).

BALEEN WHALES
Suborder Mysticeti

RORQUALS
Family Balaenopteridae

The rorquals are toothless whales that strain their food through a baleen sieving apparatus. The members of this family are easily recognized by the longitudinal folds and grooves present on the throat, the chest, and sometimes the belly. Rorquals weigh up to 32,000 pounds. These animals occur in all of the world's oceans and adjacent seas. The rorquals feed by engulfing schools of small fish or masses of crustaceans. The Balaenopteridae are known from Miocene to modern times. A single species, the humpback whale (*Megaptera novaeangliae*), is known from the Pleistocene of the Great Lakes region.

Humpback Whale
Megaptera novaeangliae
The sturdy humpback whale may be recognized by the conspicuous grooves that extend from the tip of the bottom of the snout back onto the belly almost to the base of the tail. These whales attain a weight in excess of 60,000 pounds. Humpback whales are graceful swimmers and quite acrobatic, often doing somersaults

in the water. Humpback whales mainly feed on schools of small fish, which they gulp into the mouth and strain through their baleen meshwork.

Humpback whales occur in all of the world's oceans and adjacent seas. In the Great Lakes region, they are known from the Late Wisconsinan Champlain Sea in Ontario. These whales would have been important consumers of vast numbers of small fishes in the Champlain Sea in Late Wisconsinan times.

Site. Welsh Gravel Pit near Smith Falls, Lanark County, Ontario, Rancholabrean LMA–Late Wisconsinan (Harington 1988).

BOWHEAD AND RIGHT WHALES
Family Balaenidae

The family Balaenidae consists of both the smallest and largest of the baleen whales. Balaenid whales lack the conspicuous body grooves and folds that are so conspicuous on the rorquals. Balaenids feed on zooplankton by swimming along slowly with the mouth open through the clouds of zooplankton that occur in the oceans and seas. Whales of this family occur in all ocean waters, with the exception of the south polar and tropical seas. The Balaenids are known from Early Miocene to modern times. A single species, the bowhead whale (*Balaena mysticetus*), is known from the Pleistocene of the Great Lakes region.

Bowhead Whale
Balaena mysticetus (fig. 93)
The beautiful bowhead whale is black except for the front of the lower chin, which is cream-colored. The bowhead lacks the body folds and grooves of the rorquals. Large bowhead whales attain a weight of about 140,000 pounds. Bowheads typically are found near ice floes. These gigantic animals feed on the clouds of zooplankton that occur in arctic seas by swimming along with the mouth open and filtering the food through their baleen plates.

Bowheads live chiefly in the Arctic Ocean and its tributary seas and seldom venture farther south than about 45 degrees N latitude. In the Great Lakes region, bowhead whale remains have been found in the Late Wisconsinan Champlain Sea near Ottawa, Ontario. Bowhead whales would have consumed prodigious amounts of zooplankton in the cold Late Wisconsinan Champlain Sea.

Site. Hanson's Gravel Pit, near White Lake, Renfrew County, Ontario, Rancholabrean–Late Wisconsinan (Harington 1988).

ODD-TOED UNGULATES
Order Perissodactyla

Ungulates are large, herbivorous mammals. The odd-toed ungulates are hoofed mammals with the middle toe larger than the others (fig. 94). This group includes the horses (including asses and zebras), tapirs, and rhinos. The order is native to parts of Eurasia, the East Indes, and Africa, and from southern Mexico to Argentina. Perissodactyls are known in the fossil record from the Early Eocene to modern times and trace their history back to some very strange ancient mammals called condylarths. Both the horse family (Equidae) and tapir family (Tapiridae) are known from the Pleistocene of the Great Lakes region. Pleistocene and modern horses have high crowned cheek teeth (fig. 94). Figure 95 is a location map for extinct Pleistocene perissodactyls of the Great Lakes region.

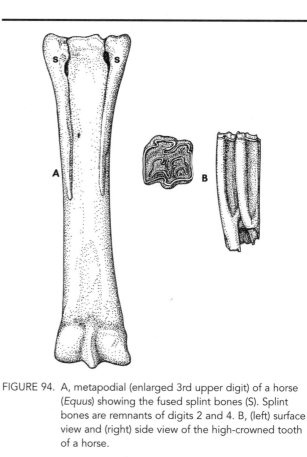

FIGURE 94. A, metapodial (enlarged 3rd upper digit) of a horse (*Equus*) showing the fused splint bones (S). Splint bones are remnants of digits 2 and 4. B, (left) surface view and (right) side view of the high-crowned tooth of a horse.

HORSES
Family Equidae

Equids are large, odd-toed ungulates with long, slender limbs adapted for running. Only the middle toe (see fig. 94), which terminates in a hoof, is functional in each limb. Wild equids weigh from about 400 to 1,000 pounds, but domestic ones have been bred much smaller and much larger. Horses and their relatives are usually social animals that tend to congregate in groups. They prefer open areas where short grasses and water is available. Presently, natural populations are found in parts of Africa and remote parts of Eurasia.

There is but a single living genus, *Equus*, in the family (figs. 96 and 97; see fig. 94). Modern horses have much more complicated and higher-crowned teeth (fig. 97; see fig. 94) than those of the tapir family (Tapiridae). Horses are known from Early Eocene to modern times. The fossil record of horses is excellent, and it demonstrates the pathways and patterns of evolution very well. The *complex-toothed horse (*Equus

complicatus) is known from the Pleistocene of the Great Lakes region.

Complex-toothed Horse
Equus complicatus (fig. 98)
The extinct complex-toothed horse was identified on the basis of the complex folds of enamel that occur on the biting surfaces of its upper cheek teeth (fig. 98). This species was the first extinct Pleistocene horse to be described in North America. It was a big horse that differs from an equally large extinct Pleistocene horse, *Equus scotti*, on the basis of the less complex enamel pattern on the teeth of *E. scotti* (fig. 98). The complex-toothed horse was probably the most common big horse in the late Pleistocene of the eastern United States. Undoubtedly, *E. complicatus* grazed on grasses, perhaps using its complex teeth on special varieties.

Equus complicatus is known from about Middle Pleistocene to Late Wisconsinan times, where it occurred from Florida to the Gulf Coast of Texas as well

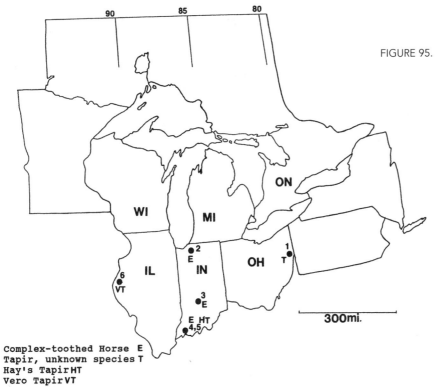

FIGURE 95. Pleistocene localities of extinct perissodactyls in the Great Lakes region. 1. Undesignated site in Columbia County, Ohio, probably Late Pleistocene. 2, Undesignated site in Laporte County, Indiana, Late Pleistocene. 3. Harrodsburg Crevice, Monroe County, Indiana, Sangamonian. 4. Undesignated Ohio River site, Vanderburg County, Indiana, probably Late Pleistocene. 5. Pigeon Creek Site. Vanderburg County, Indiana, probably Late Pleistocene. 6. Undesignated site near Carthage, Hancock County, Illinois, probably Late Pleistocene.

FIGURE 96. A generalized Pleistocene horse of the genus *Equus*.

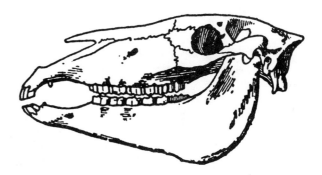

FIGURE 97. A generalized horse skull of the genus *Equus* in lateral view. From Berry (1929).

as in South Carolina, Kentucky, Indiana, and Missouri. In the Pleistocene of the Great Lakes region, horses identified as *Equus* cf. *Equus complicatus* are known from the Sangamonian of south-central Indiana and from the Late Pleistocene of northwestern and southern Indiana. The complex-toothed horse was undoubtedly an important grass-eating grazer of open regions during the Pleistocene of the Great Lakes region.

Sites. Harrodsburg Crevice, Monroe County, Indiana, Rancholabrean LMA–Sangamonian (Parmalee et al. 1978). Undesignated site, La Porte County, Indiana, Late Pleistocene (Lyon 1931). Ohio River, Vanderburg County, Indiana, Late Pleistocene (Leidy 1984–85).

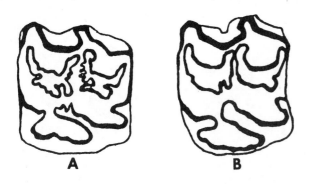

FIGURE 98. Surface view of molar teeth of extinct Equus showing the enamel patterns (heavy black lines) of A, the *complex-toothed horse (*Equus complicatus), and B, another extinct Pleistocene horse, *Equus scotti. Only the complex-toothed horse has been identified from the Pleistocene of the Great Lakes region.

TAPIRS
Family Tapiridae

Tapirs are heavy-bodied perissodactyls with short legs and a short, muscular proboscis that is formed by the upper lip and tissue around the nostrils (fig. 99). Tapirs have lower-crowned teeth with less complicated surfaces than in the horses. Tapirs weigh from about 450 to about 600 pounds. These unique animals presently inhabit wet forests and swampy areas and in some cases even wet mountain slopes. Tapirs feed upon tree leaves and tender vegetation on the forest floor as well as on aquatic plants. They are solitary, secretive animals that may have nocturnal habits from time to time. Presently, these animals occur from southern Mexico through the northern half of South America in the Western Hemisphere and from southeastern Burma and southern Thailand south through Malaya to Sumatra in the Eastern Hemisphere. The family has but a single living genus, *Tapirus*, and there are four living species. Tapirs are known from Early Oligocene to modern times. In the fossil record, the modern genus, *Tapirus*, is known from the Late Pliocene to the Late Wisconsinan. *Tapirus* probably reached North America from South America over the Central American land bridge that formed in Late Pliocene times. Both currently recognized North American Pleistocene species, *Hay's tapir (*Tapirus haysii) and the *Vero tapir (*Tapirus

veroensis), are known from the Pleistocene of the Great Lakes region.

Hay's Tapir
Tapirus haysii

Hay's tapir, named in honor of the former Indianapolis vertebrate paleontologist Oliver Perry Hay, has recently been restudied and rescued from the junk pile of discarded scientific names. Hay's tapir is a large species that occurred in the North American Pleistocene from Pennsylvania, southern Indiana, and north-central Nebraska south to Florida and Texas and west to California.

In the Pleistocene of the Great Lakes region, it has been recorded from what is probably the Late Wisconsinan of the tip of southern Indiana. A tapir jaw, now lost, was identified in 1850 as *Tapirus* sp. by several paleontologists from a site in northeast Ohio that presumably represents the Late Pleistocene. This jaw could represent either Hay's tapir or the Vero tapir. Hay's tapir would have been an important vegetarian of moist woodlands and swamps during the Pleistocene of the southern part of the Great Lakes region.

Sites. Pigeon Creek, Vanderburg County, Indiana, presumably Rancholabrean LMA–Late Wisconsinan (Ray and Sanders 1984). Columbiana County, Ohio, unknown site, probably Late Pleistocene (*Tapirus* sp., Hansen 1992).

Vero Tapir
Tapirus veroensis (fig. 99)

The Vero tapir was named for the Late Wisconsinan Vero Site in Florida, where it was found. It is considered to be a smaller animal than Hay's tapir, and its Pleistocene range is not as well documented as that of Hay's tapir, although it is known that in Florida, both species occurred in the same general area. A neglected record of the Vero tapir based on a tooth collected in 1887 from a site presumably representing the Late Pleistocene in west-central Illinois has recently been brought to light. The Vero tapir was undoubtedly a vegetarian, and it is assumed it would have existed in wet woodlands and swamps in the Pleistocene of the Great Lakes region.

Site. Carthage Tapir Site, Hancock County, Illinois, Rancholabrean LMA–Late Wisconsinan (Ray and Sanders 1984).

FIGURE 99. The *Vero tapir *(*Tapirus veroensis)* in a rocky area near the Illinois River of the Pleistocene in western-central Illinois.

EVEN-TOED UNGULATES
Order Artiodactyla

Artiodactyls are ungulates that have the body weight mainly borne by the sometimes fused third and the fourth toes of each limb (fig. 100). This group is bewilderingly diverse and includes many extinct and modern families. Artiodactyls range in size from the delicate tragulids that weigh a little over 1 pound to the hippopotamus that reaches a weight of over 9,000 pounds. Three artiodactyl families—the peccaries (Tayassuidae), deer and relatives (Cervidae), and bovids (Bovidae)—are known as Pleistocene fossils in the Great Lakes region.

PECCARIES
Family Tayassuidae

Peccaries have a piglike nose and body, but the legs are long and slim and the hooves are small. The upper canines are developed as short, very sharp, downwardly di-

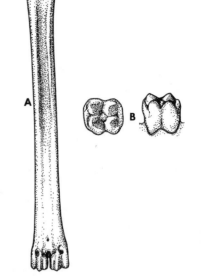

Figure 100. A, an artiodactyl (white tailed deer, *Odocoileus virginianus*) metapodial composed of fused digits 3 and 4. B, left (surface) and right (side) view of generalized cheek tooth of Pleistocene *peccary (*Platygonus*).

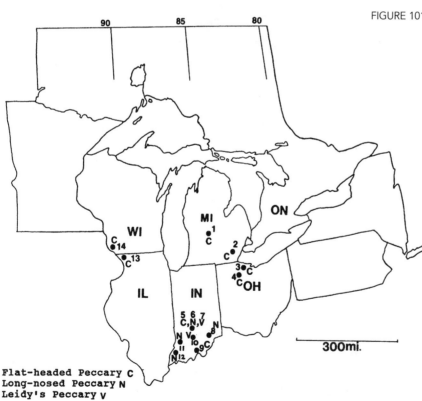

Flat-headed Peccary C
Long-nosed Peccary N
Leidy's Peccary V

FIGURE 101. Pleistocene peccary localities in the Great Lakes region. Selected sites are shown for the *flat-headed peccary (*Platygonus compressus*), *long-nosed peccary (*Mylohyus nasutus*), and *Leidy's peccary (*Platygonus vetus*). 1. Belding Peat Bog Site, Ionia County, Michigan, Late Wisconsinan. 2. Huron River Site, Washtenaw County, Michigan, Late Wisconsinan. 3. Fremont Site, Sandusky County, Ohio, Late Wisconsinan. 4. Sheriden Pit Cave Site, Wyandot County, Ohio, Late Wisconsinan. 5. Indun Rock Shelter, Monroe County, Indiana, Late Wisconsinan. 6. Undesignated site, Monroe County, Indiana, Late Pleistocene. 7. Harrodsburg Crevice, Monroe County, Indiana, Sangamonian. 8. Undesignated site, Jennings County, Indiana, Late Pleistocene. 9. Megenity Peccary Cave, Crawford County, Indiana, Late Wisconsinan. 10. Undesignated site, Lawrence County, Indiana, Sangamonian. 11. Prairie Creek D Site, Daviess County, Indiana, Late Wisconsinan. 12. Undesignated site, Gibson County, Indiana, Late Pleistocene. 13. Galena Site, Jo Daviess County, Illinois, Late Pleistocene. 14. Castle Rock Cave, Grant County, Wisconsin, Late Pleistocene.

rected tusks that are used for fighting and defense, but the other teeth are low and have blunt cusps (fig. 100). Modern peccaries weigh about 50 or 60 pounds and move in groups of about ten to twenty-five individuals. When confronted or excited, peccaries raise the hair on the neck and back and give off a strong scent from a gland on the back. It has been reported that intruders have been charged by groups of modern peccaries. Peccaries are omnivorous and feed upon invertebrates; small, slow vertebrates; and roots, bulbs, and fruits. Presently, peccaries range from the southwestern United States south to Argentina. The Tayassuidae are known from Early Oligocene to modern times. Two extinct genera, *Mylohyus and *Platygonus, are known from the Pleistocene of the Great Lakes region. Figure 101 is a location map of Pleistocene peccary sites in the Great Lakes region; only selected sites are shown for the *flat-headed peccary (*Platygonus compressus).

*Long-nosed Peccary
Mylohyus nasutus (fig. 102)
The long-nosed peccary had a long rostrum and long, slender legs. It was about the size of a small white-tailed deer. The long-nosed peccary was a swift runner, had good eyesight, and has not been found in large groups like the extinct peccary genus *Platygonus*. Thus it is believed that long-nosed peccaries were solitary animals rather than running in packs. This peccary probably inhabited open areas or forest edges.

The long-nosed peccary is known from about Middle Pleistocene to Late Wisconsinan times, where it occurred from Pennsylvania, Indiana, and Iowa south through Oklahoma and Texas to Florida. It appears to have been much less common than the extinct peccary genus *Platygonus* in most areas except Florida. In the Pleistocene of the Great Lakes region, the long-nosed peccary is known from the Late Wisconsinan of southwestern Indiana and other Late Pleistocene sites in south-central, southeastern, and the extreme southern tip of Indiana. The Gibson County site in the extreme southern tip of Indiana yielded the type specimen of the species. This species would have been an important large browser in the Late Pleistocene of Indiana.

Sites. Prairie Creek D, Daviess County, Indiana, Rancholabrean LMA–Late Wisconsinan (Tomak 1975 1982). Monroe County, Indiana, Late Pleistocene (Richards 1984b). Jennings County, Indiana, Late Pleistocene (Richards 1984b). Undesignated site, Gib-

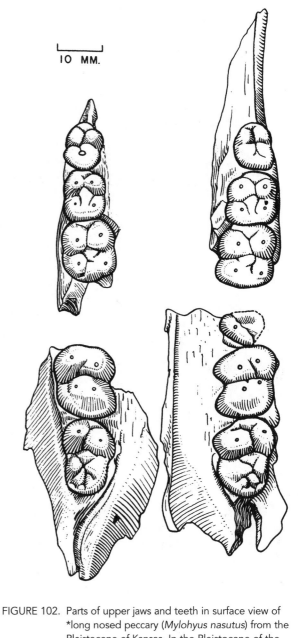

FIGURE 102. Parts of upper jaws and teeth in surface view of *long nosed peccary (*Mylohyus nasutus*) from the Pleistocene of Kansas. In the Pleistocene of the Great Lakes region, the long-nosed peccary is known only from sites in southern Indiana (see fig. 101).

son County, Indiana (type specimen), Late Pleistocene (Cope and Wortman 1885).

*Flat-headed Peccary
Platygonus compressus (figs. 103 and 104)
The flat-headed peccary was somewhat smaller than the long-nosed peccary (about the size of a modern

FIGURE 103. A group of *flat-headed peccaries *(*Platygonus compressus)* in the Late Wisconsinan of Michigan.

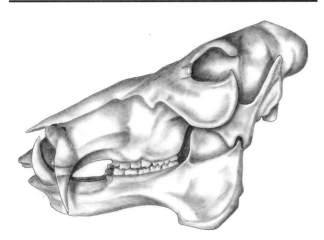

FIGURE 104. Flat-headed peccary (*Platygonus compressus*) skull from the Belding Peat Bog Site, Ionia County, Michigan, Late Wisconsinan (see fig. 101).

wild pig) and had a shorter snout. Pleistocene cave deposits, including those of the Great Lakes region, have yielded large numbers of flat-headed peccaries Flat-headed peccaries lived in small herds or packs made up of all age groups and both sexes (fig. 103). These animals had very sharp lower canines that fitted snugly in a groove in the upper jaw (fig. 104), which were present in both sexes and were used for defense against predators and for intraspecific brawls. The structure of the molar teeth indicates that flat-headed peccaries could chew tough vegetation, thus suggesting browsing food habits. It has also been suggested that this species preferred open areas or woodland edges.

Flat-headed peccaries were very widespread in the Pleistocene, occurring from New York to California and south into Mexico. They are known from Sangamonian to Late Wisconsinan times. In the Pleistocene of the Great Lakes region, they are known from the Late Pleistocene of all states but not from Canada. In Ohio alone, they have been identified from fifteen counties. Herds of flat-headed peccaries must have been important browsers in open situations and woodland edges during the Late Pleistocene of the Great Lakes region.

Selected Sites. Belding Peat Bog, Ionia County, Michigan, Rancholabrean LMA–Late Wisconsinan (Wilson 1967). Huron River, Ann Arbor, Washtenaw County, Michigan, Rancholabrean LMA–Late Wisconsinan (Eshelman et al. 1972). Sheriden Pit Cave, Wyandot County, Ohio, Rancholabrean LMA–Late Wisconsinan (McDon-

ald 1994). Fremont, Sandusky County, Ohio, Rancholabrean LMA–Late Wisconsinan (Hoare et al. 1964). Indun Rock Shelter, Monroe County, Indiana, Rancholabrean LMA–Late Wisconsinan (Richards and Munson 1988). Megenity Peccary Cave, Crawford County, Indiana, Rancholabrean LMA–Late Wisconsinan (Richards 1988a). Galena, Jo Daviess County, Illinois, Late Pleistocene (Baker 1920). Castle Rock Cave, Grant County, Wisconsin, Late Pleistocene (Palmer 1974).

*Leidy's Peccary
Platygonus vetus

Leidy's peccary is one of the largest peccaries known and further differs from other species in having massive, widely flaring cheek bones that extend as flaps down the skull to below the level of the upper tooth rows. Little is known about the habits of this peccary except that it inhabited plains and open forests. Leidy's peccary occurred in scattered localities from Appalachia and Florida well into the western states but only from about Middle Pleistocene to Sangamonian times. In the Great Lakes region, it is known from Sangamonian sites in south-central and southern Indiana.

Sites. Harrodsburg Crevice, Monroe County, Indiana, Rancholabrean LMA–Sangamonian (Parmalee et al. 1978; Munson et al. 1980; Volz 1977). Undesignated site, Lawrence County, Indiana, Rancholabrean LMA–Sangamonian (Hay 1912).

CAMELS AND LLAMAS
Family Camelidae

Camelids are hornless artiodactyls with a long, thin neck and very long limbs (fig. 105). Instead of having hooves, their toes splay out on the ground and are capped with small, padded nails. Modern camels reach a weight of 1,300 pounds, and llamas are smaller. Most wild camelids are grazers that travel in herds. Wild camels occur in the Gobi Desert and in northern China. Wild llamas occur in the Andes of South America from southern Peru to the tip of the continent. This family is known from the Eocene to the Late Pleistocene of North America. Camelids did not reach Europe, North Africa, and Asia until the Pliocene and did not reach South America until the Late Pleistocene. Only one

FIGURE 105. A, an *American camel (*Camelops). B, right maxillary fragment in surface view of an American camel (Camelops sp.) from the Pleistocene of Texas. In the Pleistocene of the Great Lakes region, Camelops is known only from an undesignated site of probable Late Pleistocene age in Harrison County, Indiana.

genus, *Camelops (fig. 105), has been recorded from the Pleistocene of the Great Lakes region.

*American Camel
*Camelops sp. (fig. 105)
Material identified only as *Camelops sp. has been recorded from the Pleistocene (presumably Late Pleistocene) of southern Indiana. The teeth of this genus are relatively uncomplicated in surface view (fig. 105).

Site. Undesignated site in Harrison County, Indiana, presumably Late Pleistocene (Richards 1984b).

DEER AND RELATIVES
Family Cervidae

The cervids comprise a very large group of artiodactyls that have crescentlike enamel ridges on their somewhat high-crowned teeth (fig. 106). Antlers are usually present, with some species that lack antlers having saberlike upper canines. Cervids weigh from about 20 to over 1,600 pounds. Presently, cervids are nearly worldwide in distribution and are absent only from permanent ice fields, Africa south of the Sahara region, Australia, and New Zealand. They are known from Oligocene to modern times. Five genera—moose (*Alces*), *Scott's moose (stag moose) (*Cervalces*), wapiti or elk (*Cervus*), white-tailed and mule deer (*Odocoileus)*, and caribou (*Rangifer*)—occur in the Pleistocene of the Great Lakes region. C. R. Harington of the Manitoba Museum of Nature, Ottawa, Ontario, has suggested in a communication to me (May 2001) that *Alces* may be the best generic name for Scott's moose

Moose
Alces alces (fig. 106)
The familiar moose, with its overhanging snout, humped shoulders, neck flap, and backwardly directed, flattened antlers, is the largest modern cervid, weighing from about 700 to over 1,000 pounds. Moose live in boreal woodlands that provide shrubs, immature trees, and adjacent aquatic habitats such as ponds and cedar swamps. These large mammals are good swimmers and can run over 30 miles per hour. Moose feed on tender tree leaves as they appear in the spring and then begin to eat plants like water lilies, arrowheads, and horsetails that occur in or near aquatic situations.

Presently, the moose lives in boreal forest regions across northern Europe, Asia, and North America. In the fossil record, it is known from the Early Wisconsinan and possibly earlier in Alaska. In the Great Lakes region, moose are known from a number of Holocene sites, but reliably dated Pleistocene sites are lacking, other than a Late Wisconsinan site in southeastern Michigan. The moose certainly would have been an important cervid herbivore during the Pleistocene of the Great Lakes region.

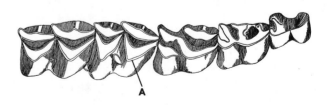

FIGURE 106. Lower dentition of a moose (Alces) in surface view showing crescentlike enamel ridges (A). From Hibbard (1940a).

FIGURE 107. A *Scott's moose (stag moose) (*Cervalces scotti*) in the Late Wisconsinan of Michigan.

Site. Shelton, Oakland County, Michigan, Rancholabrean LMA–Late Wisconsinan (Shoshani et al. 1989).

*Scott's Moose (Stag Moose)
Cervalces scotti (figs. 107 and 108)

The Scott's moose, sometimes called the stag moose, is a mooselike animal of large size that differs from the modern moose in having narrower antlers with a more complicated branching pattern (figs. 107 and 108). They also have a different enamel pattern on the teeth, and the nasal bones are longer. It is believed that Scott's moose had similar habitat requirements to the modern moose (*Alces alces*). Scott's moose is known from the Sangamonian to the Late Wisconsin, so it has been suggested that competition with the modern moose, which

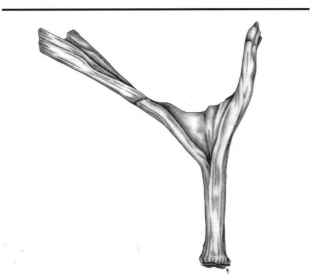

FIGURE 108. Portion of a Scott's moose (stag moose) *(Cervalces scotti)* antler from the Late Pleistocene of Michigan.

records of this extinct animal (fig. 109). In the Great Lakes region, Scott's moose is known from the Sangamonian and Early Wisconsinan of the Toronto region of Ontario and Late Wisconsinan sites in Michigan, Ohio, Indiana, and Illinois. Both Indiana and Illinois have records of the Scott's moose from at least nine sites, and there are at least six records in Ohio and at least three records in Michigan. Scott's moose must have been an important herbivore of muskegs and similar environments during the Pleistocene of the Great Lakes region.

Selected Sites. Don Formation, near Toronto, Ontario, Rancholabrean LMA–Sangamonian (Karrow et al. 1980). Pottery Road Formation, near Toronto, Ontario, Rancholabrean LMA–Early Wisconsinan (Coleman 1933). Powers Mastodont, Van Buren County, Michigan, Rancholabrean LMA–Late Wisconsinan (Garland and Cogswell 1985). Shelton, Oakland County, Michigan, Rancholabrean LMA–Late Wisconsinan (Shoshani et al. 1989). Sheriden Pit Cave, Wyandot County, Ohio, Rancholabrean LMA–Late Wisconsinan, (McDonald 1994). Carter, Darke County, Ohio, Rancholabrean LMA–Late Wisconsinan (McDonald 1994). Perkins *Cervalces*, DeKalb County, Indiana, Rancholabrean–Late Wisconsinan (Farlow et al. 1986). Shoals, Martin

arrived in the Early Wisconsinan, might have driven Scott's moose to extinction.

Records of Scott's moose are rather common in the Great Lakes region with the exception of Wisconsin, which as far as I can determine, does not have any

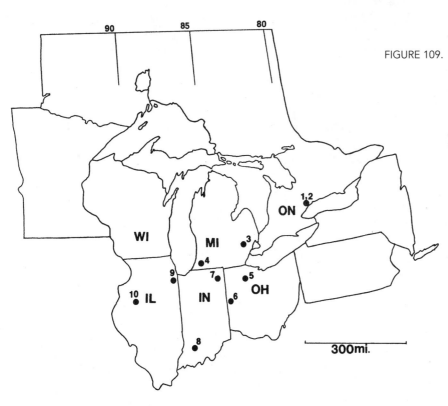

FIGURE 109. Selected Pleistocene localities of Scott's moose (stag moose) *(Cervalces scotti)* in the Great Lakes region. 1. Don Formation, near Toronto, Ontario, Sangamonian. 2. Pottery Road Formation, near Toronto, Ontario, Early Wisconsinan. 3. Shelton Mastodont Site, Oakland County, Michigan, Late Wisconsinan. 4. Powers Mastodont Site, Van Buren County, Michigan, Late Wisconsinan. 5. Sheriden Pit Cave Site, Wyandot County, Ohio, Late Wisconsinan. 6. Carter Site, Darke County, Ohio, Late Wisconsinan. 7. Perkins *Cervalces* Site, DeKalb County, Indiana, Late Pleistocene. 8. Shoals Site, Martin County, Indiana, Late Wisconsinan. 9. Beecher Site, Will County, Illinois, Late Wisconsinan. 10. Lake Wildwood Site, Fulton County, Illinois, Late Wisconsinan.

County, Indiana, Rancholabrean LMA–Late Wisconsinan (Richards 1992b). Beecher, Will County, Illinois, Rancholabrean LMA–Late Wisconsinan (Churcher and Pinsof 1987). Lake Wildwood, Fulton County, Illinois, Rancholabrean LMA–Late Wisconsinan (Strode 1976).

Wapiti (Elk)
Cervus elaphus

The wapiti, often called the elk, is a very large deer with antlers that consist of single cylindrical beams that curve back over the shoulder and bear several narrow branches. A large light patch surrounds the tail, and a shaggy mane occurs on the neck and chest. Both females and males have upper canine teeth. Wapiti weigh from about 500 to about 1,000 pounds. This large deer prefers open woods with adjoining meadows. Wapiti are social animals, and females and their young congregate in small herds in the summer. In early autumn, after intense fighting, males command a harem of up to twenty females and mate with each one as she comes into heat. After mating is finished, wapiti form winter herds that are often composed of both sexes. Wapiti feed on a variety of grasses, forbs, and mushrooms in the spring and summer but become browsers, chiefly on twigs and bark, in the fall and winter.

Wapiti occur today in scattered patches in both Eurasia and North America. In Britain they are called red deer. Within historic times, wapiti occurred over most of the continental United States and Canada but were extirpated from large areas by habitat destruction and overhunting. In the fossil record, wapiti occur from Illinoian to modern times. In the Great Lakes region, there are many Holocene records but relatively few Pleistocene records. The Pleistocene records include Late Wisconsinan deposits in southeastern Michigan, northwest Indiana, and southeastern, south-central, and southwestern Wisconsin. The wapiti was probably a widespread large herbivore of open woods and meadows during the Pleistocene of the Great Lakes region and may have become even more important after the demise of many large herbivores at the end of the Pleistocene.

Sites. Adrian Peat Bog, Lenawee County, Michigan, Rancholabrean LMA–Late Wisconsinan (Hay 1923). Duncker Muskox, LaPorte County, Indiana, Rancholabrean LMA–Late Wisconsinan (Lyon 1931). Undesignated sites in Waukesha, Dane, and Trempealeau Counties, Wisconsin, Rancholabrean LMA–Late Wisconsinan (West and Dallman 1980).

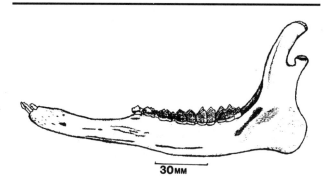

FIGURE 110. Lower jaw of a white-tailed deer (*Odocoileus virginianus*) from a modern Michigan specimen.

White-tailed Deer
Odocoileus virginianus (fig. 110)

The white-tailed deer is a relatively small cervid that may be distinguished from the much larger wapiti on the basis that the main antler beam curves forward rather than backward and that it has a white throat patch that is lacking in the wapiti. White-tailed deer weigh from about 100 to somewhat over 300 pounds, but populations of very small white-tails (such as the Florida Key deer) occur in several places. This species prefers open woodland habitats with clearings and meadows. White-tails are not as social as wapiti, but small groups of adult females, female yearlings, and new fawns tend to form. The species feeds on tender leaves, grasses, and herbs in the summer and switches to tree buds and twigs in the winter. Acorns and mushrooms are also favorite foods.

At present, white-tailed deer have a very large range from southern Canada to South America. In the fossil record, they occur from Late Pliocene to modern times. In the Pleistocene, records of this species appear to be concentrated in the eastern and central part of North America. In the Pleistocene of the Great Lakes region, white-tailed deer have been recorded from all of provinces and states. They are known from the Sangamonian and Early Wisconsinan of the Toronto region of Ontario and from the Late Sangamonian/Early Wisconsinan of southeastern Ontario. They are also recorded from the Late Wisconsinan of south-central and southeastern Michigan, northwestern and west-central Ohio, and northwestern and east-central Indiana south-central Wisconsin, and from undesignated Pleistocene localities in Illinois. The white-tailed deer must have been a very important herbivore of the woodlands and meadows during the Pleistocene of the Great Lakes region.

Selected Sites. Innerkip, Oxford County, Ontario, Rancholabrean LMA–Late Sangamonian/Early Wisconsinan, (Churcher et al. 1990). *Castoroides* Site, Eaton County, Michigan, Rancholabrean LMA–Late Wisconsinan (Holman et al. 1986). Adrian Peat Bog, Lenawee County, Michigan, Rancholabrean LMA–Late Wisconsinan (Hay 1923). Sheriden Pit Cave, Wyandot County, Ohio, Rancholabrean LMA–Late Wisconsinan (McDonald 1994). Carter, Darke County, Ohio, Rancholabrean LMA–Late Wisconsinan (McDonald 1994). Reeker Mastodont, Madison County, Indiana, Rancholabrean LMA–Late Wisconsinan (Richards 1984b). Christensen Bog, Hancock County, Indiana, Rancholabrean LMA–Late Wisconsinan (Graham et al. 1983). Undesignated site or sites, Illinois, Late Pleistocene (Kurtén and Anderson 1980). Raddatz Rock Shelter, Sauk County, Wisconsin, Rancholabrean LMA–Late Wisconsinan (Parmalee 1959).

Caribou

Rangifer tarandus (figs. 111 and 112)

Caribou, called reindeer in the Eastern Hemisphere, are the only cervids where females as well as males have antlers. The main beam of the caribou antler curves backward, then upward and forward, looking something like a question mark in side view (fig. 111). The left and the right antlers are usually asymmetrical. The legs end in very broad feet. Caribou weigh from about 200 to over 400 pounds in some large individuals. Caribou occur in both tundra and woodland situations. In fact, they are referred to as either tundra ("barren ground") or woodland caribou. Tundra caribou are very gregarious and migratory, while woodland caribou tend to have home ranges, wander much less, and exist in much smaller groups. Like other cervids, caribou tend to eat tender grasses, herbs, fruit, and mushrooms in the spring and summer and shift to browsing on twigs and bark when the snowfall comes. Caribou also eat lichens, both those that are found on the ground and on conifer trees.

The caribou occurs in both Eurasia and North America, extending in North America from the Arctic Circle south into the northern part of the Great Lakes region. Fossil caribou are known from about 1.3 to 1.6 million years B.P. to modern times. In the Pleistocene, caribou ranged as far as south Virginia and Tennessee.

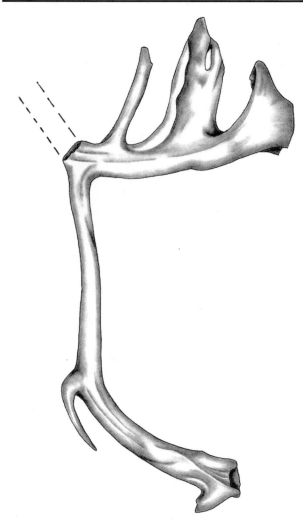

FIGURE 111. Caribou *(Rangifer tarandus)* antler from the Late Wisconsinan of Michigan.

In the Pleistocene of the Great Lakes region, caribou occur in Ontario and all of the states (fig. 113). They are known from undesignated Pleistocene sites in Ontario and from Late Wisconsinan sites in eastcentral, and southeastern Michigan; northern and northwestern Ohio; northwestern and east-central Indiana; northwestern Illinois; and southeastern Wisconsin. Caribou would have been important herbivores of ground plants, twigs, bark, and lichens in both tundra and woodland situations during the Pleistocene of the Great Lakes region.

Selected Sites. Undesignated Pleistocene sites in Ontario (Kurtén and Anderson 1980). Holcombe Beach, Macomb County, Michigan, Rancholabrean LMA–Late

FIGURE 112. A group of the woodland variety of the caribou *(Rangifer tarandus)* from the Late Wisconsinan of Michigan.

Wisconsinan (Cleland 1965). Fowlerville Caribou Site, Livingston County, Michigan, Rancholabrean LMA–Late Wisconsinan (Hibbard 1952). Sheriden Pit Cave, Wyandot County, Ohio, Rancholabrean LMA–Late Wisconsinan (McDonald 1994). Kolarik Mastodont, Starke County, Indiana, Rancholabrean LMA–Late Wisconsinan (Ellis 1982). Christensen Bog, Hancock County, Indiana, Rancholabrean LMA–Late Wisconsinan (Graham et al. 1983). Valentine Bog, Kendall County, Illinois, Rancholabrean LMA–Late Wisconsinan (Graham and Graham 1989). Wauwatosa Caribou, Milwaukee County, Wisconsin, Rancholabrean LMA–Late Wisconsinan (West 1978). Zelienka Caribou, Waushara County, Wisconsinan, Rancholabrean LMA–Late Wisconsinan (Long 1986).

BOVIDS
Family Bovidae

Bovids comprise another very large family of artiodactyls. They have crescentlike enamel ridges on the chewing surfaces of their teeth, but the teeth are higher-crowned than those of the Cervidae. Bovids have true horns that are never shed (fig. 114) rather than antlers. These horns often occur on females. Bovids range from about 6 to over 2,700 pounds. Within historic times, wild bovids occurred throughout Africa, most of Eurasia, much of temperate North America, and western Greenland. In the fossil record, the family occurred from Early Miocene to modern times. Three genera—bison (*Bison*), *woodland (helmeted) muskox (*Bootherium*),

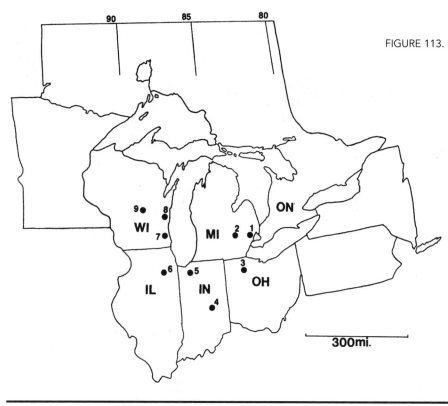

FIGURE 113. Selected Pleistocene caribou (*Rangifer tarandus*) localities in the Great Lakes region. 1. Holcombe Beach Site, Macomb County, Michigan, Late Wisconsinan. 2. Fowlerville Caribou Site, Livingston County, Michigan, Late Wisconsinan. 3. Sheriden Pit Cave Site, Wyandot County, Ohio, Late Wisconsinan. 4. Christensen Bog Mastodont Site, Hancock County, Indiana, Late Wisconsinan. 5. Kolarik Mastodont Site, Starke County, Indiana, Late Wisconsinan. 6. Valentine Bog Site, Kendall County, Illinois, Late Wisconsinan. 7. Wauwatosa Caribou Site, Milwaukee County, Wisconsin, Late Wisconsinan. 8. Oostburg Caribou Site, Sheyboygan County, Wisconsin, Late Wisconsinan. 9. Zelienka Caribou Site, Waushara County, Wisconsin, Late Wisconsinan.

and muskox (*Ovibos*)—are known from the Pleistocene of the Great Lakes region.

American Bison

Bison bison (fig. 114)

The familiar American bison, often called buffalo, has long, shaggy hair on the head, neck, and forelegs and a great hump on the back, which is larger in males than in females. The horns are short and curve upward (fig. 114). American bison weigh from about 800 to over 2,000 pounds. This species mainly subsists on grasses, which they obtain in prairies as well as in mountains and open forests. Bison tend to move about in small groups of related individuals.

Within historic times, bison occurred in western Canada and throughout most of the contiguous United States. They also may have occurred in Alaska. It was once estimated that at least 50 million bison existed in North America, but European settlers hunted the species down to fewer than 1,000 individuals by late in the nineteenth century. In the fossil record, *Bison bison* is believed to have entered North America across the

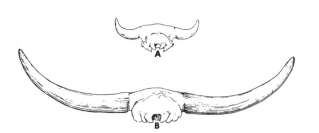

FIGURE 114. A, horns of American bison (*Bison bison*). B, horns of *giant bison (*Bison latifrons*) from the Pleistocene of the Great Plains. In the Pleistocene of the Great Lakes region, the giant bison is unquestionably known only from the Sangamonian of Ohio.

Bering Strait from Siberia in Sangamonian times. Some early *Bison bison* were larger than the modern form and were given specific and subspecific names that are now discarded. Because it is very difficult to tell bison bones and teeth from those of domestic cattle, many controversial finds of "bison" material have been recorded in literature on the Pleistocene. In the Pleistocene of the Great Lakes region, *Bison bison* has been identified from the Late Wisconsinan of southwestern Indiana and northwestern Wisconsin. Bison material identified merely as *Bison* sp. from the Late Wisconsinan of the Toronto region of Ontario probably represents *Bison bison*. The bison would have been an important grassland grazer of the grassy areas and open woodlands during the Pleistocene of the Great Lakes region.

Selected Sites. Lake Iroquois deposits near Toronto, Ontario, Rancholabrean LMA–Late Wisconsinan (*Bison* sp., Karrow et al. 1980). Knox County, Indiana, Rancholabrean LMA–Late Wisconsinan (Middleton and Moore 1900). Interstate Park Locality, Polk County, Wisconsin, Rancholabrean LMA–Late Wisconsinan (Palmer 1954).

*Giant Bison
Bison latifrons (fig. 114)

Giant bison are identified on the basis of their large size and enormous horn cores, which may be three times as long as those in the modern bison (fig. 114). Giant bison are first known from the Late Illinoian (perhaps somewhat earlier in Nebraska) and became extinct in Late Wisconsinan times. The giant bison was most widespread in North America in Sangamonian times and then declined and became confined to restricted populations in the Wisconsinan. It has been suggested that the giant bison might have become extinct because it was either out-competed or absorbed into the gene pool of (bred into) the modern bison. *Bison latifrons* has been reported from a site in southeastern Ohio that is questionably Sangamonian. Bison remains identified only as *Bison* sp. have been identified from the Sangamonian and Early Wisconsinan near Toronto, Ontario. These fossils represent either *Bison latifrons* or *Bison bison* but are tentatively included in the giant bison section because of their relatively early age. It is not known whether the giant bison ever existed in the Great Lakes region in significant numbers. Nevertheless, the individuals that were present were obviously important grazing herbivores of very large size.

Sites. Don Formation, near Toronto, Ontario, Rancholabrean LMA–Sangamonian (*Bison* sp., Karrow et al. 1980). Pottery Road Formation, near Toronto, Ontario, Rancholabrean LMA–Early Wisconsinan (*Bison* sp., Coleman 1933). Brown County, Ohio, un-

designated site(?), Rancholabrean LMA–(?) Sangamonian, Hansen 1992).

Barren Ground (Tundra) Muskox
Ovibos moschatus (figs. 115 and 116)

Barren ground (tundra) muskoxen are short, compact, shaggy bovids with a coat of coarse hair that extends down to the feet (fig. 115). They also have an insulating

FIGURE 115. Barren ground (tundra) muskoxen (*Ovibos moschatus*) forming a semicircle.

FIGURE 116. Head view of barren ground (tundra) muskox (*Ovibos moschatus*).

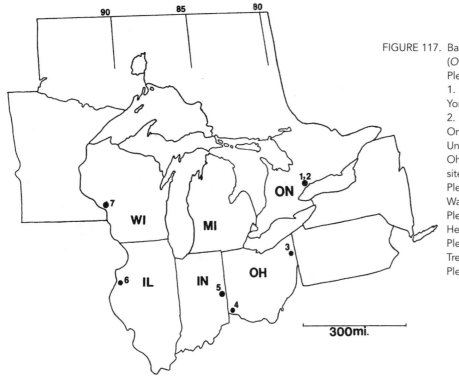

FIGURE 117. Barren ground (tundra) muskoxen (*Ovibos moschatus*) localities in the Pleistocene of the Great Lakes region. 1. Scarborough Bluff Site, near Toronto, York County, Ontario, Early Wisconsinan. 2. Lake Iroquois deposits near Toronto, Ontario, Late Wisconsinan. 3. Undesignated site, Mahoning County, Ohio, Late Pleistocene. 4. Undesignated site, Hamilton County, Ohio, Late Pleistocene. 5. Undesignated site, Wayne County, Indiana, Late Pleistocene. 6. Jinks Hollow Site, Henderson County, Illinois, Late Pleistocene. 7. Undesignated site, Trempealeau County, Wisconsin, Late Pleistocene.

inner coat of soft hair that protects them from the cold. These muskoxen have horns in both sexes that broadly meet in the middle of the head and then curve downward and outward (fig. 116). They lack the roughened basin in the area where the horns join in the middle of the head that occurs in the extinct *woodland (helmeted) muskoxen (*Bootherium bombifrons*). Barren ground muskoxen weigh from about 400 to about 800 pounds and are presently strictly arctic tundra species. They prefer moist habitats in the summer and ascend to slopes and plateaus in the winter to avoid deep snow. In the summer, the preferred food consists of grasses and sedges, and in the winter, they browse on such plants as willows, cowberry, and crowberry. These muskoxen occur in herds of about ten individuals in the summer and up to about twenty in the winter. When wolves appear, muskoxen form protective circles or semicircles around the calves (see fig. 115).

Within historic times, muskoxen ranged from northern Alaska to Hudson's Bay, on the islands in the north and west of the Canadian arctic, and in Greenland. There are now only a few significant populations (many have been reintroduced) that together are said to total about 25,000 individuals. In the fossil record, muskoxen are known from Illinoian to modern times. It has been suggested that in pre-Wisconsinan times muskoxen ranged south of the Illinoian ice sheet but that many populations retreated northward during the Sangamonian. A few lingering populations are believed to have remained in cold areas south of the Wisconsinan ice sheet. In the Pleistocene of the Great Lakes region, it is somewhat difficult to outline the history of muskoxen movements because several sites where they are found do not have firm dates. Records are from an Early and a Late Wisconsinan (*Ovibos* sp.) site near Toronto, Ontario; undesignated Late Pleistocene sites in northeastern and southwestern Ohio and east-central Indiana; a pre-Wisconsinan site in northwestern Illinois; and a Late Pleistocene (*Ovibos* sp.) site in southwestern Wisconsin. (Records cited as "*Ovibos* sp." by their original authors are considered here to be *Ovibos moschatus*.) The muskoxen would have been an important herbivore in tundra or possibly other cold communities during the Pleistocene of the Great Lakes region. Figure 117 is a location map for Pleistocene barren ground muskoxen sites in the Great Lakes region.

Sites. Scarborough Bluffs near Toronto, Ontario, Rancholabrean LMA–Early Wisconsinan (*Ovibos* sp., Churcher and Karrow 1977). Lake Iroquois deposits near Toronto, Ontario, Rancholabrean LMA–Late

FIGURE 118. A group of *woodland (helmeted) muskoxen (*Bootherium bombifrons*) in the Late Wisconsinan of Michigan.

Wisconsinan (*Ovibos* sp., Karrow et al. 1980). Undesignated sites, Mahoning and Hamilton Counties, Ohio, Late Pleistocene (McDonald 1994). Undesignated site in Wayne County, Indiana, Late Pleistocene (Richards 1984b). Jinks Hollow, Henderson County, Illinois, pre-Wisconsinan (Ray et al. 1968). Undesignated site in Trempealeau County, Wisconsin, Late Pleistocene (*Ovibos* sp., West and Dallman 1980).

Woodland (Helmeted) Muskox
Bootherium bombifrons (figs. 118 and 119)
The woodland muskox was taller and more gracile (slenderer) than the barren ground muskox. Moreover, the woodland muskox had a large, pitted, roughened basin in the horns where they join in the middle of the head (fig. 119) that does not occur in the barren ground muskox. The woodland muskox is thought to have been adapted to warmer conditions than in the modern form and to have inhabited both plains and woodlands. *Bootherium bombifrons* is one of the most common extinct animals in the Great Lakes region but is absent from Ontario. It is curious, but unexplained,

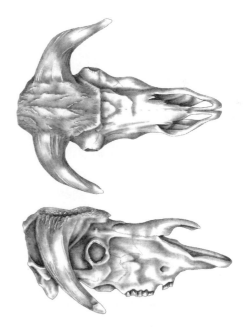

FIGURE 119. Woodland (helmeted) muskox (*Bootherium bombifrons*) skull from the Late Pleistocene of Michigan. Upper, dorsal view; lower, lateral view. The depressed, roughened area between the horn cores is visible in both views.

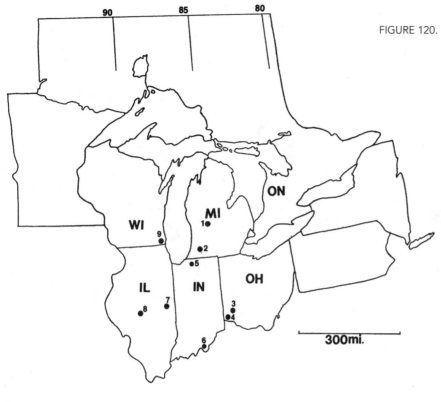

FIGURE 120. Selected woodland muskoxen (*Bootherium bombifrons*) localities in the Pleistocene of the Great Lakes region. 1. Turk Lake Site, Montcalm County, Michigan, Late Wisconsinan, 2. Coville Farm Site, Kalamazoo County, Michigan, Late Wisconsinan. 3. Foster Muskox Site, Warren County, Ohio, Late Pleistocene. 4. Hamilton Muskox Site, Hamilton County, Ohio, Late Wisconsinan. 5. North Liberty Muskox Site, St. Joseph County, Indiana, Late Wisconsinan. 6. King Leo Pit Cave, Harrison County, Indiana, Late Wisconsinan. 7. Busy Farm Site, Champaign County, Illinois, Late Wisconsinan. 8. Sangamon Sand and Gravel Company, Sangamon County, Illinois, Late Wisconsinan. 9. Undetermined site, Kenosha County, Wisconsin, Late Wisconsinan.

why there are at least ten records of the woodland muskox in the southern Lower Peninsula of Michigan and none in Ontario; yet there are records of the modern muskox in Ontario but none in Michigan. Figure 120 is a location map of selected woodland muskoxen sites in the Pleistocene of the Great Lakes region.

Woodland muskoxen occur from Illinoian to Late Wisconsinan times. Based on the fossil record, it has been suggested that they evolved in the intermontane valleys of the West and spread from there to the woodlands of the East. In the Pleistocene of the Great Lakes region, the extinct woodland muskox is much more common than the modern muskox. Woodland muskoxen are known from at least ten Late Wisconsinan sites in the southern part of the Lower Peninsula of Michigan, four in Ohio, twenty-three in Indiana, eight in Illinois, and one in Wisconsin. The woodland muskox was obviously an important herbivore of the southern part of the Great Lakes region in the Pleistocene. The reason for its extinction is obscure, as it

probably did not compete with the modern muskox for habitats.

Selected Sites. Turk Lake, Montcalm County, Michigan, Rancholabrean LMA–Late Wisconsinan (Holman 1990a). Coville Farm, Kalamazoo County, Michigan, Rancholabrean LMA–Late Wisconsinan (Hibbard and Hinds 1960). Foster Muskox, Warren County, Ohio, Rancholabrean LMA–Late Wisconsinan (McDonald and Davis 1989). Hamilton Muskox, Hamilton County, Ohio, Rancholabrean LMA–Late Wisconsinan (McDonald and Davis 1989). North Liberty Muskox, St. Joseph County, Indiana, Rancholabrean LMA–Late Wisconsinan (Lyon 1926). King Leo Pit Cave, Harrison County, Indiana, Rancholabrean LMA–Late Wisconsinan (Richards and McDonald 1991). Busy Farm, Champaign County, Illinois, Rancholabrean LMA–Late Wisconsinan (Ray et al. 1968). Sangamon Sand and Gravel Company, Sangamon County, Illinois, Rancholabrean LMA–Late Wisconsinan (Ray et al. 1968). Undesignated site, Kenosha County, Wisconsin, Rancholabrean–Late Wisconsinan (West and Dallman 1980).

RODENTS
Order Rodentia

Rodents are mainly small mammals, the largest modern ones reaching a little over 100 pounds. They are characterized by having a single pair of upper and lower incisor teeth that are separated from the cheek teeth (premolars and molars) by a space called a diastema (fig. 121). The front portions of the incisors, usually yellowish, are composed of enamel, and the back portions are composed of a softer dentine material, so that when gnawing occurs, very sharp cutting edges develop on each pair. The cheek teeth are modified for either crushing or grinding. Crushing teeth have knobby chewing surfaces, whereas grinding teeth have enamel ridges on their chewing surfaces. Rodents can either gnaw with their incisors or chew with their cheek teeth, but they are unable to do both at the same time.

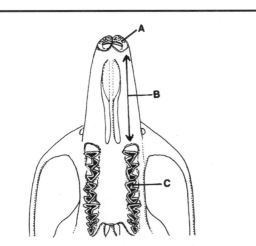

FIGURE 121. Palatal view of a generalized rodent (vole) showing A, incisor tooth; B, diastema (space between arrow points); C, cheek tooth.

Rodents occur on all of the continents and occupy an amazing variety of habitats. They are the most successful of all modern mammalian orders and comprise about 40 percent of all of the world's mammalian species. Rodents are known from Late Paleocene to modern times, the Paleocene ancestors being somewhat squirrel-like. Six families of rodents—Squirrels (Sciuridae), pocket gophers (Geomyidae), beavers, (Castoridae), rats, mice, and voles (Muridae), dipodids (Dipodidae), and New World porcupines (Erethizontidae)—occur in the Pleistocene of the Great Lakes region.

SQUIRRELS
Family Sciuridae

Unfortunately for the layperson, families of rodents are often distinguished mainly on the basis of characters found in the skull. Squirrels all have a well-developed bony process on the skull called the postorbital process, which extends from the outer edges of paired bones over the eye region (frontal bones) and projects over the eye cavity. Squirrels have four cheek teeth on each side of the lower jaw (fig. 122) and either four or five cheek teeth on each side of the upper jaw, for a total of either twenty or twenty-two teeth, counting the four incisors.

Squirrels weigh from a few ounces up to about 20 pounds in some of the marmots. Sciurids live in boreal, temperate, and tropical zones and inhabit arctic tundra, forests, grasslands, and deserts. They mainly live in trees (some are gliders) or holes in the ground. Today, they are almost worldwide in occurrence, with the exception of polar regions, the Australian region, Madagascar, and southern South America. In the fossil record, squirrels are known from Oligocene to modern times. Six genera—New World flying squirrels (*Glaucomys*), woodchucks (*Marmota*), tree squirrels (*Sciurus*),

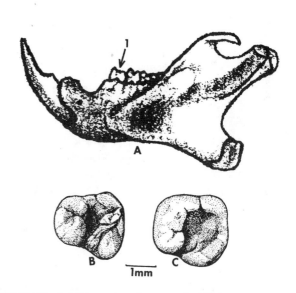

FIGURE 122. A, left lower jaw of gray squirrel (*Sciurus carolinensis*); 1 refers to the first lower premolar. B, surface view of the first lower premolar of a gray squirrel. C, surface view of the first lower premolar of fox squirrel (*Sciurus niger*).

ground squirrels (*Spermophilus*), chipmunks (*Tamias*), and red squirrels (*Tamiasciurus*)—are known from the Pleistocene of the Great Lakes region. The general lack of fossil records of tree squirrels is most likely due to their arboreal habits.

Southern Flying Squirrel
Glaucomys volans

The dainty little southern flying squirrel has very large eyes, a flat tail, and a fold of skin that runs from its wrists to its ankles. It may be distinguished from the related northern flying squirrel on the basis of having the hairs of the belly completely white rather than dark at the base like its northern relative. The southern flying squirrel weighs only about 2 to 3 ounces. These soft, furry little animals prefer open woodlands and often use tree holes as nesting sites. Flying squirrels do not fly like birds or bats but glide from one tree to another, covering between 80 and 100 feet in some instances. The favorite food of the southern flying squirrel is nuts, but outside of the nut season, they will eat other food such as twigs, bark, and lichens.

The southern flying squirrel occurs from southern Canada, the northeastern states, and the Great Lakes region south to Florida and Texas. The fossil record of the southern flying squirrel extends from Illinoian to modern times. In the Pleistocene of the Great Lakes region, it has been identified from the Late Wisconsinan of south-central Indiana. Fossil remains identified as *Glacucomys* sp. from the Late Wisconsinan of southern Indiana may well represent *Glaucomys volans*, the southern species. This little animal would have been a nut-eating and herbivorous species of open woodlands during the Pleistocene of the Great Lakes region.

Sites. Anderson Pit Cave, Monroe County, Indiana, Rancholabrean LMA–Late Wisconsinan (Graham and Lundelius 1994). King Leo Pit Cave, Harrison County, Indiana, Rancholabrean LMA–Late Wisconsinan (*Glaucomys* sp., Richards and McDonald 1991).

Woodchuck
Marmota monax

The familiar woodchuck, or groundhog, is by far the largest squirrel in the Great Lakes region today. The large size, stout body, and short, dark tail easily distinguish the woodchuck from other squirrels. Woodchucks weigh from about 5 to about 10 pounds. This species prefers open forests where it digs burrows in well-drained areas, often hillsides. Woodchucks mainly feed on grasses and herbaceous plants and on buds, twigs, and the bark of small trees. Unlike most squirrels, woodchucks do not store food for the winter but instead go into a deep hibernation during the cold season.

Presently, woodchucks occur from Alaska and Labrador across Canada and into the eastern and central United States as far south as Alabama and eastern Oklahoma. The fossil record of woodchucks is from Illinoian to modern times. In the Great Lakes region, they are known from the Sagamonian near Toronto, Ontario, and from the Late Wisconsinan of northwest and west-central Ohio and south-central and southwestern Indiana. The woodchuck would have been an important ground-dwelling herbivore of open woodlands during the Pleistocene of the Great Lakes region.

Sites. Don Formation near Toronto, Ontario, Rancholabrean LMA–Sangamonian (Karrrow et al. 1980). Sheriden Pit Cave, Wyandot County, Ohio, Rancholabrean LMA–Late Wisconsinan (Ford 1994). Carter, Darke County, Ohio, Rancholabrean LMA–Late Wisconsinan (McDonald 1994). Anderson Pit Cave, Monroe County, Indiana, Rancholabrean LMA–Late Wisconsinan (Graham and Lundelius 1994). Prairie Creek D, Daviess County, Indiana, Rancholabrean LMA–Late Wisconsinan (Richards 1992c).

Eastern Gray Squirrel
Sciurus carolinensis (fig. 122)

The common eastern gray squirrel, which often occurs as a totally black squirrel in parts of the Great Lakes region, typically has a grayish color due to the silver-tipped hairs on the body and the bushy tail. The related fox squirrel is larger, usually has a more brownish color, and lacks the silver-tipped hairs. Gray squirrels generally weigh about 1 or 2 pounds. This species prefers mature, rather dense woodlands where acorns and nuts are available.

Eastern gray squirrels occur from southern Canada and the northeastern states and Great Lakes region south to Florida and Texas. The fossil record of the eastern gray squirrel extends from Illinoian to modern times. In the Pleistocene of the Great Lakes region, this species is known from the Late Wisconsinan of south-central Indiana. Squirrel remains identified as *Sciurus* sp. from the Late Wisconsinan of northwestern Ohio and southern Indiana may well represent the eastern gray squirrel. This species would have been a nut-eating species of rather dense woodlands during the Pleistocene of the Great Lakes region.

Sites. Anderson Pit Cave, Monroe County, Indiana, Rancholabrean LMA–Late Wisconsinan (Graham and Lundelius 1994). Sheriden Pit Cave, Wyandot County, Ohio, Rancholabrean LMA–Late Wisconsinan (*Sciurus* sp., Ford 1994). King Leo Pit Cave, Harrison County, Indiana, Rancholabrean LMA–Late Wisconsinan (*Sciurus* sp., Richards and McDonald 1991).

Fox Squirrel
Sciurus niger (fig. 122)

The fox squirrel is larger than the gray squirrel, lacks the silver-tipped hairs on the body and tail, and has a brownish rather than a grayish color. Black fox squirrel populations are rarely seen, but albanistic (white) populations are known in southern parts of the Great Lakes region. Fox squirrels weigh from about 1.5 to about 3 pounds at the height of their prewinter fat-storing period. They prefer deciduous woodlands with an open understory; thus this species was uncommon in the Great Lakes region until deforestation occurred. Fox squirrels usually raise young and overwinter in large holes or hollows of mature deciduous trees. In the spring and summer, these squirrels eat the buds, flowers, and fruits of trees and later bury nuts and acorns to be eaten in the winter season.

Presently, the eastern fox squirrel occurs in about all of the eastern half of the United States. Its fossil record extends from Wisconsinan to modern times. In the Pleistocene of the Great Lakes region, it is known from the Late Wisconsinan of south-central Indiana. The eastern fox squirrel would have been an important harvester of deciduous tree products in open woodlands during the Pleistocene of the Great Lakes region, but it probably was far less abundant than the gray squirrel.

Site. Anderson Pit Cave, Monroe County, Indiana, Rancholabrean LMA–Late Wisconsinan (Graham and Lundelius 1994).

Thirteen-lined Ground Squirrel
Spermophilus tridecemlineatus

The slender-bodied thirteen-lined ground squirrel is easily recognized on the basis of the alternating dark and light stripes on the body (the dark stripes have small spots within them) and the slender tail. These little ground squirrels weigh from about 4 to 8 ounces. This species prefers open areas where the grass is short enough for them to see over it; thus they presently are much more common in the Great Lakes region than

they were in the past, favoring road shoulders, parks, and golf courses. The thirteen-lined ground squirrel is an omnivore, but its most staple food is probably seeds. They are also fond of a wide variety of insects, small vertebrates, and an occasional ground-nesting bird's egg. Carrion is also an important item in their diet.

Presently, the species is found in central southern Canada and the Great Lakes region south through the Great Plains states into Texas. It is absent from the Northeast, Southeast, and most of the Gulf Coast region. Its fossil record extends from Illinoian to modern times. In the Pleistocene of the Great Lakes region, the thirteen-lined ground squirrel is known from the Late Wisconsinan of south-central Indiana and extreme southwestern Wisconsin. Remains assigned to *Spermophilus* sp. from the Late Wisconsinan of west-central Ohio may well represent the thirteen-lined ground squirrel. This animal would have been an important seed and insect eater of very open areas during the Pleistocene of the Great Lakes region.

Sites. Anderson Pit Cave, Monroe County, Indiana, Rancholabrean LMA–Late Wisconsinan (Graham and Lundelius 1994). Moscow Fissure, Iowa County, Wisconsin, Rancholabrean LMA–Late Wisconsinan (Foley 1984). Carter, Darke County, Ohio, Rancholabrean LMA–Late Wisconsinan (*Spermophilus* sp., McDonald 1994).

Eastern Chipmunk
Tamias striatus

The little eastern chipmunk has a long, dark stripe down the middle of its back and two dark stripes along each of its sides. A white stripe fills in the space between each pair of dark stripes on the sides. Eastern chipmunks weigh between 2 and 4 ounces. The preferred habitat for this species is open deciduous forests, especially those dominated by beech and maples. The eastern chipmunk is able to climb trees but spends most of its time on the ground, where it lives in complicated burrows with separate chambers for nesting and food storage. Staple food items include fruits, seeds, and nuts, but insects, earthworms, and slugs are readily taken.

The fossil record of the eastern chipmunk occurs from Illinoian to modern times. In the Pleistocene of the Great Lakes region, this species occurs in the Late Wisconsinan near Moose Creek Ontario, northwest and west-central Ohio, and south-central and southwestern Indiana. The eastern chipmunk would have been an important little ground-dwelling omnivore in open wooded areas during the Pleistocene of the Great Lakes region.

Sites. Moose Creek (near the Champlain Sea) near Moose Creek, Ontario, Rancholabrean LMA–Late Wisconsinan (Harington 1978). Sheriden Pit Cave, Wyandot County, Ohio, Rancholabrean LMA–Late Wisconsinan (Ford 1994). Carter, Darke County, Ohio, Rancholabrean LMA–Late Wisconsinan (McDonald 1994). Anderson Pit Cave, Monroe County, Indiana, Rancholabrean LMA–Late Wisconsinan (Graham and Lundelius 1994). Prairie Creek D, Daviess County, Indiana, Rancholabrean LMA–Late Wisconsinan (Richards 1992c).

Red Squirrel
Tamiasciurus hudsonicus
The reddish color and small size of the red squirrel readily separate it from other tree squirrels. These little animals weigh from about 5 to 8 ounces. This species prefers coniferous forests but exists quite well in deciduous forest areas if a few coniferous trees are available. The red squirrel usually lives in tree holes or makes a nest of sticks and leaves in the treetops. This species jealously guards its territory, readily chasing away other squirrels and small animals from its home area. The red squirrel's preferred food is seeds extracted from the cones of coniferous trees, but it also eats buds, seeds, fruits, insects, and small vertebrates. Red squirrels commonly feed on maple sugar, which they harvest after biting through the bark of maple trees and waiting for the sap to flow.

At present, red squirrels occur from Alaska and northern Canada south through the northeastern and Great Lakes states to Appalachia in the East and to the southern Rocky Mountains in the West. The fossil record of the red squirrel spans Illinoian to modern times. In the Pleistocene of the Great Lakes region, the red squirrel has been identified from the Late Wisconsinan of northwestern Ohio; south-central, southwestern, and southern Indiana; and extreme southwestern Wisconsin. In the south-central and southern Indiana localities, the red squirrel existed somewhat south of its modern range in the state. This animal would have been an abundant small denizen of coniferous forest and mixed coniferous and deciduous forest habitats during the Pleistocene of the Great Lakes region.

Sites. Sheriden Pit Cave, Wyandot County, Ohio, Rancholabrean LMA–Late Wisconsinan (McDonald

1994). Anderson Pit Cave, Monroe County, Indiana, Rancholabrean LMA–Late Wisconsinan (Graham and Lundelius 1994). Prairie Creek D, Daviess County, Indiana, Rancholabrean LMA–Late Wisconsinan (Richards 1992c). King Leo Pit Cave, Harrison County, Indiana, Rancholabrean LMA–Late Wisconsinan (Richards and McDonald 1991). Moscow Fissure, Iowa County, Wisconsin, Rancholabrean LMA–Late Wisconsinan (Foley 1984).

POCKET GOPHERS
Family Geomyidae

Pocket gophers are rarely seen by humans, as they spend most of their time underground like moles. The animals called "gophers" by most people are actually ground squirrels. Pocket gophers have tiny eyes and ears and, during the infrequent times that they are aboveground, look disoriented and helpless. These animals have robust bodies that appear almost neckless, and the front feet have oversized claws that are used for digging underground tunnels. Pocket gophers weigh from about 2 ounces to about 2 pounds. Members of this family presently occur from Saskatchewan and British Columbia south to northern Colombia in South America. An isolated species lives in Florida. The family is known from Early Miocene to Holocene times. Two genera, eastern pocket gophers (*Geomys*) and western pocket gophers (*Thomomys*), occur in the Pleistocene of the Great Lakes region. Figure 123 is a location map for extralimital Pleistocene occurrences of pocket gophers in the Great Lakes region.

Plains Pocket Gopher
Geomys bursarius
The plains pocket gopher is distinguished from the northern pocket gopher on the basis of being larger and having two distinct grooves down the front of each incisor. It weighs from about 6 ounces to 1 pound. This underground creature prefers sandy soil in open areas, disdains clayey or gravelly soils, and it digs tunnels up to 500 feet long. Unlike moles, pocket gophers are strictly herbivorous and eat roots and bulbs as well as tender plants and grasses that they pull down into their burrows.

Presently, the plains pocket gopher occurs from southern Manitoba south through the plains states to

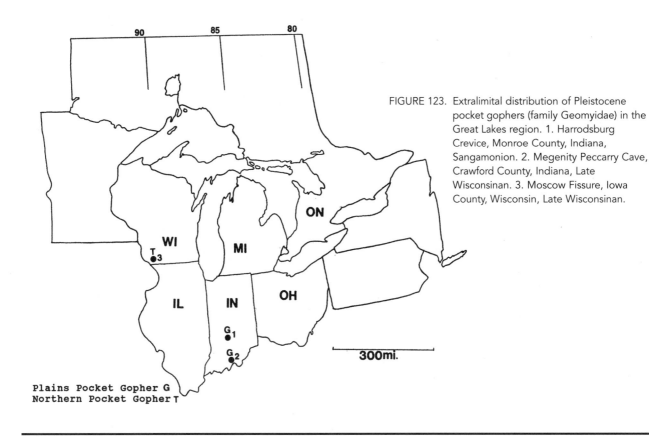

FIGURE 123. Extralimital distribution of Pleistocene pocket gophers (family Geomyidae) in the Great Lakes region. 1. Harrodsburg Crevice, Monroe County, Indiana, Sangamonion. 2. Megenity Peccarry Cave, Crawford County, Indiana, Late Wisconsinan. 3. Moscow Fissure, Iowa County, Wisconsin, Late Wisconsinan.

Texas. The fossil record of this species extends from Illinoian to modern times. In the Great Lakes region, the plains pocket gopher has been found in the Sangamonian of south-central Indiana and the Late Wisconsinan of southern Indiana. Both localities are presently out of the range of the species in Illinois and northwestern Indiana. This pocket gopher would have been an underground herbivore of open, sandy areas during the Pleistocene of the Great Lakes region.

Sites. Harrodsburg Crevice, Monroe County, Indiana, Rancholabrean LMA–Sangamonian (Parmalee et al. 1978). Megenity Peccary Cave, Crawford County, Indiana, Rancholabrean LMA–Late Wisconsinan (Richards 1988a).

Northern Pocket Gopher
Thomomys talpoides

The northern pocket gopher may be distinguished from the plains pocket gopher on the basis of being smaller and lacking the prominent grooves on the front of the incisor teeth. These pocket gophers weigh only about 3 or 4 ounces. The northern pocket gopher also spends most of its time in an underground tunnel and is uncomfortable above ground. It feeds entirely upon plant material as it excavates its burrow or pulls down plants from above. This species also prefers loose rather than clayey or gravelly soils.

Presently, northern pocket gophers range from northern Saskatchewan and Alberta south through the Dakotas and northwestern states to the mountains of New Mexico and Arizona. Northern pocket gophers occur in the fossil record from Sangamonian to modern times. In Wisconsinan times, it ranged well south and east of its modern range. In the Pleistocene of the Great Lakes region, the northern pocket gopher is known from the Late Wisconsinan of extreme southwestern Wisconsin, well east of its present range in the Dakotas. The northern pocket gopher would have been an underground herbivore in loose soils during the Pleistocene of the Great Lakes region.

Site. Moscow Fissure, Iowa County, Wisconsin, Rancholabrean LMA–Late Wisconsinan (Foley 1984).

BEAVERS
Family Castoridae

Modern beavers are the largest rodents of the temperate part of the world, and some of the extinct beavers are the largest rodents that ever lived, reaching the size of a black bear. Modern and extinct beavers are semi-aquatic animals with high-crowned cheek teeth. The modern beavers have webbed feet, and it is believed that this was also true in the extinct species. But the modern beavers have chisel-like incisors (fig. 124) for cutting down trees, while the extinct beavers had incisors that were more rounded at the tips. Beavers weigh from about 30 to probably over 300 pounds in the largest of the extinct species. The beaver family has always been restricted to the Northern Hemisphere, occurring widely in both North America and Eurasia. Two genera of beavers occur in the Pleistocene of the Great Lakes region, the modern genus, *Castor*, and the largest of the extinct beavers, *Castoroides*.

American Beaver
Castor canadensis (fig. 124)

FIGURE 124. Lateral view of the lower jaw of an American beaver (*Castor canadensis*). The arrow points to the chisel-like lower incisor.

The large body, webbed feet, and flattened tail immediately identify the modern beaver. This large rodent weighs from about 30 to about 60 pounds. Beavers prefer stream or lake habitats that are bordered by immature forests containing alder, aspen, cottonwood, and willow trees. Most beavers build lodges in which to live and often build dams across small streams to create ponds for themselves. This greatly modifies the environment in some areas. Their chisel-like incisors (fig. 124) allow them to cut down surprisingly large trees to build their lodges and dams. The American beaver feeds upon leaves, twigs, and bark of the trees in the vicinity of its lodges and dams and also eats aquatic plants, both the underwater and above-water parts.

Within historic times, the American beaver occupied almost all of North America south to northern Mexico, but clearing of the land and relentless fur trading greatly reduced beaver populations in many areas. The fossil record of the American beaver extends from Late Pliocene to modern times. In the Great Lakes region, this species is known from the Late Wisconsinan of southeastern Michigan, northwest and west-central Ohio, east-central Indiana, and southern Wisconsin. The American beaver would have been an important species of aspen-willow–lined streams, ponds, and lakes during the Pleistocene of the Great Lakes region. It would have been an important herbivore and would have impacted the environment by creating extensive wetlands.

Sites. Shelton, Oakland County, Michigan, Rancholabrean LMA–Late Wisconsinan (Shoshani et al. 1989). Sheriden Pit Cave, Wyandot County, Ohio, Rancholabrean LMA–Late Wisconsinan (McDonald 1994). Carter, Darke County, Ohio, Rancholabrean LMA–Late Wisconsinan (McDonald 1994). Dollens Mastodont, Madison County, Indiana, Rancholabrean LMA–Late Wisconsinan (Richards et al. 1987). Deerfield, Dane County, Wisconsin, Rancholabrean LMA–Late Wisconsinan (Dallman 1968).

*Giant Beaver
Castoroides ohioensis (figs. 125–27)

The giant beaver, the giant of the rodent world, is most easily identified on the basis of its huge incisor teeth that are bigger than modern horse teeth (fig. 125). These teeth are cylindrical and curved and are usually black in front and yellowish brown in back. The numerous thin grooves on the black part are unique. Otherwise, the head and body skeleton look like that of a bear-sized rodent. The giant beaver has often been characterized as somewhat of a giant edition of a modern beaver. It has been suggested that the extinct beaver could have cut down giant trees, based on the fact that it is known that modern beavers can fell large cottonwood trees. Carrying this reasoning further, it has been

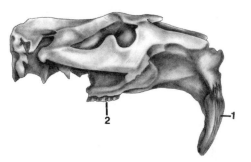

FIGURE 125. Lateral view of the skull of a *giant beaver (*Castoroides ohioensis*) from the Late Wisconsinan of Michigan. 1, incisor tooth; 2, cheek teeth.

FIGURE 126. A giant beaver (*Castoroides ohioensis*) indicating a rounded rather than a flat tail.

FIGURE 127. A pair of giant beavers (*Castoroides ohioensis*) by a pond in the Late Wisconsinan of Michigan.

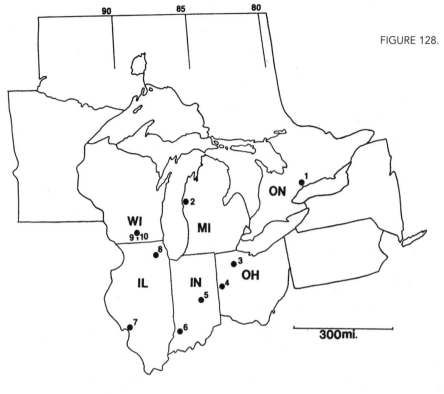

FIGURE 128. Selected Pleistocene localities of the giant beaver (*Castoroides ohioensis*). 1. Don Formation, near Toronto, Ontario, Sangamonian. 2. Ludington Beach Site, Mason County, Michigan, Late Wisconsinan. 3. Sheriden Pit Cave Site, Wyandot County, Ohio, Late Wisconsinan. 4. Carter Site, Darke County, Ohio, Late Wisconsinan. 5. Christensen Bog Mastodont Site Hancock County, Indiana, Late Wisconsinan. 6. Prairie Creek D Site, Daviess County, Indiana, Late Wisconsinan. 7. Alton Site, Madison County, Illinois, Late Pleistocene. 8. Phillips Park Site, Kane County, Illinois, Late Wisconsinan. 9. Hope Site, Dane County, Wisconsin, Late Wisconsinan. 10. Witte Farm Site, Dane County, Wisconsin, Late Wisconsinan.

proposed that giant beavers could have impounded huge bodies of water.

The other view is that the giant beaver was probably more like a giant muskrat or a South American capybara (a large aquatic rodent) than a giant modern beaver. In other words, it was suggested that the giant beaver did not fell trees or build lodges and dams but more likely, muddled about in the abundant ponds, lakes, and swamps of the Pleistocene of the Great Lakes region, eating water lilies and other common aquatic plants. Modern studies, however, show that the giant beaver lived a life somewhat in between the ones depicted in these two views.

In older accounts it was suggested that the giant beaver had a round rather than flat tail (fig. 126), but modern studies indicate that the tail was somewhat flattened but not as flat as in the modern beaver. Although the incisor teeth of the giant beaver (see fig. 125) are blunter than those of the modern beaver, which are very chisel-like (see fig. 124); the incisors of the giant form appear to be a combination of a gouge and a chisel that could cut wood and strip bark as well as allow the giant animals to root for aquatic plants. The cheek teeth of the giant beaver, however

(fig. 125), do resemble those of the capybara. Moreover, a cast from the skull of a giant beaver showed that its brain was small and smooth, indicating less complex behavior patterns than those indicated by the relatively large, wrinkled brains of the modern beaver. Although figure 127 shows round tails in giant beavers, the habitat depicted is thought to be typical of this huge rodent.

The fossil record of the giant beaver extends from the Late Pliocene to the Late Wisconsinan, when it became extinct. Although the giant beaver occurred from Alaska (represented by only one fragmentary specimen) to Florida and eastward to Nebraska, it was most abundant in the southern part of the Great Lakes region. It is believed that it became extinct because of competition with the more resourceful modern beaver and because of a reduction of its habitat at the end of the Wisconsinan. In the Pleistocene of the Great Lakes region, the giant beaver is known from the province and all the states. It is known from many sites in Ohio, Indiana, and Illinois only one in Ontario, and from sites in both Michigan and Wisconsin. Figure 128 is a location map of selected Pleistocene giant beaver sites in the Great Lakes region.

A giant beaver jaw was found by strollers on Ludington Beach in Mason County, Michigan (fig. 129). This jaw undoubtedly washed out of sediments on the bottom of Lake Michigan, probably during a storm. The jaw is not only as big as that of a black bear but is the northernmost record of *Castoroides* in the state (see fig. 128). The giant beaver must have been a very important herbivore of ponds, lakes, and swamps during the Pleistocene of the Great Lakes region.

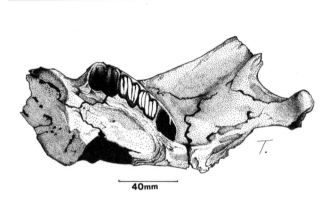

FIGURE 129. The Ludington Beach, Mason County, Michigan, Late Wisconsinan giant beaver (*Castoroides ohioensis*) right lower jaw in medial view.

Selected Sites. Don Formation, near Toronto, Ontario, Rancholabrean LMA–Sangamonian (Karrow et al. 1980). Ludington Beach, Mason County, Michigan, Rancholabrean LMA–Late Wisconsinan (Holman 1991). Sheriden Pit Cave, Wyandot County, Ohio, Rancholabrean LMA–Late Wisconsinan (McDonald 1994). Carter, Darke County, Ohio, Rancholabrean LMA–Late Wisconsinan (McDonald 1994). Christensen Bog, Hancock County, Indiana, Rancholabrean LMA–Late Wisconsinan (Graham et al. 1983). Prairie Creek D, Daviess County, Indiana, Rancholabrean LMA–Late Wisconsinan (Tomak 1975). Phillips Park, Kane County, Illinois, Rancholabrean LMA–Late Wisconsinan (Graham and Graham 1986). Alton, Madison County, Illinois, Late Pleistocene (Bagg 1909). Hope, Dane County, Wisconsin, Rancholabrean LMA–Late Wisconsinan (Dallman 1968). Witte Farm, Dane County, Wisconsin, Rancholabrean LMA–Late Wisconsinan (Dallman 1969).

RATS, MICE, AND RELATIVES
Family Muridae

The family Muridae is the largest family of mammals in the world. It is defined mainly on the basis of about twenty-five skull characters, including the position of a large opening (infraorbital foramen) in the rostrum, which important muscles run through in the side of the skull. This family contains familiar animals such as rats, mice, and muskrats, as well as voles and lemmings, which may not be as familiar to most people. The small size and ability to reproduce very rapidly have enabled murids to be very successful animals. They range in size from less than 1 ounce to about 4 pounds in a species of rat that occurs in the Philippines. They are generally herbivorous, but some are omnivorous; a few specialize on eating insects, other small invertebrates, or even small fishes. Murids occur natively in all continents. The fossil record of the Muridae spans Early Oligocene to modern times. Two large subfamilies of the Muridae occur in the Great Lakes region, the New World rats and mice (Sigmodontinae) and the voles, lemmings, and muskrats (Arvicolinae). In the Pleistocene, many northern species of murids, especially voles, existed south of their modern range.

NEW WORLD RATS AND MICE
Subfamily Sigmodontinae

Sigmodontines have rooted cheek teeth with knobby surfaces (fig. 130). These teeth are well suited for eating seeds, nuts, and insects. Voles, lemmings, and muskrats, of the subfamily Arvicolinae, usually have ever-growing incisor teeth with complex enamel patterns (fig. 131).

Eastern Woodrat
Neotoma floridana (fig. 132)
The gentle eastern woodrat is about the size of the introduced Norway rat, but it can be distinguished from this pest in that it has a softer coat, larger eyes and ears, a white belly, and a furry rather than a scaly tail. Eastern woodrats weigh from about .5 to 1 pound. Favorite habitats for these animals are rocky outcrops, crevices, and caves, usually in deciduous forests. In these rocky places, the woodrat makes a large, open nest from wood fibers. The eastern woodrat is often

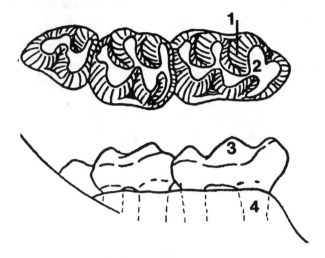

FIGURE 130. Cheek teeth from the lower jaw of a typical sigmodontine rodent (New World rats and mice). Upper, cheek teeth in surface view: 1, enamel cusp; 2, dentine exposed by wear. Lower, cheek teeth in lateral view: 3, knobby enamel cusp; 4, permanent root. Modified from Hibbard (1950).

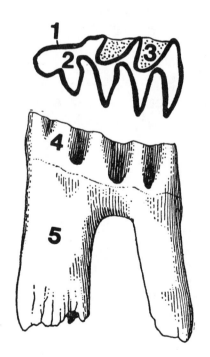

FIGURE 131. Cheek tooth from the lower jaw of a typical arvicoline rodent (voles, lemmings, and muskrats). Upper, cheek tooth in surface view: 1, enamel ridge; 2, dentine; 3, cementum. Lower, cheek tooth in lateral view: 4, beveled cusp; 5, ever-growing root. Modified from Hibbard (1950).

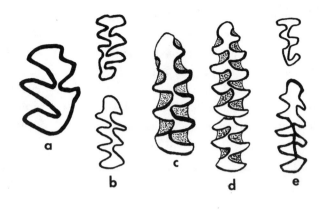

FIGURE 132. Surface views of enamel patterns of A, eastern woodrat, *Neotoma floridana* (upper left first molar); B, red-backed vole, *Clethrionomys gapperi* (top, upper right third molar; bottom, lower right first molar); C, prairie vole, *Microtus ochrogaster* (lower first molar); D, meadow vole, *Microtus pennsylvanicus* (lower first and second molar); E, woodland vole, *Microtus pinetorum* (top, upper right third molar; bottom, lower right first molar).

called the packrat because it gathers all kinds of odds and ends—such as bottle caps, pieces of glass, and other small bits of trash—around its nest. It eats a wide variety of plants and plant products, including leaves, seeds, fruits, and nuts. The populations of woodrats that occur in the Appalachian region and extend into southern Ohio, Indiana, and Illinois have sometimes been given a separate specific name (*Neotoma magister*) from the main population, but I am retaining the name *Neotoma floridana* for these animals.

Eastern woodrats presently occur mainly from the southern part of the Great Lakes region into the southeastern and south-central part of the United States to northern Florida and northeastern Texas. The fossil record of the eastern woodrat extends from Illinoian to modern times. This rat has been identified from three Late Wisconsinan sites in south-central Indiana north of the present northern limit of the species in the state (fig. 133). Another record of the species from the Late Wisconsinan of southern Indiana is within the modern range of the species. Fossils from the Late Wisconsinan of west-central Ohio assigned to *Neotoma* sp. probably represent the eastern woodrat. This Ohio record (see fig. 133) is northwest of the present range of *Neotoma* in Ohio. The eastern woodrat would have been an herbivore of rocky outcrops, crevices, and caves in the

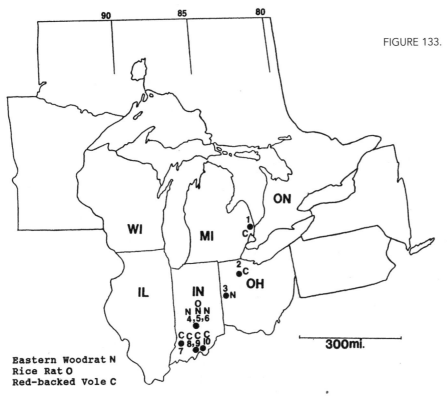

FIGURE 133. Extralimital Pleistocene localities of woodrats, rice rats, and red-backed voles. 1. Mill Creek Site, St. Clair County, Michigan, pre-Late Wisconsinan. 2. Sheriden Pit Cave Site, Wyandot County, Ohio, Late Wisconsinan. 3. Carter Site, Darke County, Ohio, Late Wisconsinan. 4. Harrodsburg Crevice, Monroe County, Indiana, Sangamonian. 5. Anderson Pit Cave, Monroe County, Indiana, Late Wisconsinan. 6. Indun Rock Shelter, Monroe County, Indiana, Late Wisconsinan. 7. Prairie Creek D Site, Daviess County, Indiana, Late Wisconsinan. 8. Alton Mammoth Site, Crawford County, Indiana, Late Wisconsinan. 9. Megenity Peccary Cave, Crawford County, Indiana, Late Wisconsinan. 10. King Leo Pit Cave, Harrison County, Indiana, Late Wisconsinan.

southern part of the Great Lakes region during the Pleistocene.

Sites. Carter, Darke County, Ohio, Rancholabrean LMA–Late Wisconsinan (*Neotoma* sp., McDonald 1994). Harrodsburg Crevice, Monroe County, Indiana, Rancholabrean LMA–Sangamonian (Richards 1987). Anderson Pit Cave, Monroe County, Indiana, Rancholabrean LMA–Late Wisconsinan (Richards 1987). Indun Rock Shelter, Monroe County, Indiana, Rancholabrean LMA–Late Wisconsinan (Richards and Munson 1988). Megenity Peccary Cave, Crawford County, Indiana, Rancholabrean LMA–Late Wisconsinan (Richards 1987).

Rice Rat
Oryzomys palustris
The rice rat may be distinguished from the eastern woodrat on the basis of its smaller size and scaly tail and from the introduced Norway rat by its smaller size and lighter underparts and tail. Rice rats weigh about 2 or 3 ounces. They live in moist habitats near water and are good swimmers and divers. Their food consists of tender parts of grasses and sedges, seeds, fruits, invertebrates, and even small fishes.

Presently, rice rats mainly occur in the southeastern United States and northern Mexico, reaching the Great Lakes region only in southern Indiana and Illinois. In the fossil record, rice rats extend from Illinoian to modern times. In the Pleistocene of the Great Lakes region, the species has been reported from the Late Wisconsinan of south-central Indiana, somewhat north of its modern range in southern Indiana (see fig. 133). The rice rat would have been an omnivorous inhabitant of moist areas near ponds, lakes, and swamps of the southern part of the Great Lakes region during the Pleistocene.

Site. Anderson Pit Cave, Monroe County, Indiana, Rancholabrean LMA–Late Wisconsinan (Richards 1980).

White-footed Mouse
Peromyscus leucopus
The white-footed mouse is very similar to its close relative, the deer mouse. Both have large, bulging, black eyes and large dark ears with white rims. Moreover,

both have brownish backs and sides with white underparts. The differences in these species are so subtle that it often takes an experienced field mammalogist to tell them apart. Among the differences are that the white-footed mouse has a lighter back, a less distinct tail tuft, and a tail that is more distinctly brown above and white below than in the deer mouse. White-footed mice weigh from about .5 to 1 ounce.

This mouse is most common in deciduous woodlands where cover such as logs or rock piles are abundant. They build small nests out of feathers, fur, grass, or other soft material that can be found in the woods. They hide these nests in holes at the bases of trees, under logs, or under rocks. White-footed mice are omnivores, feeding upon seeds, nuts, berries, grains, and grass as well as larval and mature insects.

Presently, the white-footed mouse ranges from southern, eastern, and central Canada south through about the eastern two-thirds of the United Sates and south as far as the Yucatán peninsula of Mexico. They are replaced by another species in Florida. The fossil record of this species spans Sangamonian to recent times. In the Pleistocene of the Great Lakes region, the white-footed mouse occurs in Late Wisconsinan sites in south-central Indiana. Mouse remains identified as *Peromyscus* sp. from northwest and west-central Ohio and southwest and southern Indiana represent either this species or the deer mouse. The little white-footed mouse would have been an omnivorous occupant of deciduous woodlands during the Pleistocene of the Great Lakes region.

Sites. Sheriden Pit Cave, Wyandot County, Ohio, Rancholabrean LMA–Late Wisconsinan (*Peromyscus* sp., McDonald 1994). Carter, Drake County, Ohio, Rancholabrean LMA–Late Wisconsinan (*Peromyscus* sp., McDonald 1994). Anderson Pit Cave, Monroe County, Indiana, Rancholabrean LMA–Late Wisconsinan (Graham and Lundelius 1994). Prairie Creek D, Daviess County, Indiana, Rancholabrean LMA–Late Wisconsinan (*Peromyscus* sp., Richards 1992c). King Leo Pit Cave, Harrison County, Indiana, Late Wisconsinan (*Peromyscus* sp., Richards and McDonald 1991).

Deer Mouse
Peromyscus maniculatus
Characters that distinguish the deer mouse from the white-footed mouse have been discussed in the previous account. Deer mice weigh from about .25 to about .5 of an ounce. In the Great Lakes region, some popu-

lations favor open areas, while others favor woodland habitats. Like the white-footed mouse, the deer mouse makes nests and is omnivorous.

Presently, deer mice range from northern Canada through most of the United States, except many southeastern states, and well into Mexico. The fossil range of the deer mouse is from Sangamonian to modern times. In the Pleistocene of the Great Lakes region, deer mice are known from the Late Wisconsinan of northwestern Ohio; northwestern, south-central, and southern Indiana; and extreme southwestern Wisconsin. A mouse identified as *Peromyscus* sp. from the pre-Late Wisconsinan of southeastern Michigan could well represent *Peromyscus maniculatus*. The deer mouse would have been an important little omnivore of both open and wooded areas during the Pleistocene of the Great Lakes region.

Sites. Mill Creek, St. Clair County, Michigan, Rancholabrean LMA–pre-Late Wisconsinan (*Peromyscus* sp., Karrow et al. 1997). Sheriden Pit Cave, Wyandot County, Ohio, Rancholabrean LMA–Late Wisconsinan (Ford 1994). Kolarik Mastodont Site, Starke County, Indiana, Rancholabrean LMA–Late Wisconsinan (Ellis 1982). Anderson Pit Cave, Monroe County, Indiana, Rancholabrean LMA–Late Wisconsinan (Graham and Lundelius 1994). King Leo Pit Cave, Harrison County, Indiana, Rancholabrean LMA–Late Wisconsinan (Richards and McDonald 1991). Moscow Fissure, Iowa County, Wisconsin, Rancholabrean LMA–Late Wisconsinan (Foley 1984).

VOLES, LEMMINGS, AND MUSKRATS
Subfamily Arvicolinae

Arvicolines are rodents that usually have ever-growing cheek teeth with complex enamel patterns on their surfaces (see fig. 131). These teeth are well suited for eating grasses and sedges.

Southern Red-backed Vole
Clethrionomys gapperi (fig. 134; see fig. 132)
Voles are small animals that differ from other mouse-like rodents in having much smaller eyes, smaller ears partly hidden by fur, and short tails. The southern red-backed vole may be distinguished on the basis of the broad, chestnut-colored stripe that runs along the

FIGURE 134. Red-backed vole (*Clethrionomys gapperi*).

back. The tail of this species is about one-quarter the total length of the animal. An average specimen weighs a little over .5 of an ounce. Southern red-backed voles live in both coniferous and mixed coniferous and deciduous forests. They prefer moist habitats such as cedar or tamarack swamps but may live in more upland situations if water sources are nearby. These voles feed on leaf parts and tender plant shoots in the spring, maturing berries and fruits as summer comes on, and seeds and nuts in the fall.

Presently, the southern red-backed vole occurs from northern Canada into the northern part of the Great Lakes region and southward in the Appalachian and Rocky Mountain areas. The fossil record of this species extends from Late Illinoian to modern times. In the Pleistocene of the Great Lakes region, it is known from the Late Wisconsinan of northwest Ohio, southwestern and southern Indiana, and extreme southwestern Wisconsin. *Clethrionomys* sp. is known from the pre-Late Wisconsinan of southeastern Michigan. In the Ohio and Indiana Pleistocene localities (see fig. 133), the southern red-backed vole is well south of its modern range in northern Lower Michigan and Wisconsin. In the southwestern Wisconsin Pleistocene locality, the species is a little south of its present range in the state. This vole would have been an important small herbivore of the floor of coniferous and mixed coniferous and deciduous forests during the Pleistocene of the Great Lakes region.

Sites. Mill Creek, Sinclair County, Michigan, Rancholabrean LMA–pre-Late Wisconsinan (*Clethrionomys* sp., Karrow et al. 1997). Sheriden Pit Cave, Wyandot County, Ohio, Rancholabrean LMA–Late Wisconsinan (McDonald 1994). Prairie Creek D, Daviess County, Indiana, Rancholabrean LMA–Late Wisconsinan (Richards 1992c). Alton Mammoth Site, Crawford County, Indiana, Rancholabrean LMA–Late Wisconsinan (Richards 1991). Megenity Peccary Cave, Crawford County, Indiana, Rancholabrean LMA–Late Wisconsinan (Richards 1988b). King Leo Pit Cave, Harrison County, Indiana, Rancholabrean LMA–Late Wisconsinan (Richards and McDonald 1991). Moscow Fissure, Iowa County, Wisconsin, Rancholabrean LMA–Late Wisconsinan (Foley 1984).

*Cape Deceit Vole
*Lasiopodemys deceitensis

The Cape Deceit Vole was originally described from Irvingtonian Land Mammal Age sites in Alaska and Canada. This extinct species was named on the basis of characters on its molar teeth. In the Great Lakes region, this extinct vole of the Far North is known only from an Irvingtonian II Land Mammal Age site in west-central Illinois. This is one of the oldest well-dated Pleistocene small mammal sites in the Great Lakes region.

Site. County Line Site, Hancock County, Illinois (Miller et al. 1994). A specific glacial or interglacial age was not suggested by the authors. This site is in the Irvingtonian II Land Mammal Age.

Prairie Vole
Microtus ochrogaster (see fig. 132)

The prairie vole has a yellowish belly and coarser fur than the closely related meadow vole, which has softer fur and a grayish belly. Prairie voles weigh from about 1 to 2 ounces. The prairie vole prefers grassy habitats, where its major food consists of grasses.

Presently, the prairie vole ranges from the prairie regions of central and western Canada south in the United States to New Mexico, Oklahoma, and the middle southern states. It occurs in the western and southern part of the Great Lakes region. In the fossil record, the prairie vole occurs from Late Illinoian to modern times. In the Pleistocene of the Great Lakes region, it occurs in the late Wisconsinan of southern Indiana and extreme southwestern Wisconsin. The prairie vole would have been an important grass-eating species in open areas during the Pleistocene of the Great Lakes region.

Sites. King Leo Pit Cave, Harrison County, Indiana, Rancholabrean LMA–Late Wisconsinan (Richards and McDonald 1991). Moscow Fissure, Iowa County, Wisconsin, Rancholabrean LMA–Late Wisconsinan (Foley 1984).

*Hibbard's Tundra Vole
*Microtus paroperarius

Hibbard's Tundra Vole was originally described from an Irvingtonian Land Mammal Age site in Kansas and later identified in other Pleistocene sites in the central and eastern United States. This extinct species was named on the basis of characters in its molar teeth, as is the case in many fossil voles. In the Great Lakes region, Hibbard's Tundra Vole is known only from an Irvingtonian II Land Mammal Age site in west-central Illinois. This is one of the oldest well-dated Pleistocene small mammal sites in the Great Lakes region.

Site. County Line, Hancock County, Illinois (Miller et al. 1994). A specific glacial or interglacial age was not suggested by the authors. This site is in the Irvingtonian II Land Mammal Age.

Meadow Vole
Microtus pennsylvanicus (see fig. 132)

Characters for distinguishing the meadow vole from its close relative, the prairie vole, were given in an earlier account. Meadow voles weigh from about 1 to 2 ounces. This species prefers moist fields or marshes and bogs bordered by thick grasses and sedges, where it constructs a complex of runways. It is not surprising that the chief food of this hardy rodent consists of grasses and sedges. Insects and caterpillars are eaten sparingly.

Presently, meadow voles occur from Alaska and Northern Canada south in the United States to central Georgia and northern New Mexico. The fossil record of this species extends from Illinoian to modern times. In the Pleistocene of the Great Lakes region, the meadow vole was widespread, occurring in the Late Sangamonian/Early Wisconsinan of southeastern Ontario; pre-Late Wisconsinan of southeastern Michigan; and Late Wisconsinan of southeastern and eastern-central lower Michigan; northwest and west-central Ohio; northwest, southwest, and southern Indiana; and extreme southwestern Wisconsin. The meadow vole was obviously an important grass- and sedge-eating species of the marshes and bogs during the Pleistocene of the Great Lakes region.

Selected Sites. Innerkip, Oxford County, Ontario, Rancholabrean LMA–Late Sangamonian/Early Wisconsinan (Churcher et al. 1990). Mill Creek, St. Clair County, Michigan, Rancholabrean LMA–pre-Late Wisconsinan (Karrow et al. 1997). Shelton, Oakland County, Michigan, Rancholabrean LMA–Late Wisconsinan (Shoshani et al.

1989). Charles Adams, Livingston County, Michigan, Rancholabrean LMA–Late Wisconsinan (Holman 1979). Sheriden Pit Cave, Wyandot County, Ohio, Rancholabrean LMA–Late Wisconsinan (McDonald 1994). Carter, Darke County, Ohio, Rancholabrean LMA–Late Wisconsinan (McDonald 1994). Kolarik Mastodont, Starke County, Indiana, Rancholabrean LMA–Late Wisconsinan (Ellis 1982). Prairie Creek D, Daviess County, Indiana, Rancholabrean LMA–Late Wisconsinan (Richards 1992c). King Leo Pit Cave, Harrison County, Indiana, Rancholabrean LMA–Late Wisconsinan (Richards and McDonald 1991). Moscow Fissure, Iowa County, Wisconsin, Rancholabrean LMA–Late Wisconsinan (Foley 1984).

Woodland Vole
Microtus pinetorum (see fig. 132)

The woodland vole may be distinguished on the basis of its very thick, soft, molelike fur; its reddish color; and its short tail, which is only slightly longer than its hind foot. Woodland voles normally weigh a little more than 1 ounce. The woodland vole inhabits woodland areas rather than grasslands and marshes like the vole species discussed previously. It prefers deciduous forests but will live in mixed coniferous and deciduous forests. The woodland vole excavates burrows and tunnels in sandy woodland floors and spends much of its time in a very restricted area. It feeds on underground roots and stems as well as foraging on the surface for grass, fruits, and seeds.

Presently, the woodland vole occurs in southeastern Ontario and about the eastern third of the United States, from the northeastern states and the southern Great Lakes region south to Florida and Texas. The fossil record of this species extends from Sangamonian to modern times. In the Great Lakes region, it has been identified from the Late Wisconsinan of southern Indiana. Remains of voles thought to be either the pine vole or the prairie vole have also been identified from the Late Wisconsinan of southern Indiana. The woodland vole would have been an important herbivore of burrows and the forest floor in woodlands during the Pleistocene of the Great Lakes region.

Sites. King Leo Pit Cave, Harrison County, Indiana, Rancholabrean LMA–Late Wisconsinan (Richards and McDonald 1991). Megenity Peccary Cave, Crawford County, Indiana, Rancholabrean LMA–Late Wisconsinan (*Microtus pinetorum* or *ochrogaster*, Richards 1988b).

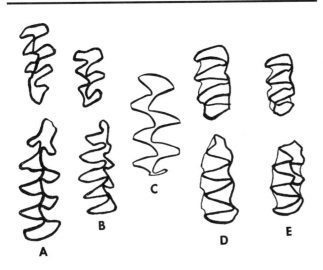

FIGURE 135. A, Upper right third molar (top) and lower right first molar (bottom) of yellow-cheeked vole (*Microtus xanthognathus*). B, Upper right third molar (top) and lower right first molar (bottom) of heather vole (*Phenacomys intermedius*). C, Upper third molar of collared lemming (*Dicrostonyx torquatus*). D, Upper right third molar (top) and lower right first molar (bottom) of northern bog lemming (*Synaptomys borealis*). E, Upper right third molar (top) and lower right first molar (bottom) of southern bog lemming (*Synaptomys cooperi*).

Yellow-cheeked Vole
Microtus xanthognathus (fig. 135)

The yellow-cheeked vole may be distinguished by its large size, yellow rostrum (cheek area), and relatively long tail, which is about 2 inches long. This vole is one of the largest species in this large genus and weighs from about 4 to 6 ounces. The yellow-cheeked vole inhabits both forests and bogs. This species is a strict herbivore, eating grasses and sedges as well as a variety of other plant species.

Presently, the yellow-cheeked vole is restricted to the Far North, ranging from Alaska through the northern Canadian territories south to the northern parts of Manitoba, Saskatchewan, and Alberta. In the Pleistocene, however, the yellow-cheeked vole ranged as far south as Virginia, Tennessee, and Arkansas. The fossil record of this species extends from Illinoian to modern times. In the Pleistocene of the Great Lakes region, the yellow-cheeked vole has been identified from the pre-Late Wisconsinan of southeastern Michigan and from the Late Wisconsinan of northwest Ohio, southwest and southern Indiana, southwest Illinois, and extreme southwestern Wisconsin. All of these localities are far south of the present range of the species in Alaska and Canada (fig. 136). This big vole would have been an

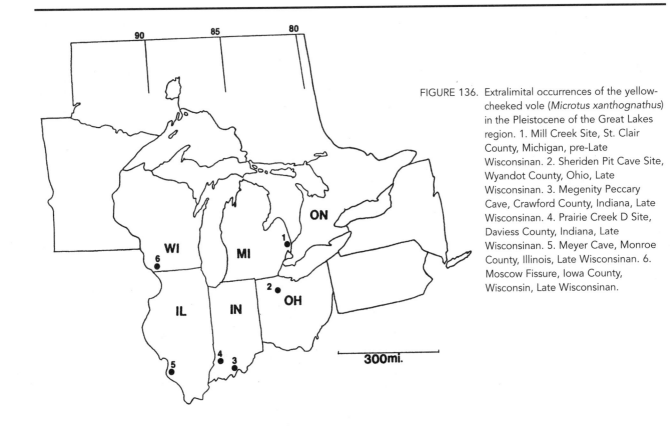

FIGURE 136. Extralimital occurrences of the yellow-cheeked vole (*Microtus xanthognathus*) in the Pleistocene of the Great Lakes region. 1. Mill Creek Site, St. Clair County, Michigan, pre-Late Wisconsinan. 2. Sheriden Pit Cave Site, Wyandot County, Ohio, Late Wisconsinan. 3. Megenity Peccary Cave, Crawford County, Indiana, Late Wisconsinan. 4. Prairie Creek D Site, Daviess County, Indiana, Late Wisconsinan. 5. Meyer Cave, Monroe County, Illinois, Late Wisconsinan. 6. Moscow Fissure, Iowa County, Wisconsin, Late Wisconsinan.

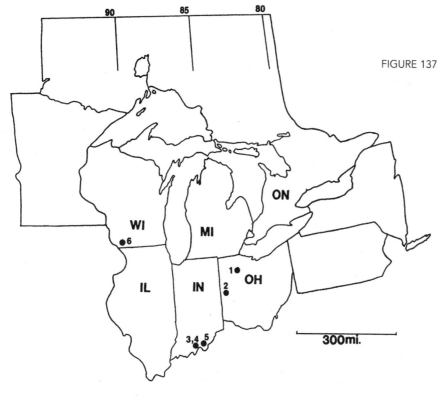

FIGURE 137. Extralimital occurrences of the heather vole (*Phenacomys intermedius*) in the Pleistocene of the Great Lakes region. 1. Sheriden Pit Cave Site, Wyandot County, Ohio, Late Wisconsinan. 2. Carter Site, Darke County, Ohio, Late Wisconsinan. 3. Alton Mammoth Site, Crawford County, Indiana, Late Wisconsinan. 4. Megenity Peccary Cave, Crawford County, Indiana, Late Wisconsinan. 5. King Leo Pit Cave, Harrison County, Indiana, Late Wisconsinan. 6. Moscow Fissure, Iowa County, Wisconsin, Late Wisconsinan.

important herbivore of forest and bog habitats during the Pleistocene of the Great Lakes region.

Sites. Mill Creek, St. Clair County, Michigan, Rancholabrean LMA–pre-Late Wisconsinan (Karrow et al. 1997). Sheriden Pit Cave, Wyandot County, Ohio, Rancholabrean LMA–Late Wisconsinan (McDonald 1994). Prairie Creek D, Daviess County, Indiana, Rancholabrean LMA–Late Wisconsinan (Richards 1992c). Megenity Peccary Cave, Crawford County, Indiana, Rancholabrean LMA–Late Wisconsinan (Richards 1988b). Meyer Cave, Monroe County, Illinois, Rancholabrean LMA–Late Wisconsinan (Parmalee 1967). Moscow Fissure, Iowa County, Wisconsin, Rancholabrean LMA–Late Wisconsinan (Foley 1984).

Heather Vole

Phenacomys intermedius (see fig. 135)

The little heather vole differs from the voles of the genus *Microtus* in having rooted rather than ever-growing cheek teeth. It has long, fine hair and a grayish brown color. These voles weigh only 1 or 2 ounces. The heather vole occurs in a wide range of habitats, from sea level to the edge of the tree line in the mountains. It

prefers open coniferous forests with an understory of heath or shrubby areas in open situations. Heather voles feed upon heaths, berries, seeds, and the buds and bark of trees.

Presently, the heather vole occurs in most of subarctic Canada, northern Minnesota, and the mountainous areas of the western United States. In the Pleistocene, this species ranged as far south as Virginia, Tennessee, and Arkansas. The fossil record of the heather vole is restricted to Wisconsinan and modern times. In the Pleistocene of the Great Lakes region, this species occurred in the Late Wisconsinan of northwest and west-central Ohio, southern Indiana, and extreme southwestern Wisconsin. These are all southern extralimital records (fig. 137). The heather vole would have been a small herbivore of coniferous forests and open, shrubby areas during the Pleistocene of the Great Lakes region.

Sites. Sheriden Pit Cave, Wyandot County, Ohio, Rancholabrean LMA–Late Wisconsinan (Ford 1994). Carter, Darke County, Ohio, Rancholabrean LMA–Late Wisconsinan (McDonald 1994). Alton Mammoth Site, Crawford County, Indiana, Rancholabrean LMA–Late Wisconsinan (Richards 1991). Megenity Peccary Cave,

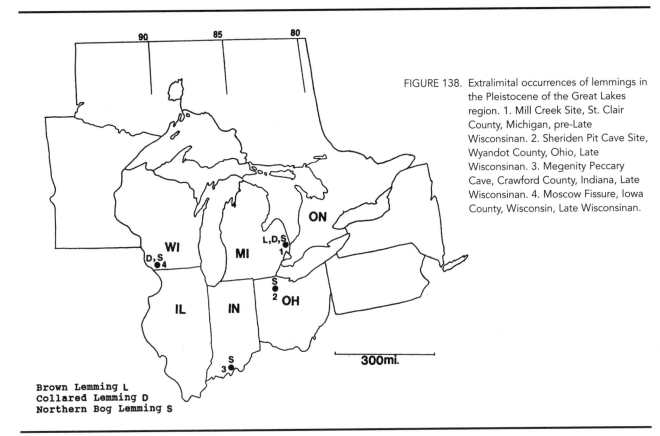

FIGURE 138. Extralimital occurrences of lemmings in the Pleistocene of the Great Lakes region. 1. Mill Creek Site, St. Clair County, Michigan, pre-Late Wisconsinan. 2. Sheriden Pit Cave Site, Wyandot County, Ohio, Late Wisconsinan. 3. Megenity Peccary Cave, Crawford County, Indiana, Late Wisconsinan. 4. Moscow Fissure, Iowa County, Wisconsin, Late Wisconsinan.

Brown Lemming L
Collared Lemming D
Northern Bog Lemming S

Crawford County, Indiana, Rancholabrean LMA–Late Wisconsinan (Richards 1988b). King Leo Pit Cave, Harrison County, Indiana, Rancholabrean LMA–Late Wisconsinan (Richards and McDonald 1991). Moscow Fissure, Iowa County, Wisconsin, Rancholabrean LMA–Late Wisconsinan (Foley 1988).

Collared Lemming

Dicrostonyx torquatus (figs. 138 and 139; see fig. 135) The robust collared lemming may be recognized by the very large third and fourth claws on the front feet, which are highly adapted for digging. In the winter, each claw develops into a double claw for even more efficient digging. This lemming is the only rodent that turns white during the winter season. Collared lemmings weigh from about 1 to 4 ounces. This is an animal of the treeless tundra regions of the arctic. It prefers dry areas with a substrate of sand or gravel with ample plant cover. In the summer, collared lemmings feed on grasses and sedges and in the winter eat bark, twigs, and buds.

Presently, this lemming occurs in the tundra regions of Alaska, northwestern and north-central

FIGURE 139. Left, collared lemming, *Dicrostonyx torquatus*; right, northern bog lemming, *Synaptomys borealis*.

Canada, and Greenland. A related species, *Dicrostonyx hudsonius*, occurs east of Hudson's Bay. The fossil record of the collard lemming is restricted to Wisconsinan and modern times. In the Pleistocene of the Great Lakes region, *Dicrostonyx* sp. has been identified from the pre-Late Wisconsinan of southeastern Michigan, and *D. torquatus* has been identified from the Late Wisconsinan of extreme southwestern Wisconsin. Both of these Pleistocene localities are very far south of the range of the species in Alaska, northern Canada, and Greenland (see fig. 138). This lemming would have been an herbivore of open, shrubby areas during the Pleistocene of the Great Lakes region.

Sites. Mill Creek, St. Clair County, Michigan, Rancholabrean LMA–pre-Late Wisconsinan (*Dicrostonyx* sp., Karrow et al 1997). Moscow Fissure, Iowa County, Wisconsin, Rancholabrean LMA–Late Wisconsinan (Foley 1984).

Brown Lemming
Lemmus sp.

The genus *Lemmus* is a Palearctic taxon characterized by a robust skull and ungrooved incisors. This brownish lemming occurs in part of Alaska and in northern Canada where it inhabits tundra and alpine meadows and is adapted to moister conditions than the collared lemming *Dicrostonyx*. In the Pleistocene of the Great Lakes region, the brown lemming genus, unidentified as to species, was found only in a pre-Late Wisconsinan site in southeastern Michigan (see fig. 138).

Site. Mill Creek, St. Clair County, Michigan, Rancholabrean LMA–pre-Late Wisconsinan (Karrow et al. 1997).

Northern Bog Lemming
Synaptomys borealis (fig. 139; see fig. 135)

The northern bog lemming and its relative, the southern bog lemming, are distinguished from voles, which they closely resemble, by having each of the incisors grooved on the outer edge. The northern and the southern species are very similar to one another except for the fact that there are four pairs of mammary glands in the female northern bog lemming and only three pairs in the female southern bog lemming. The northern bog lemming weighs from a little under 1 ounce to almost 2 ounces. This lemming inhabits wet meadows and bogs where mats of vegetation exist. It feeds mainly on low vegetation but will temper this diet with small invertebrates such as slugs and snails.

Presently, the northern bog lemming occurs from Alaska and Canada south of the tundra and barely enters the United Sates in northern Minnesota and the far

northwestern states. The fossil record of the northern bog lemming ranges from Wisconsinan to modern times. In Wisconsinan times, this lemming occurred as far south as Virginia, Tennessee, and Arkansas. In the Pleistocene of the Great Lakes region, the northern bog lemming has been identified from the pre-Late Wisconsinan of southeastern Michigan and the Late Wisconsinan of northwestern Ohio, southern Indiana, and extreme southwestern Wisconsin. All of these localities are far south of the species' present range (see fig. 138). The northern bog lemming would have been a small herbivore of wet meadows and bogs during the Pleistocene of the Great Lakes region.

Sites. Mill Creek, St. Clair County, Michigan, Rancholabrean LMA–pre-Late Wisconsinan (Karrow et al. 1997). Sheriden Pit Cave, Wyandot County, Ohio, Rancholabrean LMA–Late Wisconsinan (McDonald 1994). Megenity Peccary Cave, Crawford County, Indiana, Rancholabrean LMA–Late Wisconsinan (Richards 1988b). Moscow Fissure, Iowa County, Rancholabrean LMA–Wisconsin, Late Wisconsinan (Foley 1984).

Southern Bog Lemming
Synaptomys cooperi (see fig. 135)

Distinguishing the southern from the northern bog lemming was discussed in the preceding account. Southern bog lemmings weigh from about 1 ounce to a little less than 2 ounces. The southern bog lemming is an inhabitant of wet meadows and bogs with mats of vegetation. It feeds mainly on low plants and occasionally supplements its diet with small invertebrates.

At present, the southern bog lemming ranges from southeastern Canada to North Carolina, Tennessee, and Missouri. The fossil record of this species extends from Illinoian to modern times. In the Pleistocene of the Great Lakes region, the southern bog lemming has been found in the Late Wisconsinan of south-central and southern Indiana and extreme southwestern Wisconsin. Fossils identified as *Synaptomys* sp. from the Late Wisconsinan of southwestern Indiana may represent this species or possibly the northern bog lemming. The southern bog lemming would have been a small herbivore of wet meadows and bogs during the Pleistocene of the Great Lakes region. It appears that the southern bog lemming and the northern bog lemming may have been thrown into competition when the northern form was displaced southward by the Wisconsinan glaciation.

Sites. Anderson Pit Cave, Monroe County, Indiana,

Rancholabrean LMA–Late Wisconsinan (Richards, pers. com.). Megenity Peccary Cave, Crawford County, Indiana, Rancholabrean LMA–Late Wisconsinan (Richards 1988b). Prairie Creek D, Daviess County, Indiana, Rancholabrean LMA–Late Wisconsinan (*Synaptomys* sp., Richards 1992c). Moscow Fissure, Iowa County, Wisconsin, Rancholabrean LMA–Late Wisconsinan (Foley 1984).

Muskrat
Ondatra zibethicus (fig. 140)

The large, semiaquatic muskrat is easily recognized on the basis of its naked, scaly tail that is compressed from side to side and used as a rudder (rather than flattened from top to bottom like the beaver) and its partially webbed hind feet. Muskrats weigh about 2 or 3 pounds and are the largest members of the family Muridae in North America. Muskrats prefer shallow, very slowly flowing or still water; thus marshes are a favorite habitat. These animals either make a den in the bank above the water level but with underwater entrances, or they build houses composed of mud and emergent aquatic plants. Their main food consists of the lower parts of large aquatic plants such as cattails or water lilies, but they also eat small invertebrates and vertebrates from time to time.

At present, muskrats range from Alaska and northern Canada south through the United States to the Florida panhandle, Texas, southern California, and northern Mexico. In the fossil record, muskrats are known from Late Illinoian to modern times. In the Pleistocene of the Great Lakes region, muskrats are known from the Middle Wisconsinan of southeast Ontario and the probable Late Wisconsinan of southeast Ontario; the Late Wisconsinan of southeastern Michigan, northwest and west-central Ohio, northwest, east-central, and southwest Indiana, and southwest Wisconsin. The muskrat would have been an important large rodent herbivore of marshes and shallow ponds during the Pleistocene of the Great Lakes region.

Selected Sites. Innerkip, Oxford County, Ontario, Rancholabrean LMA–Middle Wisconsinan (Churcher et al. 1990). Kelso Cave, Halton County Ontario, Rancholabrean LMA–probably Late Wisconsinan (Churcher and Dods 1979; Mead and Grady 1996). Shelton, Oakland County, Michigan, Rancholabrean LMA–Late Wisconsinan (Shoshani et al. 1989). Sheriden Pit Cave, Wyandot County, Ohio, Rancholabrean LMA–Late Wisconsinan (McDonald 1994). Carter, Darke County, Ohio, Rancholabrean LMA–Late Wisconsinan (McDonald 1994). Kolarik Mastodont, Starke County, Indiana, Rancholabrean LMA–Late Wisconsinan (Ellis 1982). Christensen Bog, Hancock County, Indiana, Rancholabrean LMA–Late Wisconsinan (Graham et al. 1983). Prairie Creek D, Daviess County, Indiana, Rancholabrean LMA–Late Wisconsinan (Richards 1992c). Moscow Fissure, Iowa County, Wisconsin, Rancholabrean LMA–Late Wisconsinan (Foley 1984).

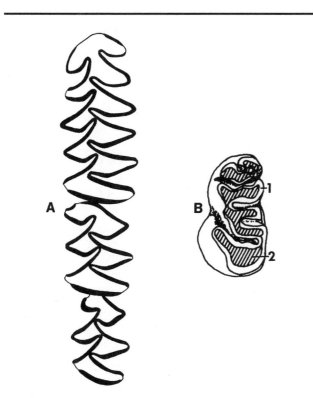

FIGURE 140. A, surface view of lower molars 1 (top) through 3 (bottom) of muskrat (*Ondatra zibethicus*); modified from Hibbard (1950). B, surface view of first lower molar of meadow jumping mouse (*Zapus hudsonius*): 1, enamel ridge; 2, exposed dentine.

JUMPING MICE AND RELATIVES
Family Dipodidae

Dipodids may be recognized by their large, kangaroo-like hind feet, hopping locomotion, and long tail that acts as a balance. They are small to medium-sized rodents that weigh from 1 or 2 ounces up to 6 or 7 ounces. Dipodids are omnivorous but usually eat more plant than animal food, and seeds are a staple diet for most of them. The members of this family undergo true hibernation. Dipodids occur in North America, temperate Eurasia, and northern Africa. In the fossil record, dipodids are known from Late Eocene to Holocene times. Only one genus of jumping mouse, *Zapus*, occurs in the Pleistocene of the Great Lakes region.

Meadow Jumping Mouse
Zapus hudsonius (fig. 140)

Jumping mice of the genus *Zapus* have large, kangaroo-like hind feet and almost hairless tails that are much longer than the body. Observable differences between the meadow jumping mouse and its relative, the western jumping mouse, are that the meadow jumping mouse is smaller and has an olive-colored body whereas the western jumping mouse is larger and has a dark back with yellowish sides. Meadow jumping mice weigh about .75 of an ounce. The meadow jumping mouse prefers damp, grassy sites where seeds are a staple food. They also eat other food items such as fungi, fruit, and even insects from time to time. Meadow jumping mice spend the winter in underground chambers where they undergo true hibernation, with the body temperature lowered to between 35 and 40 degrees F.

Presently, the meadow jumping mouse occurs from Alaska and northern Canada south to Alabama and Georgia. It is absent from Texas and most of the states westward. In the fossil record, the meadow jumping mouse occurs from Late Illinoian to modern times. In the Pleistocene of the Great Lakes region, the meadow jumping mouse has been identified from the Late Wisconsinan of southwestern Indiana. Fossils identified as *Zapus* sp. from a pre-Late Wisconsinan site in southeastern Michigan may represent this species. In the Pleistocene of the Great Lakes region, the Meadow jumping mouse would have been a seed-eating occupant of damp, grassy, meadows.

Sites. Mill Creek, St. Clair County, Michigan, Rancholabrean LMA–pre-Late Wisconsinan (*Zapus* sp., Karrow et al. 1997). Prairie Creek D, Daviess County, Indiana, Rancholabrean LMA–Late Wisconsinan (Richards 1992c).

Western Jumping Mouse
Zapus princeps

Distinguishing the western jumping mouse from the meadow jumping mouse was discussed in the preceding account. The western jumping mouse typically weighs a little over 1 ounce. This mouse mainly occurs in mountain regions where it occupies open, moist areas and meadows. Feeding habits appear to be generally similar to those of the meadow jumping mouse.

Presently, the western jumping mouse occurs from Alaska south to South Dakota and through the mountains of the southwestern United States. In the fossil record, this species is known from Wisconsinan to Holocene times. In the Pleistocene of the Great Lakes region, this species has been tentatively identified from the Late Wisconsinan of extreme southwestern Wisconsin, very much east of its modern range. This mouse would have also been a seed-eating, meadowland inhabitant during the Pleistocene of the Great Lakes region.

Site. Moscow Fissure, Iowa County, Wisconsin, Rancholabrean LMA–Late Wisconsinan (*Zapus* tentatively referred to *Zapus princeps*, Foley 1984).

NEW WORLD PORCUPINES
Family Erethizontidae

New World porcupines may be recognized by their stout bodies; short legs with strong, curved claws; and spines on various parts of the bodies. All of them are excellent climbers that feed, at least in part, on tree products. New World porcupines occur from Alaska and northern Canada south through much of the United States, except for the Southeast; through Mexico and Central America; and into about the northern two-thirds of South America. In the fossil record, they are known from the Oligocene to the Holocene. A single genus, the common porcupine (*Erithizon*), is known from North America and from the Pleistocene of the Great Lakes region.

Common Porcupine
Erithizon dorsatum (fig. 141)

The common porcupine is easily recognized on the basis of its stout body that is covered with long spines. Common porcupines weigh from about 10 to 30 pounds. These animals prefer deciduous or coniferous forests, and the diet consists of tree products such as pine needles, bark, and buds, as well as leaves, roots, flowers, and seeds in the summer season. The easily detached, penetrating spines protect the common porcupine from most predators but the fisher.

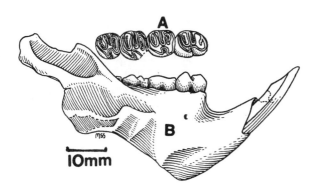

FIGURE 141. A, surface view of dentition of right lower jaw of a porcupine (*Erithizon dorsatum*). B, lateral view of right lower jaw of the same porcupine.

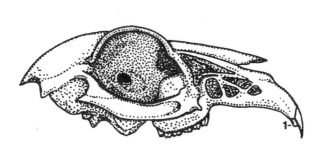

FIGURE 142. Skull of young cottontail rabbit (*Sylvilagus floridanus*) showing the latticework of openings on the snout and the extra pair of incisors (1).

Presently, the common porcupine occurs from Alaska and northern Canada south into the Great Lakes region. It also occurs in the western mountain states and south into Mexico. In the Pleistocene, however, the common porcupine occurred as far south as Florida and Texas. In the Pleistocene of the Great Lakes region, it is known from the Late Wisconsinan of west-central Ohio, south of its modern range, and from extreme southwestern Wisconsian. The common porcupine would have been an important woodland herbivore during the Pleistocene of the Great Lakes region.

Sites. Carter, Darke County, Ohio, Rancholabrean LMA–Late Wisconsinan (McDonald 1994). Moscow Fissure, Iowa County, Wisconsin, Rancholabrean LMA–Late Wisconsinan (Foley 1984).

HARES, JACKRABBITS, RABBITS, AND PIKAS
Order Lagomorpha

Lagomorphs share a common ancestry with rodents far back in the fossil record. The skull (fig. 142) is rodent-like in that it lacks canine teeth and has a diastema between the incisor teeth and cheek teeth and in that the cheek teeth are high-crowned and rootless. But the lagomorph skull differs, among other characters, in having a second pair of small, upper incisors (fig. 142) behind the enlarged first pair in the upper jaw and in having a large opening or a latticework of openings on the rostrum in front of the eye.

Presently, lagomorphs are nearly worldwide in distribution, with the exception of the Philippines, Madagascar, and many oceanic islands. They have been widely introduced into Australia and New Zealand in historic times. Lagomorphs are a much less diverse group than are the rodents.

Lagomorphs are known from Late Paleocene to modern times. Oddly, remains of lagomorphs are rare in the Pleistocene of the Great Lakes region, and many identifications within the order are very tentative (for example, bones identified merely as Lagomorpha sp. have been found atthe Late Wisconsinan of the Sheriden Pit Cave in northwest Ohio; Ford 1994). Two families, the Ochotonidae and the Leporidae, have been identified from the Pleistocene of the Great Lakes region. Figure 143 is a location map of extralimital Pleistocene occurrences of lagomorphs in the Great Lakes region.

PIKAS
Family Ochotonidae

The short-eared pikas, with their nonvisible tails, are readily distinguished from their long-eared relatives, the hares, jackrabbits, and rabbits of the family Leporidae. The family Ochotonidae is composed of a single living genus, *Ochotona*. Ochotonids weigh from about 4 ounces to a little under 1 pound and range in habitats from deserts, plains, and steppes to the timberline areas of high mountains. Pikas eat a wide variety of plant material, and most species construct hay piles to feed upon in the winter months. The pika family ranges from the steppes of the former Soviet Union through

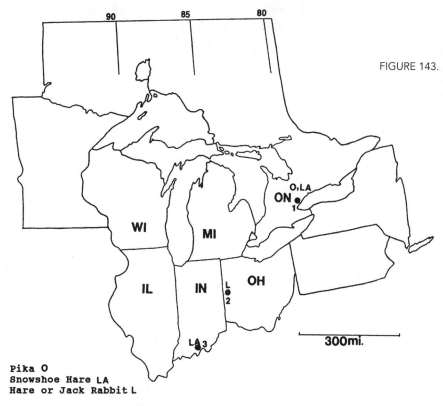

FIGURE 143. Extralimital Pleistocene localities of lagomorphs (pikas, hares, rabbits) in the Great Lakes region. 1. Kelso Cave, Halton County, Ontario, Rancholabrean LMA, probably Late Wisconsinan. 2. Carter Site, Darke County, Ohio, Late Wisconsinan. 3. Megenity Peccary Cave, Crawford County, Indiana, Late Wisconsinan.

Pika O
Snowshoe Hare LA
Hare or Jack Rabbit L

Asia and into the mountains of Alaska, Canada, and the northwestern United States. Ochotonids are known from Oligocene to modern times.

Pika
Ochotona sp.

Remains of a very large pika of the type that occurs in some parts of Asia are known from a single cave site in southeastern Ontario (fig. 143).

Site. Kelso Cave, Halton County, Ontario, Rancholabrean LMA, probably Late Wisconsinan (Churcher and Dods 1979; Mead and Grady 1996).

HARES, JACKRABBITS, AND RABBITS
Family Leporidae

The long-eared hares, jackrabbits, and rabbits are easily distinguished from their short-eared relatives, the pikas of the family Ochotonidae. Leporids weigh from about .5 of a pound to about 15 pounds. Leporids oc-

cupy a tremendous variety of habitats including arctic tundra, forests, shrubby country, grasslands, and deserts. Leporids are herbivores that reswallow and redigest their firstly eliminated soft green fecal pellets for essential proteins and vitamins. The secondly eliminated dry brown pellets are left alone. The geographic distribution of leporids today is the same as for the order Lagomorpha. In the fossil record, this family is known from Late Paleocene to modern times. Two genera, hares and jackrabbits (*Lepus*) and cottontail rabbits (*Sylvilagus*), are known from the Pleistocene of the Great Lakes region.

Bones identified merely as Lagomorpha sp. have been identified from the Late Wisconsinan of northwest Ohio. Leporidae sp. Fossils have been found in the pre-Late Wisconsinan of southeastern Michigan and from the Late Wisconsinan of southern Indiana and extreme southwestern Wisconsin.

Sites. Sheriden Pit Cave, Wyandot County, Ohio, Rancholabrean LMA–Late Wisconsinan (Lagomorpha sp., Ford 1994). Mill Creek, St. Clair County, Michigan, Rancholabrean LMA–pre-Late Wisconsinan (Leporidae sp., Karrow et al. 1997). Alton Mammoth Site,

Crawford County, Indiana, Rancholabrean LMA–Late Wisconsinan (Leporidae sp., Richards 1991a). Moscow Fissure, Iowa County, Wisconsinan, Rancholabrean LMA–Late Wisconsinan (Leporidae sp., Foley 1984).

Snowshoe Hare

Lepus americanus

The snowshoe hare may be immediately recognized in the winter, when it is white except for the black tips of its ears. In the summer, it may be distinguished from the cottontail by its larger size, much larger hind foot, and the lack of reddish brown fur on the neck. The snowshoe hare favors closed forests with a dense understory and prefers coniferous and mixed coniferous/deciduous woodlands to deciduous wooded areas. This hare is perfectly adapted to life in the Far North, as its large hind feet have a coating of stiff hairs that function as snowshoes, which allows the animal to travel easily over the top of the snow. In the summer, the snowshoe hare prefers to eat grasses and clover and in the winter switches to bark, twigs, and buds of deciduous trees and conifer needles.

At present, the snowshoe hare ranges from Alaska and northern Canada to the northern Great Lakes region and south in the Appalachian and western mountains. In the Pleistocene of the Great Lakes region, it has been found extralimitally in the probable Late Wisconsinan of southeastern Ontario and in the Late Wisconsinan of southern Indiana (see fig. 143). This hare would have been an important herbivore of coniferous and mixed coniferous/deciduous forests during the Pleistocene of the Great Lakes region.

Sites. Kelso Cave, Halton County, Ontario, Rancholabrean LMA, probably Late Wisconsinan (Churcher and Dodds 1979; Mead and Grady 1996). Megenity Peccary Cave, Crawford County, Indiana, Rancholabrean LMA–Late Wisconsinan (Richards and Whitaker 1997).

Hare or Jackrabbit

Lepus sp.

Bones that represent either hare or jackrabbit remains have been identified from the Late Wisconsinan of west-central Ohio (see fig. 143). At present, hares and jackrabbits do not occur in Ohio, as hares range to the north and jackrabbits to the west of the state. Bones referred to *Lepus* or *Sylvilagus* from the Late Wisconsinan of southern Indiana may represent a hare or jackrabbit (*Lepus*) or a cottontail (*Sylvilagus*).

Sites. Carter Site, Darke County, Ohio, Rancholabrean LMA–Late Wisconsinan (*Lepus* sp., McDonald 1994). King Leo Pit Cave, Harrison County, Indiana, Rancholabrean LMA–Late Wisconsinan (*Lepus* or *Sylvilagus*, Richards and McDonald 1991).

Eastern Cottontail

Sylvilagus floridanus (fig. 144; see fig. 142)

The familiar, bouncy, eastern cottontail may be distinguished from similar leporids (such as the snowshoe hair in its summer coat) on the basis of its smaller size, much smaller hind foot, and the reddish brown fur on the neck. Eastern cottontails weigh from about 2 to about 4 pounds. Throughout its vast range, the eastern cottontail prefers somewhat open habitats where thickets exist for hiding places. Deep forests and grasslands are avoided. Eastern cottontails are herbivores, with grasses forming the bulk of the diet.

Presently, the eastern cottontail occurs from southeastern and south-central Canada south through the United States to Argentina and Paraguay. In the fossil record, this species is known from Late Illinoian to Holocene times, and there are tentative reports of earlier occurrences. In the Pleistocene of the Great Lakes region, the eastern cottontail has been identified from the probable Late Wisconsinan of southeastern Ontario and was tentatively identified from the Late Wisconsinan of southwestern Indiana. The eastern cottontail would have been an important grass-eating species of the brushy habitats of the Pleistocene of the Great Lakes region.

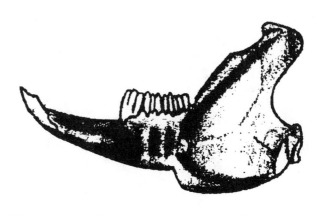

FIGURE 144. Left dentary in lateral view of an adult cottontail rabbit (*Sylvilagus floridanus*).

Sites. Kelso Cave, Halton County, Ontario, Rancholabrean LMA, probably Late Wisconsinan (Churcher and Dods 1979; Mead and Grady 1996). Prairie Creek D, Daviess County, Indiana, Rancholabrean LMA–Late Wisconsinan (*Sylvilagus* tentatively identified as *Sylvilagus* cf. *floridanus*, Richards 1992c).

PROBOSCIDEANS
Order Proboscidea

This order contains, among other groups, the mastodonts and mammoths. These giant, extinct mammals are the most important fossil vertebrates in the Great Lakes region. Their role as megaherbivores (dominant giant herbivores) will be discussed in detail later in the book. Definitive characters of the Proboscidea are the presence of a trunk or proboscis composed of the extended upper lip and nose, a much shortened skull, ever-growing incisor teeth specialized as tusks and composed of dentine only, and cheek teeth that are replaced horizontally as they are worn down. Modern proboscideans, the elephants, occur natively in Africa south of the Sahara (African elephant, *Loxodontia*) and south of the Himalayas in India, Malaya, Sri Lanka, and Sumatra (Indian elephant, *Elephas*). The fossil record shows that proboscideans were a dominant group in the Tertiary and the Pleistocene. The order comprised ancient groups such as gomphotheres and stegodonts as well as the more familiar mastodonts, mammoths, and elephants.

Fossil proboscideans originated in the Early Eocene and spread from North Africa into all continents except for Australia and Antarctica. They occurred from tropical rain forests to the arctic tundra. By the end of the Pleistocene, all but African and Indian elephants became extinct. Evolutionary tendencies in the proboscidea included an increase in size, lengthening of the limbs and proboscis (trunk), and specializations in the teeth. Two genera of proboscideans, *Mammut* (mastodonts) and *Mammuthus* (mammoths), occurred in the Pleistocene of the Great Lakes region. Each of these genera occupies a separate family.

FIGURE 145. Skeleton (left) and surface view of tooth (right) of an *American mastodont (*Mammut americanum*).

MASTODONTS
Family Mammutidae

Many people are confused about the terms mastodont and mammoth and think that the names apply to a single Ice Age mammal that looks something like a long-haired elephant. Actually, these animals have had a long, separate evolutionary history and are as different from one another as horses are from rhinos. Mastodonts are relatively primitive animals that occupy the family Mammutidae. Mammoths are more advanced animals that occupy the family Elephantidae along with the modern elephants. Some of the differences between mastodonts and mammoths are summarized in the following paragraphs.

In the mastodonts (fig. 145), the skull is flat on top, and both the skull and the sheaths that contain the bases of the tusks extend almost straight out. In the mammoths (fig. 146), the skull is domed on top, and both the skull and the tusk sheaths are oriented almost vertically. The tusks of the mastodont exit the skull horizontally, then curve outward and downward, whereas the tusks of mammoths exit the skull downward and curve outward. Some mastodonts grow an

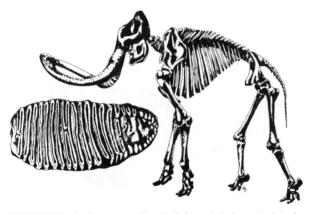

FIGURE 146. Surface view of tooth (left) and skeleton (right) of a *Jefferson mammoth (*Mammuthus jeffersonii*).

additional pair of smaller lower tusks, but mammoths never grow these.

The cheek teeth of mastodonts and mammoths are very different. Mastodont cheek teeth are relatively primitive in having their chewing surfaces composed of a series of large knobs, usually arranged in two rows (fig. 147; see fig. 145). These knobby teeth gave rise to the name mastodont, which technically means "nipple tooth."

Mammoths, on the other hand, have more advanced cheek teeth in having their chewing surfaces composed of transverse rows of thin enamel plates (fig. 147; see fig. 146). The jaws, muscles, and teeth of mastodonts and mammoths work differently to produce two kinds of chewing motions. In mastodonts, the lower jaw moves up and down against the upper jaw, producing a crunching type of chewing between the knobby surfaces of the teeth. In mammoths, the lower jaw moves backward and forward on the upper jaw, producing a grinding type of chewing along the surface rows of enamel on the teeth.

These differences in the jaws, muscles, and teeth of mastodonts and mammoths reflect the food habits of these proboscideans. Mastodonts crushed up mast and other types of browse between the bumpy surfaces of their teeth, whereas mammoths ground up grasses between the transverse enamel ridges of their teeth. In the Pleistocene, mastodonts were more abundant in woodlands where trees, shrubs, and mast were common items to browse upon, whereas mammoths were more abundant in prairie states where grasses were more abundant.

The body plan of the two proboscideans is also quite different. Mastodonts have a more low-slung, piglike body with shorter legs and with the back end at about the same level as the front end (see fig. 145). The mammoth has a more arched body, with longer legs and a sloping back (see fig. 146). The mastodont also tends to have a more robust skeletal frame to its body than the mammoth, which is more lightly built.

It has been suggested from time to time that the mastodont was not as intelligent an animal as the mammoth. Part of this is based on the relative size of the brains of the two animals, and it also might be based on the realization that mammoths are much more similar to modern elephants than are mastodonts.

The Pleistocene mastodont is composed of a single species, *Mammut americanum*. This species is by far the most abundant mammalian fossil identified from the Pleistocene of the Great Lakes region.

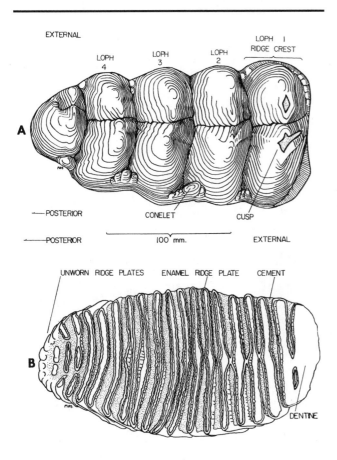

FIGURE 147. Detailed terminology used in the identification of the teeth of A, American mastodont (*Mammut americanum*), and B, Jefferson mammoth (*Mammuthus jeffersonii*). Both teeth are upper third molars in surface view.

*American Mastodont
*Mammut americanum (figs. 148 and 149)

Characters of the American mastodont compared with the mammoth have been discussed earlier. Mastodonts ranged from about 8 to 10 feet high at the shoulders and weighed from about 4 to 6 tons. The preferred habitat of mastodonts appears to have been open spruce woodlands, spruce forests, and open pine forests in the northeast and Great Lakes region, but in the Great Plains, Florida, and Texas they are believed to have lived in valleys, lowlands, and swamps. An analysis of undigested food remains inside of the rib cages of several mastodonts indicated the presence of twigs, conifer cones, leaves, tough grasses, marsh and swamp plants, and even moss.

The American mastodont lived from the Late

FIGURE 148. A group of American mastodonts (*Mammut americanum*) in the Late Wisconsinan of Michigan.

FIGURE 149. American mastodonts (*Mammut americanum*) digging for salt in a shallow pond in southern Michigan during the Late Wisconsinan.

Pliocene to the end of the Wisconsinan. It ranged throughout Alaska, the Yukon, and Prairie provinces, as well as New Brunswick, Quebec, Nova Scotia, and southern Ontario and most of the continental United States but was most common in the regions south of the Great Lakes and along the Atlantic coast. Specimens have been found by commercial anglers from over forty sites under the sea along the continental shelf off the northeastern North American coast. Some finds were made more than 200 miles from shore. It is believed these animals lived on the shelf during glacial times, when the sea level was lower and the area would have been covered by coniferous forest.

There are many more records of American mastodonts than of any other vertebrates in the Pleistocene of the Great Lakes region, and they were unusually abundant in the southern half of Lower Michigan. In the Late Wisconsinan alone, there are at least 61 mastodont sites in Ontario, 211 in Michigan, 136 in Ohio, 23 in Indiana, 40 in Illinois, and 20 in Wisconsin. The taphonomy, paleoecology, and possible interaction of mastodonts with humans are discussed in chapter 9.

Selected Sites. Hamilton Mastodont, Wentworth County, Ontario, Rancholabrean LMA–Late Wisconsinan (McAndrews and Jackson 1988). Saint Catharines Mastodont, York County, Ontario, Rancholabrean LMA–Late Wisconsinan (McAndrews and Jackson 1988). Powers Mastodont, Van Buren County, Michigan, Rancholabrean LMA–Late Wisconsinan (Garland and Cogswell 1985). Shelton, Oakland County, Michigan, Rancholabrean LMA–Late Wisconsinan (Shoshani et al. 1989). Charles Adams Mastodont, Livingston County, Michigan, Rancholabrean LMA–Late Wisconsinan (Holman 1979). Carter, Darke County, Ohio, Rancholabrean LMA–Late Wisconsinan (McDonald 1994). Burning Tree Mastodont, Licking County, Ohio, Rancholabrean LMA–Late Wisconsinan (Fisher et al. 1994). Kolarik Mastodont, Starke County, Indiana, Rancholabrean LMA–Late Wisconsinan (Ellis 1982). Christensen Bog, Hancock County, Indiana, Rancholabrean LMA–Late Wisconsinan (Graham et al. 1983). New Milford Mastodont Site, Winnebago County, Illinois, Rancholabrean LMA–Late Wisconsinan (Anderson 1905). Maple Park Mastodont, Kane County, Illinois, Rancholabrean LMA–Late Wisconsinan (Graham and Graham 1986). Boaz Mastodont, Richland County, Wisconsin, Rancholabrean LMA–Late Wisconsinan (Palmer and Stoltman 1976). Deerfield Mastodont, Dane County, Wisconsin, Rancholabrean LMA–Late Wisconsinan (Dallman 1968).

MAMMOTHS AND ELEPHANTS
Family Elephantidae

The mammoth weighed several tonne
and when his short lineage was done,
his huge bones remained but have not yet explained
if his species were three, two, or one

The family Elephantidae is readily separated from the proboscidean family Mammutidae (mastodonts) in that both elephants and mammoths have cheek teeth that are formed by a series of up-and-down oriented, thin plates (see figs. 146 and 147). Each plate is composed of relatively soft dentine on the inside and harder enamel on the outside, so that wear on the plates forms a series of enamel ridges. Mastodonts have bumpy cusps on the surface of their teeth (see figs. 145 and 147).

There has been very much confusion and argumentation in scientific circles about specific names (see the preceding limerick) for mammoths (*Mammuthus*). Here I will follow the concept that there are two species of mammoths in the Great Lakes region: a *woolly mammoth (*Mammuthus primigenius*), which generally inhabited the Far North but occasionally reached areas in southern Canada and the Great Lakes region during the Pleistocene, and the *Jefferson mammoth (*Mammuthus jeffersonii*), which had a more southern distribution. It is often difficult to identify these species on the basis of the fragmentary remains that one usually finds at mammoth sites. Thus some fossils identified as the Jefferson mammoth may actually be woolly mammoths and vice versa.

Jefferson Mammoth

Mammuthus jeffersonii (figs. 150 and 151; see fig. 146) Characters that separate mammoths from mastodonts have been given in the account on the family Mammutidae (mastodonts). The Jefferson mammoth, named for Thomas Jefferson, may be distinguished from the woolly mammoth on the basis of being larger and having less complicated teeth. But misidentification of the two mammoths has probably been made from time to time. The preferred habitat of the Jefferson mammoth was grasslands and prairies.

In Michigan, where woodlands dominated grasslands during the Pleistocene, mammoths were outnumbered by mastodonts by more than four to one. But in the Great Plains region, where grasslands were dominant during the Pleistocene, mammoths greatly outnumbered mastodonts. I have already discussed the fact that the complicated enamel ridges on the teeth of mammoths indicate that they were grassland grazers.

Jefferson mammoths were hunted by humans, and several obvious human kill sites have been discovered in the West. The postulated relationships between humans and mastodonts and mammoths in the Great Lakes region will be discussed in chapter 9.

The Jefferson mammoth occurred from Sangamonian to Late Wisconsinan times, when it is thought to have become extinct about 11,000 years ago. Jefferson mammoths have been identified from sites in Ontario; the western, central, and southern United States; and Mexico. In the Late Pleistocene alone (some records, however, could be misidentified woolly mammoths), at least 27 mammoth sites are known from Ontario, 49 from southern Lower Michigan, 57 from Ohio, 13 from Indiana, 11 from Illinois, and 28 from Wisconsin.

Selected Sites. Scarborough Bluffs near Toronto, Ontario, Rancholabrean LMA–Early Wisconsinan (Karrow et al. 1980). Rostock, Perth County, Ontario, Rancholabrean LMA–Late Wisconsinan (McAndrews and Jackson 1988). Mead, Ingham County, Michigan, Rancholabrean LMA–Late Wisconsinan (Abraczinskas 1993). Marion, Marion County, Ohio, Rancholabrean LMA–Late Wisconsinan (McDonald 1994). Alton

FIGURE 150. A group of Jefferson mammoths (*Mammuthus jeffersonii*) grazing on grass in Subregion I during the Late Wisconsinan.

FIGURE 151. Jefferson mammoths (*Mammuthus jeffersonii*) soaking in a Late Wisconsinan pond in the Great Lakes region. A group of flat-headed peccaries (*Platygonus compressus*) moves in the foreground.

Mammoth, Crawford County, Indiana, Rancholabrean LMA–Late Wisconsinan (Richards 1991). Cairo Mammoth, Alexander County, Illinois, Rancholabrean LMA–Late Wisconsinan (Crook 1927). Schaefer, Kenosha County, Wisconsin, Rancholabrean LMA–Late Wisconsinan (West and Dallman 1980).

Woolly Mammoth
Mammuthus primigenius

The woolly mammoth was a smaller animal with more complicated enamel ridges on the teeth than the Jefferson mammoth and had a more northerly distribution. I have already commented on the difficulty of distinguishing the two mammoths based on the fragmentary remains that one usually finds at mammoth sites. Nevertheless, it appears that from time to time, the woolly mammoth invaded the Great Lakes region, even reaching as far as southeastern Ohio.

Fortunately, because of the many cave paintings and engravings in Eurasia as well as the thousands of bones, tusks, and even frozen carcasses in Siberia, we know more about the woolly mammoth than any other Western Hemisphere extinct mammal. This mammoth had the very general appearance of a shaggy elephant, but there were several important differences. It was relatively small, reaching a maximum height of about 10 feet at the shoulders. The ears were quite small, very much unlike the large, floppy ears of modern elephants. Also, the trunk was short compared to extant elephants, and there were two fingerlike projections on the trunk, one on the front and one on the back. The African elephant has a finger on the front of the trunk and a thick thumb on the back. The Indian elephant has a single finger on the front of the trunk. The tusks of woolly mammoths became so twisted in some individuals that they sometimes pointed inward at one another.

The shaggy coat of woolly mammoths has been the subject of several artistic Paleolithic cave paintings in Europe. Carcasses have shown that this coat consisted of an outer coat of long hair and an inner coat of short hair. Both reddish and dark-colored hairs of mammoths have been found, and it has been suggested that they had a change of hair color at the onset of the summer season, as presently happens in many arctic mammals. It has also been suggested that the hair became bleached during the animals' long burial and that the outer hair was black.

Woolly mammoths lived from Sangamonian to Late Wisconsinan times, and they also are thought to have become extinct about 11,000 years ago. The woolly mammoth entered North America across the Bering Strait in Sangamonian times. Woolly mammoths have been reported from the Pleistocene of Ontario, Michigan, and Ohio. A complete skull was recovered in southeastern Ohio.

Selected Sites. Woodbridge, York County, Ontario, Rancholabrean LMA–Middle Wisconsinan (Churcher 1968). Lake Iroquois Deposits, near Toronto, Ontario, Rancholabrean LMA–Late Wisconsinan (Karrow et al. 1980). Glacial Lake Mogodore, Cass County, Michigan, Late Pleistocene (Case et al. 1935). Whiskey Run, near Mt. Healthy, Hamilton County, Ohio, Rancholabrean LMA–Late Wisconsinan (complete skull, McDonald 1994).

8

Important Pleistocene Vertebrate Sites in the Great Lakes Region

This chapter describes some important Pleistocene vertebrate sites in the Great Lakes region. Sites will be presented in chronological order from older to younger Pleistocene sequences. Emphasis will be on the larger sites that have yielded large vertebrate faunas and/or especially important species. For instance, the Late Wisconsinan Sheriden Pit Cave Site in northwestern Ohio will be featured because of its extensive faunal list as well as its important species. On the other hand, the Sangamonian Hopwood Farm Site in south-central Illinois will be discussed mainly because of the presence of a single important species, the giant land tortoise (*Hesperotestudo crassiscutata*).

ONTARIO

The most important Pleistocene sites in Ontario occur in the southern part of the province near Ottawa and Toronto.

In the Toronto region, one finds the most extensive superimposed sequence of Pleistocene vertebrate faunas anywhere in the Great Lakes region. Here, superimposed Sangamonian, Early Wisconsinan, Middle Wisconsinan, and Late Wisconsinan sites occur, often with glacial tills above and below them. The most important

Ontario Pleistocene vertebrates are the widespread mastodont and mammoth fossils and the large marine mammals of the Champlain Sea deposits near Ottawa. The most disappointing aspect of the Ice Age vertebrate fauna of Ontario is how few Pleistocene records of amphibian and reptiles exist.

Sangamonian Site (Rancholabrean)

Don Formation, Don Valley Brickyard Site, Toronto, York County, Ontario. The Don Valley Brickyard Site in Toronto is the type locality for the Sangamonian Don Formation, which is underlain by the York Till of Illinoian age and overlain by the Scarborough Formation of Early Wisconsinan age. The fossils from the Don Valley Brickyard were collected from a 25-foot thick layer of stratified, cross-bedded clay and sand. Cross-bedding indicates that these relatively fine sediments were deposited by relatively slow currents that moved over a shallow drop-off.

Mollusk shells, wood, and leaf fossils, as well as pollen and diatoms (tiny protists with silica shells), suggest that the sediments of the Don Formation were deposited in a freshwater bay of a very large Pleistocene lake that was at least 60 feet higher than Lake Ontario is today. Pollen taken from sediments at the bottom of the Don Formation suggests that the plants existed in a temperate climate that was as much as 5 degrees F warmer than at present. But pollen and faunal evidence from the upper part of the Don Formation beds indicates that a cooling trend was in progress.

The vertebrate fauna consists of the following animals: trout and whitefish (Salmonidae), pike or muskellunge (*Esox* sp.), shiner (cf. *Notropis* sp.), channel catfish (*Ictalurus punctatus*), burbot (*Lota lota*), yellow perch (*Perca flavescens*), Probable freshwater drum (cf. *Aplodinotus grunniens*, a member of the sculpin family (Cottidae), brown bear (*Ursus arctos*), Scott's moose (stag moose) (*Cervalces scotti*), white-tailed deer (*Odocoileus* sp.), woodchuck (*Marmota monax*), and giant beaver (*Castoroides ohioensis*). Since Sangamonian-age fossils are rare in the Great Lakes region, these Don Formation fossils are of considerable importance, especially since they occur in such a northern area. This is one of the earliest records of the giant beaver. It is surprising that there have been no reptile fossils, especially turtle shell elements, recovered from the Don Formation, particularly from the lower beds where temperatures warmer than at present were indicated. The lack of turtle bones might indicate that the freshwater bay lacked the dense aquatic vegetation that is preferred by most aquatic turtle species.

References. Coleman (1933), Karrow (1969), Harington (1978, 1990), Karrow et al. (1980), Eyles and Williams (1992).

Late Sangamonian or
Early Wisconsinan Site (Rancholabrean)

Innerkip Site, Oxford County, Ontario. The Innerkip Site lies about 2 miles from the village of Innerkip about 30 miles northeast of London on a tributary of the Thames River. This site, unlike the Don Valley Brickyard Site in Toronto, is not a part of an extensive stratigraphic sequence. The fossils were recovered from an isolated peat deposit that represents a Pleistocene bog. The peat rests on silt and clay and is overlain by organic muds, followed in sequence by clay, silt, and glacial tills. Based on a radiocarbon date of over 50,000 B.P. on black organic muds and floral and faunal evidence, a Late Sangamonian or Early Wisconsinan age is suggested.

Plants identified from the Innerkip Site include pollen, mosses, wood fragments, and seeds. Insects from the site include ants, true bugs, caddisflys, and beetles. The plant remains suggest that the dominant vegetation of the area was a broadleaf forest, and the plant and beetle remains together suggest a temperate climate with moderate rainfall. The fact that the bog had a rich flora and fauna indicates that it was not formed in the vicinity of glacial ice.

The vertebrate fauna consists of the following animals: Blanding's turtle (*Emydoidea blandingii*), white-tailed deer (*Odocoileus virginianus*), meadow vole (*Microtus pennsylvanicus*), muskrat (*Ondatra zibethicus*), and ermine, long-tailed weasel, or mink (*Mustela* sp.). The vertebrate fauna forms an assemblage that might presently be found at the edge of a small pond in southern Ontario and, like the plant and insect remains, suggests a moderate climate. The white-tailed deer is sensitive to deep snow and low temperatures, and its occurrence in the Innerkip fauna may be indicate limitations of snowfall and low temperature.

References. Pilny and Morgan (1987), Churcher et al. (1990), Harington (1990).

Early Wisconsinan Sites (Rancholabrean)

Pottery Road Formation Sites at Scarborough Bluffs and the Don Valley Brickyard, Toronto, Ontario. The sediments of the Pottery Road Formation lie above the Sangamonian beds of the Toronto area and below glacial tills that indicate progressive colder conditions. The Scarborough Bluffs Site lies in bluffs on the north shore of Lake Ontario, east of Toronto, and has yielded vertebrate fossils eroding out of the Early Wisconsinan Pottery Road Formation. The Don Valley Brickyard Site in Toronto also contains Pottery Road Formation sediments that have produced vertebrate fossils. Pollen, mollusk shells, and vertebrate fossils from the Pottery Road Formation indicate boreal forest conditions, probably with open forests and grasslands.

The combined vertebrate faunas of these Pottery Road Formation sites consist of the following vertebrates: grizzly bear (*Ursus arctos*), Scott's moose (stag moose) (*Cervalces scotti*), white-tailed deer (*Odocoileus virginianus*), tundra muskox (*Ovibos* sp.), bison (*Bison* sp.), and mammoth (*Mammuthus* sp.). The combination of the grizzly bear and tundra muskox indicate cold conditions. On the other hand, the presence of the white-tailed deer indicates a lack of deep snow and savagely cold temperatures. The bison prefers areas where grasses are available. Thus a boreal climate with open woodlands and grasses is indicated.

References. Coleman (1933), Churcher and Karrow (1977), Karrow et al. (1980).

Middle Wisconsinan Site (Rancholabrean)

Woodbridge Borrow-pit, York County, Ontario. The fossils from the Woodbridge Site near Toronto came from glacial deposits exposed in the floor of a borrow-pit used in the construction of an embankment. Stratigraphic studies at the pit showed that the site was underlain by Early Wisconsinan till and overlain by tills of Middle Wisconsinan age, and thus the site was interpreted as representing an interstadial within the Middle Wisconsinan. The age of the site is considered to be between about 50,000 to 40,000 B.P. Peat, wood, pollen, and mollusks have been recovered from the site.

The vertebrate fauna consists of only two species, but they are very important ones. A grizzly bear (*Ursus arctos*) reported from the site represents one of the earliest records of this large bear in North America and is one of the rare reports of this species in the Great Lakes region. A woolly mammoth (*Mammuthus primigenius*) from Woodbridge is also one of the few records of the species from the region. The occurrence of the two species together indicates a relatively cold climate at the time.

References. Churcher (1968), Churcher and Morgan (1976), Karrow et al. (1980).

Probable Late Wisconsinan Site (Rancholabrean)

Kelso Cave, near Milton, Ontario. Kelso Cave west of Milton, Ontario, lies on the north face of a large promontory (projecting ridge) of the Niagara Escarpment. The cave was quarried for limestone at one time and then abandoned. Vertebrate fossils and pollen were taken from the cemented dolomite breccia (cemented mass of cave flowstone) in the cave.

Pollen from Kelso Cave shows that spruce, pine, birch, oak, elm, lime, and black walnut coexisted with the vertebrate fauna. These plant species indicate a somewhat moist climate, but not a subarctic one. An important vertebrate fossil that was collected at the site was a very large pika (*Ochotona* sp.) that represented a species about as large as some of the giant pikas from the Pleistocene of the Bering Straits region and in Asia. Because these large pikas were once thought to be restricted to Illinoian age sites in eastern North America, it was suggested that the Kelso Cave fauna might represent the Illinoian. It has recently been shown, however, that large pikas lived well into the Late Wisconsinan, and because other Kelso Cave species are typical of this age, its fauna is now considered to be of probable Late Wisconsinan age.

Other fossil vertebrates from Kelso Cave are the American toad (*Bufo americanus*), grouse (referred to the ruffed grouse, *Bonasa umbellus*), little brown bat (*Myotis lucifugus*), striped skunk (*Mephitis mephitis*), deer mouse (*Peromyscus maniculatus*), muskrat (*Ondatra zibethicus*), snowshoe hare (*Lepus americanus*), and eastern cottontail (*Sylvilagus floridanus*). All of the Kelso Cave vertebrate species, with the exception of the large pika, presently occur in southern Ontario, thus the fauna and the plant species together suggest a climate similar to that which occurs in the area today.

References. Churcher and Dods (1979), Mead and Grady (1996).

Late Wisconsinan Sites (Rancholabrean)

Glacial Lake Iroquois Deposits, Toronto Region, Ontario. Fossils from sediments derived from Glacial Lake Iroquois have yielded remains of wood, pollen, mollusks, ostracods, and large vertebrates. These deposits lie over glacial tills and are followed by a sequence of Holocene sediments. The vertebrate-bearing sediments are considered to represent a time of about 12,000 B.P., thus representing very Late Wisconsinan times.

The vertebrate fauna includes the following large mammals: bison (*Bison* sp.), tundra muskox (*Ovibos* sp.), and mammoth (*Mammuthus*). At one time, both the woolly mammoth and the Jefferson mammoth were recorded from the Lake Iroquois sediments, but it now seems probable that only a single species was represented or, if the woolly mammoth and the Jefferson mammoth were actually subspecies of a single species, that an intermediate form between the two subspecies was represented. The muskox indicates cold conditions, and the bison suggests the presence of grasslands in the area during the Late Wisconsinan.

References. Bensley (1923), Karrow (1967), Karrow et al. (1980).

Champlain Sea Sites near Ottawa, Ontario. Several sites near Ottawa, Ontario, have yielded remains of Late Wisconsinan vertebrates associated with sediments of the Champlain Sea This sea was a significant Late Wisconsinan event that affected Quebec as well as Ontario. When the Wisconsinan ice mass retreated to the north, the Atlantic Ocean flooded the St. Lawrence River valley. The sea reached its maximum size about 11,500 B.P., when it covered at least 20,000 square miles (fig. 152). This coverage included the lower Ot-

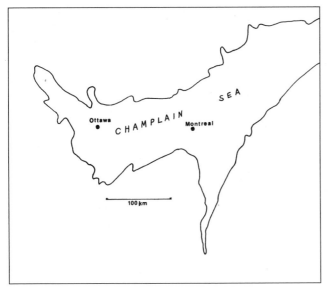

FIGURE 152. General outline of the Champlain Sea near its maximum in the Late Wisconsinan.

tawa River and Lake Champlain valleys as well as the region between Brockville and Quebec City.

Radiocarbon dates indicate that fossil vertebrates, including large marine mammals, occupied the Champlain Sea from about 11,500 to 10,500 B.P. Based on evidence from shells of marine mollusks, it has been estimated that the early Champlain Sea environment was probably similar to that of the modern Gulf of St. Lawrence, and in fact, most of the Pleistocene species can presently be found there. But during later Wisconsinan times, the Champlain Sea became warmer and less salty and finally reached a freshwater phase at about the end of the Pleistocene. In Holocene times, the freshwater lake drained and the landscape became as it is presently.

Fossil fishes identified from sediments of the Champlain Sea include: lake trout (*Salvelinus namaycush*), capelin (*Mallotus villosus*), rainbow smelt (*Osmerus mordax*), Atlantic tomcod (*Microgadus tomcod*), threespine stickleback (*Gasterosteus aculeatus*), ninespine stickleback (*Pungitius* sp.), hook-eared sculpin (*Artediellus uncinatus*), and lumpfish (*Cyclopterus lumpus*). These fishes came from nodules from the Green Creek Site, Carlton County, near Ottawa. The fishes of this site suggest that a cool marine habitat was fed by streams from a nearby deep lake near Ottawa.

Nonmarine mammals from the Champlain Sea sed-

iments include the American marten (*Martes americanus*) from a calcareous nodule at the Green Creek Site and an eastern chipmunk (*Tamias striatus*) from Champlain Sea age deposits near Moose Creek, Ontario. The most interesting mammals from the Champlain Sea sites, however, are the seals and whales.

A bearded seal (*Erignathus barbatus*) is known from a site radiocarbon-dated at about 11,000 B.P. near Finch, Stormont County, Ontario, and a ringed seal (*Phoca hispida*) is known from a site near the Ottawa International Airport. There is no radiocarbon date for the ringed seal. A harp seal (*Phoca groenlandica*) with no radiocarbon date is known from the Green Creek Site. These seals are all adapted to breeding on fast ice.

White whales (*Delphinapterus leucas*) are known from sites radiocarbon-dated at about 10,500 B.P. from Ottawa and from Pakenham in Lanark County, Ontario. The bowhead whale (*Balaena mysticetus*) is known from a site radiocarbon-dated at about 11,500 B.P. from near White Lake, Renfrew County, Ontario, and an undated humpback whale (*Megaptera novaeangliae*) is known from near Smith Falls, Lanark County, Ontario. All of these whales are adapted to cool, nearshore conditions in modern seas.

References. Wagner (1970, 1984), Harington (1977, 1978, 1988), McAllister et al. (1981).

Late Wisconsinan Proboscidean Sites. At least sixty-one mastodonts and twenty-seven mammoths are known from postglacial Wisconsinan (Rancholabrean) sites in Ontario ranging from about 12,000 to 10,000 B.P. Most of the mastodont sites are found in the southern part of the province in lake plains areas, which suggests the animals lived in or near wetland habitats. On the other hand, mammoth sites are found mainly northward in uplands, especially on ancient lake beaches.

Most mastodont finds are concentrated in the area southwest of Lake Ontario. The northernmost mastodont record is well south of Lake Simcoe in Dufferin County, near Laurel. Mammoth sites are mainly concentrated around the eastern end of Lake Ontario, and the northernmost verified record is just south of Lake Simcoe north of Zephyr. The northernmost records of mastodonts and mammoths in Ontario are nearly identical in latitude to the northernmost records of mastodonts and mammoths in Michigan. The northern distribution of mastodonts and mammoths in the Great Lakes region is discussed in detail in chapter 9.

Probably the most important proboscidean site, relative to the ecological changes that might have affected proboscideans in Ontario, is the Rostock Mammoth Site in Perth County, near Stratford. This site provides a pollen record of vegetational changes correlated with radiocarbon dates. From 14,500 to 13,000 B.P. the Rostock area was a wasteland very near the glacier. From 13,000 to 12,000 B.P. the area was a tundra woodland. From 12,000 to 10,000 B.P. the area was a boreal woodland.

Human hunters were contemporaneous with proboscideans in Ontario, as shown by the existence there of fluted spear points of the type used to kill proboscideans in other parts of North America. Nevertheless, there is no clear evidence of the association of these spear points with mastodonts and mammoths. Thus it is possible that in Ontario, at least, extinction of mastodonts and mammoths may be linked with a change to a warmer, drier climate that eventually eliminated the boreal forest habitat.

Reference. McAndrews and Jackson (1988).

MICHIGAN

Most Michigan Pleistocene vertebrate records are based on isolated finds of mastodonts and mammoths. In fact, substantiated Michigan Ice Age proboscidean sites are far more numerous than such substantiated sites in any other province or state in the Great Lakes region. Most of these Pleistocene vertebrate sites consist of infilled kettle bogs or shallow basins, and all are confined to southern Lower Michigan south of the Mason-Quimby Line (see fig. 10).

Only one Michigan Pleistocene vertebrate site is older than Late Wisconsinan, and almost all represent the time span between about 12,000 and 10,000 B.P. Nevertheless, two Michigan sites show that there was a withdrawal of the Wisconsinan ice about 24,000 years ago. As in Ontario, Pleistocene fish, amphibian, and reptile fossils are rare. Also as in Ontario, there is no direct association of human lithic (stone) tools or artifacts with proboscidean remains.

Pre-Late Wisconsinan Site (Rancholabrean)

Mill Creek Site, near Port Huron, St. Clair County, Michigan. The Mill Creek Site is exposed in an eroded stream bank northwest of Port Huron, Michigan, near the Canadian border. This is the only Michigan Pleistocene vertebrate fauna that is older than Late Wisconsinan. The fossil zone at Mill Creek is overlain by two Late Wisconsinan tills (Mill Creek Till and Fisher Road Till) and contained pollen, fossil wood, plant debris, mollusks, ostracods, and small vertebrate fossils. The stratigraphy of the site indicates a Middle Wisconsinan (Rancholabrean) age, but conflicting radiocarbon and amino acid dates have led to the designation of the site merely as "pre-Late Wisconsinan."

The local vegetation of the time was mainly a pine-spruce association. Thirty-nine taxa of mollusks and six ostracods were identified. The vertebrate fauna consists of the following animals: pike or musky (*Esox* sp.), Cyprinidae (minnow), Catostomidae (sucker), ninespine stickleback (*Pungitius* sp.), freshwater drum (*Aplodinotus grunniens*), pygmy shrew (*Sorex hoyi*), ermine (*Mustela erminea*), deer mouse (*Peromyscus* sp.), lemming (*Dicrostonyx* sp.), brown lemming (*Lemmus* sp.), northern bog lemming (*Synaptomys borealis*), red-backed vole (*Clethrionomys* sp.), meadow vole (*Microtus pennsylvanicus*), yellow-cheeked vole (*Microtus xanthognathus*), jumping mouse (*Zapus sp.*), and rabbit or hare (Leporidae sp.). Seven of the eleven mammals of the site have boreal or tundra affinities, but some gastropods, fishes, and other mammals indicate somewhat warmer conditions. More equable winters than presently occur in boreal and tundra situations might have allowed for such an assemblage.

References. Benninghof et al. (1977), Karrow et al. (1997).

Late Wisconsinan Sites (Rancholabrean)

Casnovia Duck Site, Muskegon County, Michigan. The Casnovia Duck Site is one of two Michigan Pleistocene vertebrate sites that indicate that the Late Wisconsinan glacier temporarily withdrew from southern Lower Michigan about 25,000 to 24,000 B.P. The second site is described in the following account. A fossil duck bone (see fig. 67) was taken from a fossil wood layer near Casnovia, Muskegon County, Michigan. The 3-foot wood layer was underlain by 1 foot of organic clay and overlain by 144 feet of glacial till that represented the return of the Late Wisconsinan glacier. A radiocarbon date of about 25,000 B.P. was obtained on the fossil wood.

The duck bone represented a lesser scaup duck, *Aythya affinis* (see fig. 66), a species that is often seen flying through southwestern Michigan today on its way north to its breeding grounds. But it is quite possible that the Casnovia duck was part of a resident breeding population during this stage of the Pleistocene, for wood and pollen from the fossiliferous layer indicate that the deposit was formed in a wet depression dominated by coniferous trees, possibly a tamarack-spruce swamp, and that the regional vegetation was a coniferous forest dominated by spruce and pine.

References. Holman (1976), Kapp (1978).

Bailer Mammoth Site, Midland County, Michigan.
The Bailer Mammoth Site is the second Michigan Pleistocene vertebrate site that indicates that the Late Wisconsinan glacier temporarily withdrew from the southern Lower Peninsula of Michigan about 25,000 to 24,000 B.P. The site is near Coleman in the northwest corner of Midland County, Michigan. Here a Jefferson mammoth tooth was taken during the dragline excavation of a farm pond. A radiocarbon date of 24,000 B.P. was obtained on the mammoth tooth.

The tooth was taken from a layer of sandy gravel under a sequence of sand (with a leaf lens in it), organic silt, and peat (fig. 153). It is believed that the bone was transported by water a short distance from the original site of deposition. This is one of the very few cases of water-transported fossils from the Pleistocene of Michigan, as most sites are in still-water situations where little or no transport takes place. Unfortunately, no plant or pollen remains could be correlated with the tooth to determine what kind of vegetation existed at the time.

Reference. Kapp (1970).

Shelton Mastodont Site, Oakland County, Michigan.
The Shelton Mastodont Site in southeastern Lower Michigan lies between Oxford and Ortonville in Oakland County (fig. 154). This site has provided the most diversified vertebrate fauna of any Late Wisconsinan site in the state. The bones of the Shelton Site were derived from sediments that formed at the edge of a lake. The Pleistocene lake was bordered by a woodland. Radiocarbon dates from the fossiliferous sediments indicate that the bone-bearing strata were deposited from about 12,300 to about 11,700 B.P. Plant remains associated with the Pleistocene vertebrates suggest a conifer forest dominated by spruce in the lower strata, giving way in the upper strata to a mixture of trees (mainly pine), sedges, and grasses. Twenty-five diatom genera and ten mollusk genera suggest that the lake was moderately organically rich.

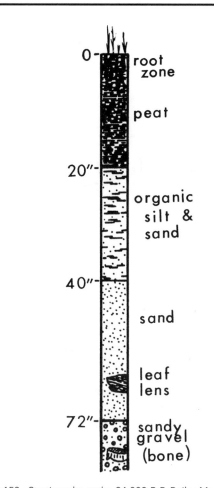

FIGURE 153. Stratigraphy at the 24,000 B.P. Bailer Mammoth Site, Midland County, Michigan. The basal gravels and sands are cross-bedded stream deposits. It is thought that deposition was discontinuous prior to the accumulation of the overlying peat.

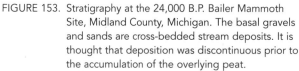

The Pleistocene vertebrate fauna consists of the following animals: northern pike (*Esox lucius*); yellow perch (*Perca flavescens*), somewhat questionably; the bullfrog (*Rana catesbeiana*); green frog (*Rana clamitans*); turkey (*Meleagris gallopavo*); meadow vole (*Microtus pennsylvanicus*); muskrat (*Ondatra zibethicus*); American beaver (*Castor canadensis*); canid (*Canis* sp.); moose (*Alces alces*); Scott's moose (stag moose) (*Cervalces scotti*); and mastodont (*Mammut americanum*).

The fishes and frogs represent the first records of these species from the Pleistocene of Michigan. The Scott's moose teeth are the first teeth of this species to have been found in Michigan. The mastodont was unusual in that it had only developed a single right tusk. Based on the stage of eruption and wear on the

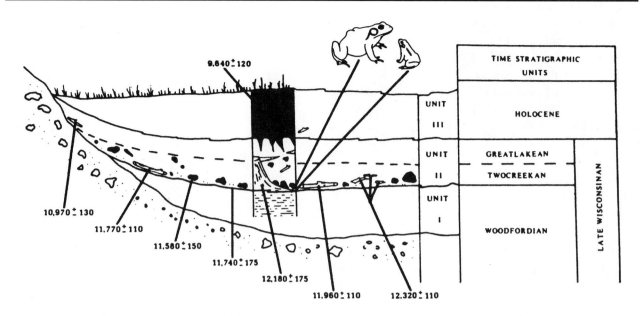

FIGURE 154. Stratigraphic diagram of the Late Wisconsinan Shelton Mastodont Site, Oakland County, Michigan.

mastodont's teeth, it was estimated that the animal was between thirteen and seventeen years old when it died.

References. Stoermer et al. (1988), Shoshani et al. (1989), DeFauw and Shoshani (1991).

Charles Adams Mastodont Site, Livingston County, Michigan. The Charles Adams Mastodont site lies in east-central Lower Michigan near Fowlerville in Livingston County. The site represents an infilled shallow basin interpreted to be of Late Wisconsinan age on the basis of its faunal content. The most striking fossil of the site was the partial skull of a mastodont with all of its upper teeth erupted. About 500 pounds of sediments surrounding the skull were taken to the lab and picked through and screened for smaller fossils. Plant remains, mollusk shells, and three kinds of small vertebrates were recovered. Plant fossils consisted of wood and plant debris. About 1,300 snail shells and 500 clam valves were collected.

The vertebrate fauna of the Charles Adams Site consisted white sucker (*Catostomus commersoni*), crappie (*Pomoxis* sp.), meadow vole (*Microtus pennsylvanicus*; fig. 155), and mastodont (*Mammut americanum*). The presence of the white sucker and crappie indicates a permanent body of water with enough oxygen to support these fishes. The meadow vole probably lived in the grassy margin of the pond or shallow lake.

References. Holman (1979), Holman et al. (1986).

Powers Mastodont Site, Van Buren County, Michigan. The Powers Mastodont Site lies in southwestern Lower Michigan near Decatur in Van Buren County. The site is of considerable interest because it was meticulously excavated, using constrained archaeological techniques (fig. 156), and because much was learned about the life of the mastodont itself. Radiocarbon dates on

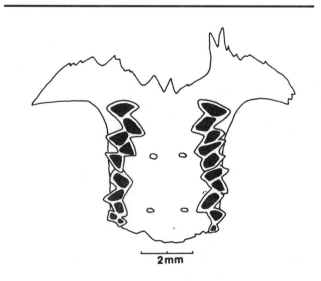

FIGURE 155. Palate of a meadow vole (*Microtus pennsylvanicus*) from the Late Wisconsinan Charles Adams Mastodont Site, Livingston County, Michigan.

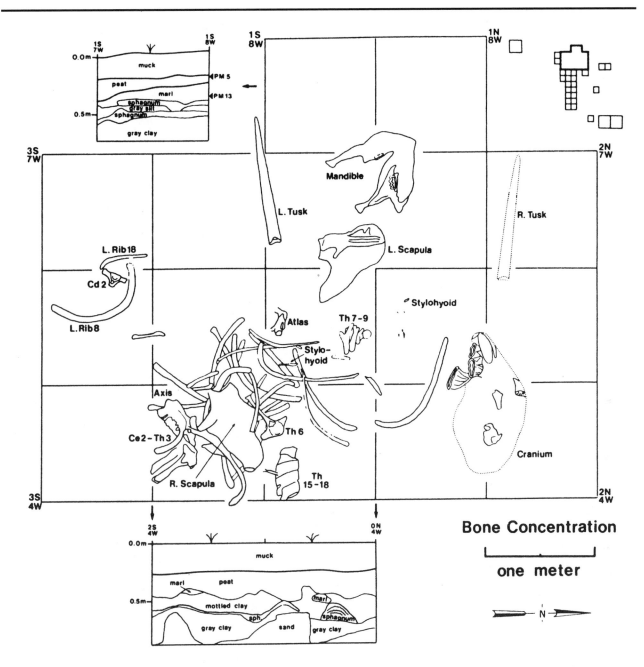

FIGURE 156. Detailed map of bone concentration in the Late Wisconsinan Powers Mastodont Site, Van Buren County, Michigan.

mastodont bone indicate a date of about 11,500 B.P. for the site. The mastodont was found in a peat deposit that formed a tamarack bog or swamp in the Late Wisconsinan. The only other vertebrate remains recovered at the site were those of the Scott's moose (*Cervalces scotti*).

Although almost every parcel of material at the Powers Site was examined, no evidence of human tools or even a flint chip was found. But a detailed study of the mastodont skeleton revealed much about the life of the individual. By counting the annual growth rings in the tusks, it was determined that the mastodont was about thirty to thirty-five years old when death occurred. The size of the tusks and dimensions of other skeletal material indicate that the animal was a female. The female had teeth that were poorly occluded, and the condition of her vertebrae shows that she had a bad back. Sometime during her life she sustained a puncture wound to her shoulder blade, and at about age twenty-two she cracked one of her tusks clear through. Minute tusk rings indicate that she died in the early spring.

Reference. Garland and Cogswell (1985).

I-96 Site, Eaton County, Michigan. The I-96 Site lies in south-central Lower Michigan along the present I-96 by-pass of Lansing in Eaton County, Michigan. The site was unearthed during the construction of the I-96 bypass and unfortunately was destroyed by the construction process. A rather diverse Pleistocene vertebrate assemblage apparently occurred in the area, as swamp or lake as well as woodland habitats are indicated by the three vertebrate species that were identified.

Thousands of freshwater snails and clams as well as wood, cones, and other plant debris were associated with colloidal marl and peat at the site. Pleistocene vertebrates include giant beaver (*Castoroides ohioensis*), white-tailed deer (*Odocoileus virginianus*), and mastodont or mammoth (Proboscidea sp.). The deer and proboscidean (probably mastodont) indicate the presence of woodlands, or at least shrubby vegetation, and the giant beaver indicates the presence of a swamp or lake.

Reference. Holman et al. (1986).

New Hudson Mastodont Site, Oakland County, Michigan. The New Hudson Mastodont Site in southeastern Lower Michigan near New Hudson in Oakland County is important in that it has yielded a mastodont (*Mammut americanum*) that shows indirect evidence of human butchery and has produced the only reptile remains in Michigan of definite Pleistocene age. The vertebrate bones of the New Hudson Site were recovered from the uppermost part of a point sand bar of a Pleistocene stream. This sandy layer was overlain by peat, indicating that the stream subsided and that a period of peat deposition followed. The site is believed to represent the Late Wisconsinan based on the presence of the mastodont.

The mastodont at the site is one of several in the state that are thought by some paleontologists to have been butchered by humans, even though no human artifacts were found at the site. In these cases, evidence such as supposed cut marks on bone, burned bone, and bone tools thought to have been fashioned from the carcass itself and an analysis of the position of the bones in the sediments have been used to suggest human butchery. The problem of indirect evidence of human butchery of mastodonts in Michigan is discussed in chapter 9.

Shell bones of two painted turtles (*Chrysemys picta*) were found in association with the mastodont bones at the New Hudson Site. Annual growth rings on a bone of the lower shell of a small individual thought to be a male showed that the turtle died when it was eight years old. The rather rapid growth pattern in the male suggests a moderately enriched aquatic situation. A large shell bone from a female could not be aged. The presence of painted turtle remains in the site is important from a paleoclimatic standpoint because they indicate that summers were warm enough for painted turtle eggs to hatch during Late Wisconsinan times in Michigan. It has previously been determined that modern painted turtles probably need a mean July temperature of at least 68 degrees F for their eggs to hatch.

Reference. Holman and Fisher (1993).

Late Wisconsinan Proboscidean Sites. An unusually large number of proboscidean sites that range from about 12,000 to about 10,000 B.P. are known from southern Lower Michigan. One study shows that at least 211 mastodonts, 48 mammoths, and 11 indeterminate proboscidean (either mastodont or mammoth) sites occur in Michigan. Moreover, several additional mastodont sites have come to light since that study. These are far more substantiated records than occur in any other Great Lakes region province or state.

All of the Michigan records occur south of the Mason-Quimby Line (fig. 157). This line roughly occurs between the present mixed coniferous and deciduous forests to the north and deciduous forests to the south. The region north of the Mason-Quimby Line has been more recently glaciated than the southern portion, and there is little evidence of Ice Age human occupation in this area. The Mason-Quimby line might well be extended across Ontario (fig. 158), as the northern extent of mastodonts and mammoths in Michigan and Ontario are almost identical.

Most mastodont and mammoth finds in Michigan have been made in Pleistocene kettle bogs or shallow basins where the bones have been found in colloidal marl or peat or between such layers. It has been suggested that these huge animals fell through mats of surface vegetation in quaking bogs and either drowned or became stuck in the marl and were killed by humans. The many pollen studies associated with Michigan mastodonts and mammoths indicate that these animals lived in the vicinity of coniferous forests with signifi-

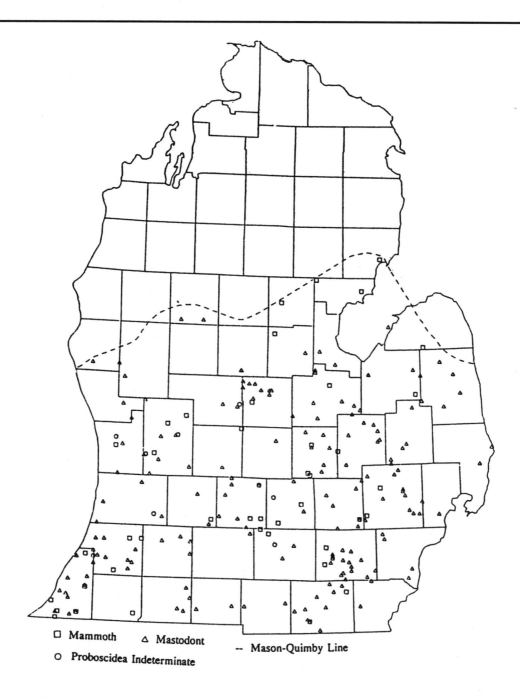

□ Mammoth   △ Mastodont

○ Proboscidea Indeterminate   -- Mason-Quimby Line

FIGURE 157. Mastodont and mammoth sites in Michigan relative to the Mason-Quimby Line (broken line above sites).

cant presence of spruce. In Michigan, mammoths extend somewhat farther north than mastodonts.

It has recently been proposed that the reason for the presence of such a large number of proboscideans in Michigan has to do with the widespread availability of salt seeps and shallow saline water in Pleistocene times. African elephants are unable to exist without a salt supply, and it has been suggested that mastodonts and mammoths made annual trips from other areas in the Great Lakes region to Michigan to replenish their salt supply. This subject is treated in detail in chapter 9.

References. Holman et al. (1986), Abraczinskas (1993).

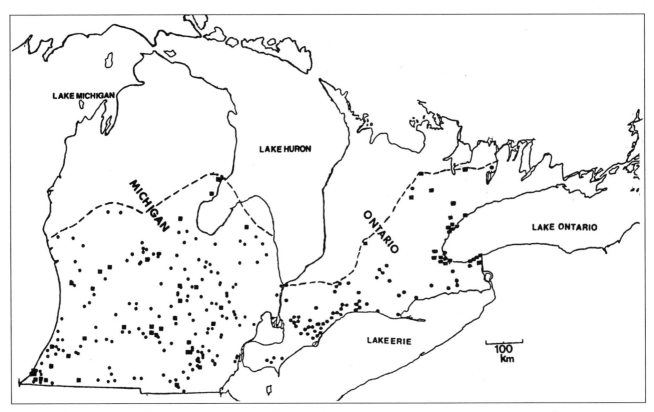

FIGURE 158. Mastodont and mammoth sites in Michigan and Ontario (circles = mastodont sites, squares = mammoth sites).

OHIO

Because it lies farther south, Ohio has a more diverse Pleistocene vertebrate fauna than either Ontario or Michigan. Yet surprisingly few individually diverse sites occur in the state, with the exception of the Late Wisconsinan Sheriden Pit Cave Site in Wyandot County in northwest Ohio and the Late Wisconsinan Carter Site in Darke County in west-central Ohio. Almost all other Ohio sites are also Late Wisconsinan in age, with the possible exception of a site in Brown County that yielded the remains of a giant bison (*Bison latrifrons*) and may represent the Sangamonian age. As in Ontario and Michigan, however, the most numerous occurrences of Pleistocene vertebrates are those of mastodonts and mammoths, although there are fewer records in Ohio than there are in Michigan. Some striking Ohio Ice Age mammals that do not occur in Ontario and Michigan are the giant ground sloth (*Megalonyx jeffersonii*), predatory bear (*Arctodus simus*), tapir (*Tapirus* sp.), and giant bison (*Bison latifrons*). Of special interest is the fact that several small mammals, both carnivores and her-

bivores, extended from the Far North into Ohio during the Pleistocene.

Sangamonian Site (Rancholabrean)
Extinct Bison Site, Brown County, Ohio. The possible Sangamonian age of the Extinct Bison Site in Ohio is not recognized by all paleontologists. In 1869 fossil bison material thought to represent the extinct, giant, longhorned species, *Bison latifrons*, was collected in Brown County in southeastern Ohio. The suggestion that the site might be of Sangamonian age is based on the fact that this extinct species was common in Illinoian and Sangamonian times in the East, with the only records for Late Wisconsinan times being in the western United States.

Reference. Hansen (1992).

Late Wisconsinan Sites
Sheriden Pit Cave Site, Wyandot County, Ohio. The Sheriden Pit Cave Site has yielded the most diverse Pleistocene vertebrate fauna known from Ohio and one of the most diverse known in the Great Lakes region.

Moreover, a spear point made by Pleistocene humans from an antler, found in association with extinct vertebrate fossils in the cave, makes the Sheriden Pit Cave Site unique. At least sixty-two species of Pleistocene vertebrates occur here, including at least ten species of fishes, eight species of amphibians, ten species of reptiles, one bird, and thirty-three species of mammals. Sheriden Pit is located among a series of caverns forming a commercial enterprise about 15 miles east of Findlay in Wyandot County, northwest Ohio. The cave system here developed in a local dolomitic ridge that is covered by a relatively thin layer of glacially derived sediments. Dolomite is a carbonate rock whose dissolution may form caves and other similar structures. Stratified sediments in the cave indicate several different depositional environments in the Late Wisconsinan.

A series of radiocarbon dates from the cave sediments indicate that the majority of the recovered bones are about 11,700 years old. A striking difference between the different groups of vertebrates excavated from Sheriden Pit is that all of the fish, amphibian, and reptile species identified from the cave may be found living in northwest Ohio at present; but several large mammals are extinct, and several small extant mammals are present that invaded Ohio from the Far North during Late Wisconsinan times.

The fish fauna (fig. 159) from Sheriden Pit represents the largest Pleistocene freshwater fish fauna known from the Great Lakes region and is one of the largest Pleistocene cave fish faunas known from North America. Pleistocene fishes from Sheriden Pit include: central stoneroller (*Campostoma anomalum*), silver chub (*Macrhybopsis storeriana*), hornyhead chub (*Nocomis biguttatus*), golden shiner (*Notemigonus chrysoleucas*), creek chub (*Semotilus atromaculatus*), shorthead redhorse (*Moxostoma macrolepidotum*), redhorse (*Moxostoma* sp.), brown bullhead (*Ameiurus nebulosus*), channel catfish (*Ictalurus punctatus*), and smallmouth bass (*Micropterus dolomieu*). It has been suggested that the fish remains entered the cave in the form of the scats (droppings) or caches (food storage places) of mink that retreated to the cave after they had caught fish from a nearby pond or stream. None of the Sheriden Pit fishes is extinct, and all presently occur in northwest Ohio.

The eighteen species of amphibians and reptiles identified from Sheriden Pit are the largest Pleistocene herpetofauna known from the Great Lakes basin. These amphibian and reptile species include: Blue-spotted salamander complex (*Ambystoma laterale* com-

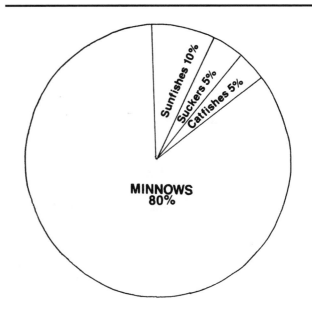

FIGURE 159. Frequency of occurrence of fish fossils in the Late Wisconsinan Sheriden Pit Cave Site, Wyandot County, Ohio.

plex), American toad (*Bufo americanus*), Fowler's toad (*Bufo fowleri*), chorus frog (*Pseudacris triseriata*), bullfrog (*Rana catesbeiana*), green frog (*Rana clamitans*), leopard frog (*Rana pipiens*), wood frog (*Rana sylvatica*), snapping turtle (*Chelydra serpentina*), painted turtle (*Chrysemys picta*), Blanding's turtle (*Emydoidea blandingii*), racer (*Coluber constrictor*), fox snake (*Elaphe vulpina*), milk snake (*Lampropeltis triangulum*), northern watersnake (*Nerodia sipedon*), smooth green snake (*Opheodrys vernalis*), queen snake (*Regina septemvittata*), and common garter snake (*Thamnophis sirtalis*). None of these snakes is extinct, and all are found within the general area of the cave today, with the exception of Blanding's turtle and the fox snake, which presently occur north of the site in the marshlands near Lake Erie. The herpetofauna suggest the presence of a pond or slow-moving stream, meadowland, and rather open woods near the cave site.

One bird and at least thirty-three mammal species have been identified from the Pleistocene strata of Sheriden Pit. The only bird positively identified was the turkey (*Meleagris gallopavo*). Shrews and bats include the short-tailed shrew (*Blarina brevicauda*), masked shrew (*Sorex cinereus*), pygmy shrew (*Sorex hoyi*), and northern bat (*Myotis septentrionalis*).

Carnivores and artiodactyls include: red fox (*Vulpes vulpes*), short-faced bear (*Arctodus simus*), black bear

(*Ursus americanus*), raccoon (*Procyon lotor*), American marten (*Martes americana*), fisher (*Martes pennanti*), ermine (*Mustela erminea*), mink (*Mustela vison*), striped skunk (*Mephitis mephitis*), northern river otter (*Lutra canadensis*), extinct flat-headed peccary (*Platygonus compressus*), Scott's moose (stag moose) (*Cervalces scotti*), white-tailed deer (*Odocoileus virginianus*), and caribou (*Rangifer tarandus*).

Rodents and lagomorphs include: American beaver (*Castor canadensis*), giant beaver (*Castoroides ohioensis*), woodchuck (*Marmota monax*), gray or fox squirrel (*Sciurus* sp.), eastern chipmunk (*Tamias striatus*), red squirrel (*Tamiastriatus hudsonicus*), deer mouse (*Peromyscus maniculatus*), redback vole (*Clethrionomys gapperi*), meadow vole (*Microtus pennsylvanicus*), yellow-cheeked vole (*Microtus xanthognathus*), muskrat (*Ondatra zibethicus*), heather vole (*Phenacomys intermedius*), northern bog lemming (*Synaptomys borealis*), porcupine (*Erythizon dorsatum*), and hare or rabbit (Lagomorpha sp.).

Extinct large mammals of Sheriden Pit include the short-faced bear, an important huge predator that probably dragged other rather large animals into the cave to eat; the Scott's moose, which is thought to have been a woodland form; the flat-headed peccary, which occurred in very large numbers in the cave fauna; and the huge black bear–sized giant beaver. In historic times, humans entering caves have sometimes been confronted by black bears utilizing the cave for a den. It must have been a frightening experience indeed for prehistoric humans to be confronted by a giant short-faced bear, standing on its hind legs in the passageways of the ancient northwestern Ohio cavern system.

The fact that humans had actually accessed the pit cave is attested to by the fact that a beautiful spear point fashioned from an antler (probably from a white-tailed deer) was found in association with the Pleistocene vertebrates at the Sheriden Pit Cave Site.

Large and small mammals, both carnivores and herbivores, found in the Sheriden Pit Pleistocene fauna had invaded northwestern Ohio from the north or Far North. These animals include the pygmy shrew, caribou, ermine, fisher, American marten, heather vole, yellow-cheeked vole, northern bog lemming, and porcupine. It is remarkable that several of these animals coexisted in the Pleistocene with more southern amphibians, reptiles, birds, and mammals at Sheriden Pit. The significance of such mixed Pleistocene faunas is discussed in chapter 9.

References. McDonald (1994), Ford (1994), Ford et al. (1996), Holman (1997).

Carter Site, Darke County, Ohio. The Carter Site in Darke County in west-central Ohio is second only in importance to the Sheriden Pit Cave Site in the diversity of its fauna. It is richer in bird species but somewhat poorer in all of the other vertebrate groups. The Carter Site fauna occurred in a Pleistocene bog setting, although the fauna as a whole indicates that a permanent aquatic situation must have been available there from time to time. The Carter Site is by far the most diverse Pleistocene bog site fauna known in the Great Lakes region. Several radiocarbon dates are available from the Carter Site, and these suggest that the fauna might have accumulated here for a considerable period of time in the postglacial Wisconsinan. Nevertheless, it appears that the majority of vertebrate species accumulated here from about 12,000 to about 10,000 B.P.

Fishes of the Carter Site include: unidentified minnows (Cyprinidae sp.), redhorse (*Moxostoma* sp.), brown bullhead (*Ameiurus nebulosus*), pike (*Esox* sp.), muskellunge (*Esox masquinongy*), mudminnow (*Umbra limi*), largemouth bass (*Micropterus salmoides*), and yellow perch (*Perca flavescens*).

Reptile bones identified include snapping turtle (*Chelydra serpentina*) and painted turtle (*Chrysemys picta*) remains. Birds from the Carter Site include five ducks and a raven: shoveler duck (*Anas clypeata*), mallard (*Anas platyrhynchos*), dabbling duck (*Anas* sp.), northern ringnecked duck (*Aythya collaris*), diving duck (*Aythya* sp.), and raven (*Corvus corax*). The fishes, turtles, and ducks indicate a permanent body of water that was able to support a diversity of aquatic and semiaquatic vertebrates.

Among mammals, xenarthrans, carnivores, and artiodactyls include: Jefferson ground sloth (*Megalonyx jeffersonii*), long-tailed weasel (*Mustela frenata*), fisher (*Mustela pennanti*), mink (*Mustela vison*), white-tailed deer (*Odocoileus virginianus*), and Scott's moose (stag moose) (*Cervalces scotti*; fig. 160). Rodents and lagomorphs include: American beaver (*Castor canadensis*), giant beaver (*Castoroides ohioensis*), woodchuck (*Marmota monax*), ground squirrel (*Spermophilus* sp.), eastern chipmunk (*Tamias striatus*), meadow vole (*Microtus pennsylvanicus*), woodrat (*Neotoma* sp.), muskrat (*Ondatra zibethicus*), deer or white-footed mouse (*Peromyscus* sp.), heather vole (*Phenacomys intermedius*), and hare or jackrabbit (*Lepus* sp.). Pro-

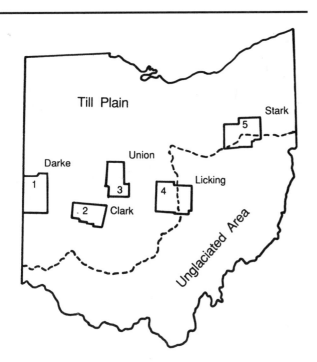

FIGURE 160. Distribution of Scott's moose (stag moose) *Cervalces scotti*, by counties in the Pleistocene of Ohio.

boscideans are represented by the American mastodont (*Mammut americanum*).

Four large, extinct mammals are known from the Carter Site, the same number that occurred at Sheriden Pit. The Carter Site extinct mammals include the giant Jefferson ground sloth and American mastodont, species that are absent from Sheriden Pit, and the Scott's moose and giant beaver, which occurred at both Ohio sites. The extinct bear, absent at the Carter Site, was present at Sheriden Pit. Northern invaders are fewer at the Carter Site than at Sheriden Pit and include only the fisher and the heather vole.

References. Mills (1975), Mills and Guilday (1972), Todd (1973), Holman (1986), and McDonald (1994).

Burning Tree Mastodont Site, Newark, Licking County, Ohio. The Burning Tree Mastodont Site in Newark in Licking County, central Ohio, produced the skeleton of an American mastodont (*Mammut americanum*) that died about 11,500 B.P. based on an average of several radiocarbon dates. The site is significant from a variety of aspects. First, it has produced one of the best preserved recently discovered American mastodont skeletons. But more important, some daring hypotheses

have been made about the conditions of death and preservation of the specimen as well as its intestinal contents.

The Burning Tree mastodont was found preserved in peat in three concentrated areas near the former margin of a Late Wisconsinan pond. Each concentration of mastodont bones included mixed elements from different regions of the body. Although no human artifacts were associated with the site, it has been suggested that human hunters killed and butchered the animal and that units of the carcass were put in the pond, possibly for storage in the winter months, as growth increments on the skeleton indicate that the animal died in the fall.

Cut marks occur on several of the bones, but the most prominent marks are shallow gouges at the origin of several major muscles and tendons (fig. 161). These marks, along with the fact that paired elements had cut

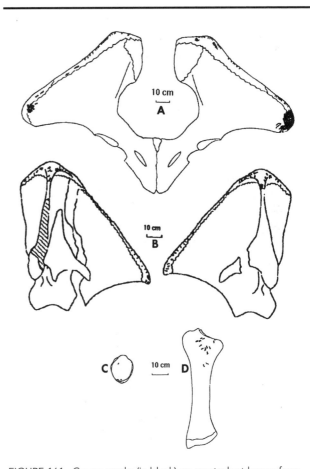

FIGURE 161. Gouge marks (in black) on mastodont bones from the Late Wisconsinan Burning Tree Mastodont Site, Licking County, Ohio. A, pelvic girdle; B, scapulae; C, right patella; D, left tibia.

marks in similar places, led to the hypothesis that the animal had been butchered. Moreover, thin striations on a series of the mastodont ribs were interpreted as the result of humans dragging the ribs or ribs attached to vertebrae across coarse gravel or sand; sand grains that stuck in the gouged surfaces of the bones were also interpreted as being derived from the dragging process. Both of these features led to the hypothesis that the mastodont parts were taken from an original setting to the Burning Tree Site pond for winter storage. Finally, material found within the rib cage, interpreted as the gut contents of the mastodont, were found to contain living intestinal bacteria that were thought to have survived since the Late Wisconsinan. Such bold hypotheses are of great interest and if verified by additional evidence will provide exciting new information about the activities of Pleistocene human hunters, mastodonts, and even bacteria that might be living fossils in the true sense of the word.

References. Lepper et al. (1991), Fisher et al. (1994).

Late Wisconsinan Proboscidean Sites. Ohio has produced at least 136 mastodont and 57 mammoth records for a total of 193, finishing second in the Great Lakes region to Michigan, which has a total of 270 Late Wisconsinan proboscidean sites. Ohio, however, has 57 to Michigan's 48 mammoth records.

As in Michigan, most mastodont remains have been recovered from bog or shallow basin sites. Despite the large number of mastodont remains discovered in Ohio, relatively little information has come from these sites, other than the Burning Tree Mastodont Site. Nevertheless, possible associations between mastodonts and humans in Ohio have been suggested previous to the discovery of the Burning Tree specimen.

A possible association of a mastodont with a broken Paleo-Indian fluted point was reported from Hardin County in 1983, and in 1978 and 1983 sites in Huron and Ashtabula Counties were suggested to be human kill sites. Mastodonts are thought to have preferred spruce forests in Ohio, and spruce, pine, and low herbaceous plants are thought to have been important in their diets.

Both the Jefferson mammoth (*Mammuthus jeffersonii*) and the woolly mammoth (*Mammuthus primigenius*) have been reported from the Late Wisconsinan of Ohio, with the Jefferson mammoth being by far the most numerous species. One of the most reliable records of the woolly mammoth comes from a nearly complete skull that was found near Mount Healthy in Hamilton County, southeastern Ohio. This skull indicates that the woolly mammoth ranged into the southern portion of the state, probably during the maximum extent of the Wisconsinan glaciation. Mammoth remains in Ohio are usually found in gravel deposits and lake beds, unlike mastodont remains, which are usually found in bog and shallow basin deposits. There is no reason to believe that grass was not an important part of the diet of Ohio mammoths.

References. Falquet and Hanebert (1978), Hansen et al. (1978), Lepper (1983a), Murphy (1983), Hansen (1992), Fisher et al. (1994).

INDIANA

Indiana has produced the largest number of diverse Pleistocene vertebrate faunas of any province or state in the Great Lakes region. This is not only because of its more southerly location but also because of the extensive network of caves in the southern part of the state. Caves are natural traps for vertebrates that fall into them by accident and are sources for the bones of animals brought in by large carnivores. Moreover, the bones of small mammals in pellets regurgitated by cave-roosting owls often become part of cave fossil faunas.

Sangamonian Site (Rancholabrean)

Harrodsburg Crevice near Bloomington, Monroe County, Indiana. The Harrodsburg Crevice resulted from the collapse of a limestone cave in the unglaciated, hilly region of south-central Indiana. The vertebrates were collected from the former floor of this cave either in clay or in cemented limestone detritus. The cemented fossils were very hard to extract, and specialized techniques were required to expose them for identification.

It is believed that the former cave served as a den or shelter for animals rather than as a natural trap and that the bones were derived either from animals that died within the cave, from portions of kills that were brought into the cave by predators, or from the fecal contents of predators. The lack of small mammals fossils at the site indicates that owls were not an important agent in the accumulation of the fossils. The fauna as a whole suggest a warmer and drier –than present shortgrass prairie/forest edge setting of Sangamonian age.

The only reptile bones in the site were fragments from the upper shell of a turtle believed to be a box turtle (*Terrapene* sp.). Mammalian insectivores and carnivores from the site include: hairy-tailed mole (*Parascalops breweri*), dire wolf (*Canis dirus*), coyote (*Canis latrans*), gray fox (*Urocyon cinereoargenteus*), black bear (*Ursus americanus*), small mustelid (*Mustela* sp.), striped skunk (*Mephitis mephitis*), spotted skunk (*Spilogale putorius*), bobcat (*Lynx rufus*), a subspecies of jaguar (*Panthera onca augusta*), and sabertooth cat (*Smilodon fatalis*).

Perissodactyls, artiodactyls, rodents, and lagomorphs include: complex-toothed horse (*Equus complicatus*), Leidy's peccary (*Platygonus vetus*), white-tailed deer (*Odocoileus virginianus*), plains pocket gopher (*Geomys bursarius*), woodchuck (*Marmota monax*), eastern woodrat (*Neotoma floridana*), prairie vole (*Microtus ochrogaster*), southern bog lemming (*Synaptomys cooperi*), and cottontail or snowshoe hare (*Sylvilagus* or *Lepus*).

Two mammals, the jaguar and sabertooth cat, are found no place else in the Great Lakes region, and for this reason alone the Harrodsburg Crevice is an important site. Leidy's peccary either had become extinct or had evolved into the flat-headed peccary before the Wisconsinan; thus at least a Sangamonian age is indicated for the deposit. Three extant species of mammals—the hairy-tailed mole, plains pocket gopher, and eastern woodrat—do not presently occur in Monroe County, Indiana.

References. Parmalee et al. (1978), Munson et al. (1980), Richards (1982a, 1984b, 1985, 1987).

Probable Sangamonian Site (Rancholabrean)

Megenity Peccary Cave, near Taswell, Crawford County, Indiana–Deep Deposits. The deep deposits in the Megenity Peccary Cave extend beyond the range of radiocarbon dating. These deposits have yielded the remains of a giant land tortoise (*Hesperotestudo* sp.). Giant land tortoises are very important in the interpretation of Pleistocene climates. In climates with cold winters, modern giant land tortoises are not able to digest the large amounts of plant food that they eat. The fermenting plant material causes a condition called enteritis (inflammation of the digestive tract) that leads to the death of these animals. Thus the presence of the giant land tortoise suggests that winters must have been mainly frost free during this interglacial time when the tortoise lived in Indiana.

Reference. Richards and Whitaker (1997).

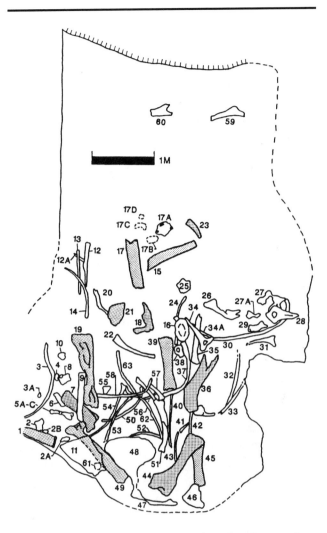

FIGURE 162. Diagram showing packed condition of bones of a Jefferson mammoth (*Mammuthus jeffersonii*) from the Late Wisconsinan Alton Mammoth Site, Crawford County, Indiana. Tusks, mandible, and major limb bones (stippled) indicate the animal was preserved resting on its sternum with the limbs flexed up under the body.

Possible Middle Wisconsinan Site (Rancholabrean)

Alton Mammoth Site, near the Ohio River, Crawford County, Indiana. It has been cautiously suggested that the Alton Mammoth Site near the Ohio River in southern Indiana represents the Middle Wisconsinan based on some somewhat conflicting stratigraphic and sedimentary evidence. Nevertheless, although the exact age is not known, there seems to be little doubt that the site represents some part of the Wisconsinan glacial age. A mammoth carcass (fig. 162), as well as several smaller vertebrates, was buried in this site, which occurred in a small terrace of the Ohio River.

The vertebrae fauna from the Alton Mammoth Site includes: unidentified bat remains, southern red-backed vole (*Clethrionomys gapperi*), meadow vole (*Microtus pennsylvanicus*), heather vole (*Phenacomys intermedius*), an indeterminate mouse, cottontail rabbit or hare (Leporidae sp.), and Jefferson mammoth (*Mammuthus jeffersonii*). The red-backed vole and heather vole are northern invaders in the fauna.

Reference. Richards (1991).

Middle to Late Wisconsinan Site (Rancholabrean)

Megenity Peccary Cave, near Taswell Crawford County, Indiana. The Megenity Peccary Cave has yielded the largest number of bones ever excavated in Indiana. The bone bed is situated about 160 feet inside of the cave, and procuring the fossils requires crawling and squeezing through a long horizontal passageway and then descending to the bottom of a 15-foot-deep pit. As implied by its name, the cave is known for its excellently preserved flat-headed peccary (*Platygonus compressus*) remains, but many species of small Pleistocene mammals also occur here.

Older, deep sediments in the cave beyond the range of radiocarbon dating yielded the giant land tortoise discussed earlier. Younger sediments in the cave have yielded radiocarbon dates from about 51,000 to about 14,000 B.P., with the majority of the identified vertebrate bones coming from a Late Wisconsinan interval ranging from about 31,000 to about 14,000 B.P.

Amphibians in the Megenity Peccary Cave are represented by numerous unidentified frog bones, and reptiles are represented by unidentified snakes. Mammalian insectivores, carnivores, and artiodactyls from the site include: short-tailed shrew (*Blarina brevicauda*), arctic shrew (*Sorex arcticus*), masked shrew (*Sorex cinereus*), smoky shrew (*Sorex fumeus*), dire wolf (*Canis dirus*), coyote (*Canis latrans*), black bear (*Ursus americanus*), fisher (*Martes pennanti*), river otter (*Lutra canadensis*), and flat-headed peccary (*Platygonus compressus*).

The cave is especially well represented by Pleistocene rodents and lagomorphs which include: plains pocket gopher (*Geomys bursarius*), eastern woodrat (*Neotoma floridana*), southern red-backed vole (*Clethrionomys gapperi*), meadow vole (*Microtus pennsylvanicus*), pine vole or prairie vole (*Microtus pinetorum* or *ochrogaster*), yellow-cheeked vole (*Microtus xanthognathus*), heather vole (*Phenacomys in-*

termedius), northern bog lemming (*Synaptomys borealis*), southern bog lemming (*Synaptomys cooperi*), and snowshoe hare (*Lepus americanus*).

A number of Megenity Peccary Cave mammals represent invaders from far to the north. These include arctic shrew, fisher, southern red-backed vole, yellow-cheeked vole, heather vole, and snowshoe hare. Mammals that are locally absent from the cave area today are plains pocket gopher and woodrat.

References. Richards (1988a, 1988b), Richards and Whitaker (1997).

Late Wisconsinan Sites (Rancholabrean)

King Leo Pit Cave, near Depauw, Harrison County, Indiana. The King Leo Pit Cave in the extreme southern part of Indiana yielded the remains of an extinct woodland muskox (*Bootherium bombifrons*) as well as fossils representing many species of small vertebrates. To retrieve the bones, the collectors had to descend into a 45-foot pit and then inch their way through a 200-foot long crawlway. The muskox bones were on top of and continued through a layer about 8 inches deep on the cave floor. When the silty sediments from this layer were washed and screened, a large number of bones of smaller vertebrates were recovered. Together, the extinct muskox and contemporaneous small vertebrate fauna indicate a Late Wisconsinan age for the deposit.

Fish fossils from the sedimentary zone containing the muskox bones were unidentified. Amphibians were represented by the cave salamander (*Eurycea lucifuga*). Snake bones belonging to a rat snake, or milk snake, or both, represented the reptiles. If a rat snake is represented by the snake bones, it may have entered the cave to prey upon bats roosting there.

Among the mammals, insectivores and bat remains were numerous and include: short-tailed shrew (*Blarina brevicauda*), least shrew (*Cryptotis parva*), masked shrew (*Sorex cinereus*), smoky shrew (*Sorex fumeus*), pygmy shrew (*Sorex hoyi*), gray bat or northern bat (*Myotis grisescens* or *septentrionalis*), small-footed myotis (*Myotis leibii*), eastern pipistrelle bat (*Pipistrellus subflavus*), big-eared bat (*Plecotus* sp.), and a myotis bat that could have been one of three species, little brown, Indiana, or southeastern myotis.

Carnivores, artiodactyls, rodents, and lagomorphs from the muskox zone include: fisher (*Martes pennanti*), woodland (helmeted) muskox (*Bootherium bombifrons*), flying squirrel (*Glaucomys* sp.), gray or fox squirrel (*Sciurus* sp.), red squirrel (*Tamiasciurus hudsonicus*), white-footed mouse (*Peromyscus leucopus*), deer mouse (*Peromyscus maniculatus*), southern red-backed vole (*Clethrionomys gapperi*), meadow vole (*Microtus pinetorum*), heather vole (*Phenacomys intermedius*), and cottontail rabbit or hare (Leporidae sp.).

The fisher, southern red-backed vole, and heather vole were all invaders from the north. The King Leo Pit Cave faunas suggest an environmental mosaic during the time of the deposition of the bones. Deciduous or coniferous forest habitats are indicated, as are dry, open conifer environments. This interpretation is in line with the plaid (vegetational mosaic) hypothesis for communities well south of the ice sheet, as discussed earlier in the book.

Reference. Richards and McDonald (1991).

Prairie Creek D Site, Daviess County, Indiana. The Prairie Creek D Site in southwest Indiana is not a cave site but represents the sediments of Lake Prairie Creek, which occurred in the Late Wisconsinan from about 16,000 to about 13,000 B.P. Radiocarbon dates show that the fossils vertebrates recovered from sedimentary zone D represent a time of about 14,000 B.P. Since other fossiliferous zones at the Prairie Creek locality represent mixed Pleistocene and Holocene as well as Holocene vertebrates, the faunas from these zones are not included here.

All vertebrate classes, including fishes, amphibians, reptiles, birds, and mammals, were recovered from the Prairie Creek D fossil zones, but only the amphibians, reptiles, and mammals have been published and are included here. The Prairie Creek D fauna includes not only several large, extinct mammals but also several invading vertebrate species from the north.

Pleistocene amphibians and reptiles from Prairie Creek D include: mole salamander (*Ambystoma* sp.), American toad or Fowler's toad (*Bufo* sp.), bullfrog (*Rana catesbeiana*), leopard frog group (*Rana pipiens* complex of species), wood frog (*Rana sylvatica*), snapping turtle (*Chelydra serpentina*), common musk turtle (*Sternotherus odoratus*), painted turtle (*Chrysemys picta*), Blanding's turtle (*Emydoidea blandingii*), map turtle (*Graptemys* sp.), cooter or red-bellied slider turtle (*Pseudemys* sp.), spiny softshell turtle (*Apalone spinifer*), water snake (*Nerodia* sp.), plain-bellied water snake (*Nerodia erythrogaster*), and garter snake or ribbon snake (*Thamnophis* sp.).

Among the mammals, xenarthrans, insectivores, and carnivores include: Jefferson's ground sloth (*Megalonyx jeffersonii*), beautiful armadillo (*Dasypus bellus*),

short-tailed shrew (*Blarina brevicauda*), least shrew (*Cryptotis parva*), masked shrew (*Sorex cinereus*), star-nosed mole (*Condylura cristata*), eastern mole (*Scalopus aquaticus*), fisher (*Martes pennanti*), long-tailed weasel (*Mustela frenata*), and mink (*Mustela vison*). It is interesting, but not unexpected in this noncave fauna, that no bats were present.

Artiodactyls, rodents, and lagomorphs from Prairie Creek D include: flat-headed peccary (*Platygonus compressus*), long-nosed peccary (*Mylohyus nasutus*), woodchuck (*Marmota monax*), eastern chipmunk (*Tamias striatus*), red squirrel (*Tamiasciurus hudsonicus*), giant beaver (*Castoroides ohioensis*), deer or white-footed mouse (*Peromyscus* sp.), southern red-backed vole (*Clethrionomys gapperi*), meadow vole (*Microtus pennsylvanicus*), yellow-cheeked vole (*Microtus xanthognatus*), bog lemming (*Synaptomys* sp.), muskrat (*Ondatra zibethicus*), meadow jumping mouse (*Zapus hudsonius*), and cottontail rabbit (*Sylvilagus* cf. *floridanus*).

Five mammals in the Prairie Creek D fauna are extinct. These include Jefferson's ground sloth, beautiful armadillo, flat-headed peccary, long-nosed peccary, and giant beaver. The appearance of two distinct species of peccaries probably indicates the varied nature of the habitat that existed at the time.

Invading mammalian species from the north include the star-nosed mole, fisher, red squirrel, southern red-backed vole, and yellow-cheeked vole. Aside from Blanding's turtle, which occurs somewhat north of the Prairie Creek D Site in Indiana today, the herpetofauna is similar to one that would be presently in the area. Thus the Prairie Creek D fauna, like the King Leo Pit Cave fauna, suggests a mosaic of habitats, including coniferous stands, mixed deciduous and coniferous stands of forest, possibly some pure deciduous stands of forest, and grassy areas. This fauna fits in perfectly with the climatic equability hypothesis and the plaid (vegetational mosaic) hypothesis previously discussed.

References. Tomak (1975, 1982), Richards (1984b, 1992c), Holman and Richards (1993).

Christensen Bog Site, near Greenfield, Hancock County, Indiana. The Christensen Bog Site, 17 miles east of Indianapolis, is important in that it not only yielded more than 290 bones representing at least two mastodonts (*Mammut americanum*) but that it produced other important vertebrate species as well, including amphibians, reptiles, birds, and mammals. Christensen Bog is an oval-shaped depression that

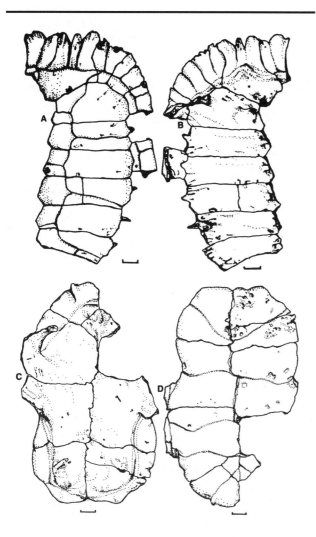

FIGURE 163. Excellently preserved painted turtle (*Chrysemys picta*) shell remains from the Late Wisconsinan Christensen Bog Mastodont Site, Hancock County, Indiana. A, carapace in dorsal view; B, carapace in ventral view; C, plastron in dorsal view; D, plastron in ventral view. Each line = 10 mm.

formed originally as a kettle lake created by wasting glacial ice that was retreating to the north.

Bones were first discovered in the bog when it was being dredged to form an artificial lake. The species listed here come from a fossil unit designated Qmp3, which, on the basis of radiocarbon dates, existed from about 13,000 to about 12,000 B.P.

The vertebrate fauna from the site includes: frogs of the northern leopard frog group (*Rana pipiens* complex), softshell turtle (*Apalone* sp.), snapping turtle (*Chelydra serpentina*), especially well preserved painted turtle (*Chrysemys picta*) specimens (fig. 163),

surface-feeding duck (*Anas* sp.), turkey (*Meleagris gallopavo*), white-tailed deer (*Odocoileus virginianus*), caribou (*Rangifer tarandus*), giant beaver (*Castoroides ohioensis*), muskrat (*Ondatra zibethicus*), and two American mastodonts (*Mammut americanum*). The bones were spread out over a relatively large area (fig. 164).

The vertebrate fauna and pollen studies at the site indicate that the vertebrate assemblage existed near a sluggish body of water surrounded by forest vegetation with a significant component of spruce. Summers, however, had to be warm enough to support the turtle population, and it has been suggested that a mean July temperature in excess of 68 degrees F must have occurred in order for the turtle eggs to hatch successfully.

Reference. Graham et al. (1983).

Kolarik Mastodont Site, near Bass Lake, Starke County, Indiana. The Kolarik Mastodont Site in northwestern Indiana produced much of a postcranial skeleton of an American mastodont plus a small vertebrate fauna. The skeleton was discovered during the excavation of a pond in a peat bog area. The partially articulated skeleton lacked the skull and jaws and was found in sandy muck that was above a blue calcareous clay that contained remains of large plants, mollusk shells, and fish bones.

The vertebrate fauna of the Kolarik Site includes: fish remains, frogs of the northern leopard frog group (*Rana pipiens* complex), caribou (*Rangifer tarandus*), deer mouse (*Peromyscus maniculatus*), meadow vole (*Microtus pennsylvanicus*), and American mastodont (*Mammut americanum*). The fauna accumulated in with a shallow kettle lake or depression.

References. Ellis (1981, 1982), Richards (1984b).

Dollens Mastodont Site, near Alexandria, Madison County, Indiana. The Dollens Mastodont Site in east-central Indiana provided skeletal elements of a young mastodont and an associated vertebrate fauna. The bones came from a peat deposit. Pollen from the site indicates that the bones were deposited in a grass-sedge fen that held standing water for a large part of the year. The pollen of tree species indicates that spruce was abundant but that several genera of hardwoods were present as well. A radiocarbon date from peat that was directly under the mastodont bones yielded a date of about 12,200 B.P.

The vertebrate fauna of the Dollens Site includes fishes, a frog, a turtle, and a few mammals. The fishes are unidentified, and the frog has been identified only

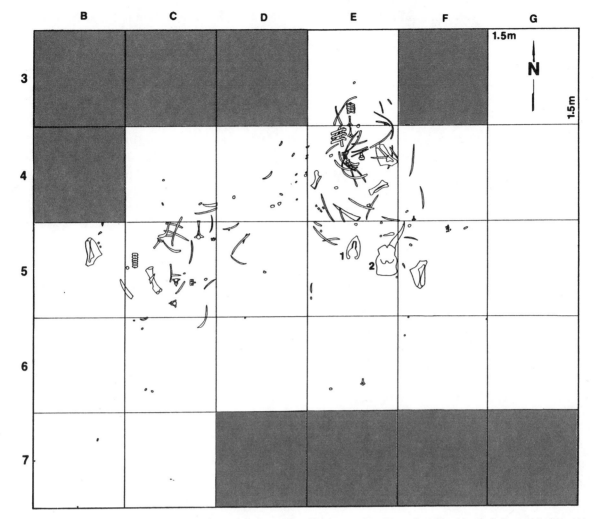

FIGURE 164. Plan view of bone distribution at the Late Wisconsinan Christensen Bog Mastodont Site, Hancock County, Indiana. Note the mastodont mandible (1) and skull (2) in square 5-E. The bones were scattered over a relatively wide area in a bog pond.

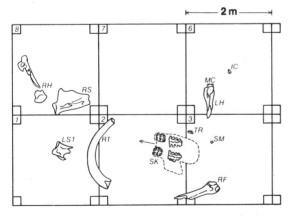

FIGURE 165. The distribution of American mastodont (*Mammut americanum*) bones at the Late Wisconsinan Dollens Mastodont Site, Madison County, Indiana. Some important bones are a left humerus (LH), right tusk (RT), skull with four molars (SK), and right femur (RF).

as *Rana* sp. The single reptile species consists of a snapping turtle (*Chelydra serpentina*). Mammalian fossils at the Dollens Site include tree squirrel (*Sciurus* sp.), beaver (*Castor canadensis*), muskrat (*Ondatra zibethicus*), and American mastodont (*Mammut americanum*; fig. 165).

Evidence from the sediments, plants, and vertebrates demonstrates that the Dollens mastodont was deposited in a vegetation-choked marshy basin that occurred during the final infilling stage of a shallow kettle lake that probably became seasonally dry. Evidence of beaver-gnawed wood and collapsed beaver and muskrat lodges in the sediments enhance the picture of the shallow marsh in east-central Indiana some 12,000 years ago.

Reference. Richards et al. (1987).

Rochester Site, Fulton County, Indiana. Only two records of the great, predatory, extinct short-faced bear (*Arctodus simus*) are known from the Great Lakes region. One find, from the Sheriden Pit Cave Site, Wyandot County, northwest Ohio, consists of relatively incomplete remains. The other discovery, from near Rochester, Fulton County, northern Indiana, consists of one the most complete skeletons known of this important predator. A life-sized restoration in the form of a mural in the National Museum of Natural History, Smithsonian Institution, Washington, D.C., is based on this Indiana fossil.

The Fulton County bones were unearthed by a backhoe operator while digging a pipeline trench. Although the exact site is uncertain, calcareous clays that adhered to the short-faced bear bones contained aquatic snails that indicate that the bones were deposited in a shallow kettle lake. The skeleton was so complete that the short-faced bear could be assigned to the giant subspecies *Arctodus simus yukonensis.*

Bones from the bear provided a radiocarbon date of about 11,000 B.P. Most of the other known skeletons of short-faced bears have been very incomplete. Thus the Indiana specimen has provided much of our knowledge about the species, including, among other important details, the verification that the skull was very short and wide compared to the elongated limbs.

Reference. Richards and Turnbull (1995).

Pigeon Creek Site, near Evansville, Vanderberg County, Indiana. The Pigeon Creek Site near Evansville in extreme southwestern Indiana occurs as silty deposit transported by water (alluvium). The site is important because it provides the only record of the extinct Hay's tapir (*Tapirus haysii*) in the Great Lakes region. This is one of the few records of this species in the northeastern quadrant of the United States.

The identification is based on a single tooth that was found in the middle 1800s. A radiocarbon date of 9,400 B.P. that was made in the 1950s on the alluvial stratum that produced the Hay's tapir is probably several hundred years too young.

References. Leidy (1884–1885), Ray and Sanders (1984).

Late Wisconsinan Proboscidean Sites. Records at the Indiana State Museum in Indianapolis document the occurrence of at least seventeen mastodont and eleven mammoth sites in Indiana. Most of these represent Late Wisconsinan occurrences. As far as I am aware, there has been no comprehensive summary of mastodont and mammoth sites in Indiana, and many more records undoubtedly exist.

The mastodont records are all those of the American mastodont (*Mammut americanum*); it is suggested that two mammoths, the Jefferson mammoth (*Mammuthus jeffersonii*) and the woolly mammoth (*Mammuthus primigenius*), were present in the state. The rather high proportion of mammoths to mastodonts based on the Indiana State Museum records is unusual compared to other records in Ontario and eastern Great Lakes states such as Michigan and Ohio, where mastodonts greatly outnumber mammoths.

References. Richards (1984b, 1991).

ILLINOIS

Illinois has yielded many individual records of large, extinct Pleistocene mammals but, compared to Indiana, has yielded few diverse Pleistocene vertebrate faunas. This may be due to the lack of paleontological collecting in the caves and fissures of Illinois. A noteworthy occurrence in the Illinois Pleistocene is a Middle Pleistocene (Irvingtonian II) site, possibly the oldest well-dated Pleistocene vertebrate site in the Great Lakes region. This site has yielded the only extinct species of small rodents (two voles) presently known in the Great Lakes region. Another noteworthy Illinois Pleistocene vertebrate fauna has yielded the spectacular extinct giant land tortoise species (*Hesperotestudo crassiscutata*), which has not been reported from any other Great Lakes province or state, although a giant tortoise identified only as *Hesperotestudo* sp. is known from probable Sangamonian sediments in a cave in southern Indiana (see above).

Middle Pleistocene (Pre-Illinoian) Site (Rancholabrean II)
County Line Locality, Hancock County, Illinois. The County Line Locality in Hancock County in west-central Illinois yielded a diverse assemblage of fossils from sediments along a cut bank near the county line between Adams and Hancock Counties. The fossil assemblage is thought to have accumulated in a depression, most likely a fissure or sinkhole. Plant remains, ostracods, mollusks, insects, and mammals were yielded by the silty deposits at the site. Two extinct voles in the

fauna represent the only extinct small rodents presently known in the Great Lakes region.

The dating of the fossils was based on the reversed magnetic signatures of the sediments and the temporal ranges of the two extinct voles. This dating was supplemented by aminostratigraphic dating (dating based on the constant conversion of specific amino acids from one type to another over long time intervals) in two mollusk genera from the site. Combined, these methods constrained the age of the fossil assemblage to between about 830,000 and 730,000 B.P.

The vertebrate fauna from the site consisted of six small mammals, only two of which could be identified to the specific level. These mammals were two shrews (*Sorex* sp. and *Blarina* sp.), a rabbit (*Sylvilagus* sp.), a deer mouse (*Peromyscus* sp.), and two extinct voles, the Cape Deceit vole (*Lasiopodomys deceitensis*) and Hibbard's Tundra vole (*Microtus paroperarius*). The two voles are boreal faunal elements.

In addition to the boreal voles, the fossil assemblage includes several species of snails, insects, and plants with modern ranges well to the north of the fossil site, but they occur in association with other fossil species that would have required relatively warm climates. It has been suggested that local conditions at the site were probably similar to those presently found in northeastern Iowa, where blocks, fissures, and joints in the carbonate bedrock provide shade and are kept cool and moist during the hot summer months by cold-air drainage and groundwater seepage. This type of local northern microclimate would possibly explain the survival of the boreal plants and animals of the County Line fauna.

Reference: Miller et al. (1994).

Sangamonian Site (Rancholabrean)

Hopwood Farm Site, near Fillmore, Montgomery County, Illinois. The Hopwood Farm Site in south-central Illinois developed in a kettle basin that formed during the Illinoian glacial age. Fossils from the kettle basin are from both Illinoian- and Sangamonian-age sediments. Pollen from these sediments shows a transition from Illinoian-age conifer communities, dominated by pine and spruce, to deciduous forest, to grasses and herbaceous plants, and finally back to deciduous forest again. The grass and herb zone reflects a warm, dry spell in the sequence.

The Sangamonian Zone at the Hopwood Farm Site has produced the only published record of the extinct giant land tortoise species (*Hesperotestudo crassiscutata*) in the Great Lakes region (*Hesperotestudo* sp. was reported from a Pleistocene cave in southern Indiana; see above) and is one of the northernmost records of this species in the United States.

The giant land tortoise is a most important indicator of climate in that it suggests above-freezing temperatures in south-central Illinois throughout the year, at least during the portion of the Sangamonian age represented by the Hopwood Farm Site. The reasoning behind this interpretation of the Sangamonian climate lies in the fact that modern giant land tortoises that live in the Aldabra and Galapagos Islands are unable to survive in areas with even intermittently cold winter temperatures. The large amounts of vegetation that they must eat ferments in their stomach during cold snaps, and the animals die of enteritis. In fact, it has been shown that giant tortoise cannot survive winters outdoors in climates as warm as southern Florida.

Other vertebrate fauna associated with the giant land tortoise at the Hopwood Farm Site includes: unidentified fishes, amphibians, and reptiles, as well as raccoon (*Procyon* sp.), bovid artiodactyls, American beaver (*Castor* sp.), giant beaver (*Castoroides ohioensis*), vole or lemming, muskrat (*Ondatra* sp.), and American mastodont (*Mammut americanum*).

Below the Sangamonian Zone at the Hopwood Farm Site are earlier vertebrate fossils that remain to be identified. These include fishes, amphibians, reptiles, birds, and mammals representing the Illinoian to Sangamonian transition. Undoubtedly these unidentified fossils will become the subject of one of the most important publications on Pleistocene vertebrates in the Great Lakes region.

Reference. King and Saunders (1986).

Late Wisconsinan Sites (Rancholabrean)

Clear Lake Site, near Springfield, Sangamon County, Illinois. The Clear Lake Site in central Illinois is important in that it yielded three species of turtles and large, extinct Pleistocene vertebrates. Turtles are particularly important climatic indicators, as they need rather warm temperatures in the summer for their eggs to hatch. The Clear Lake Site bones were collected during the excavation of gravels at the locality, which lies about 4 miles east of Springfield. The bones accumulated in fluvial (river-transported) sands and gravels and are considered to represent the Late Wisconsinan based on the local stratigraphy and vertebral faunal composition.

Vertebrate faunal remains from the Clear Lake Site include: snapping turtle (*Chelydra serpentina*), slider turtle (*Trachemys scripta*), softshell turtle (*Apalone spinifer*), giant beaver (*Castoroides ohioensis*), Scott's moose (stag moose) (*Cervalces scotti*), and American mastodont (*Mammut americanum*). The turtles and giant beaver indicate the presence of a permanent, shallow lake, and the Scott's moose indicates the presence of muskegs. The three species of turtles suggest that there were enough warm summer days for their eggs to hatch. This small, mosaic fauna provides yet another example of a climate that is not typical of any area in the present Great Lakes region.

References. Holman (1966), Parmalee (1967).

Polecat Creek Site, near Oakland, Coles County, Illinois.
The Polecat Creek Site in southeastern Illinois is another site that yielded turtles and a typical Late Wisconsinan fauna of large, extinct vertebrates. Several workers have examined materials from this site, and sometimes one wonders if it is, in fact, a unit fauna, in other words a fauna whose members all lived contemporaneously.

The reported fauna of the Polecat Creek Site includes: painted turtle (*Chrysemys picta*), eastern box turtle (*Terrapene carolina*), slider turtle (*Trachemys scripta*), Jefferson's ground sloth (*Megalonyx jeffersonii*), Scott's moose (stag moose) (*Cervalces scotti*), woodland muskox (*Bootherium bombifrons*), giant beaver (*Castoroides ohioensis*), and American mastodont (*Mammut americanum*).

This is another mosaic assemblage. The painted turtle, slider turtle, and giant beaver indicate a permanent aquatic situation, and the Scott's moose indicateds the presence of muskegs. Nevertheless, the turtles, especially the eastern box turtle, indicate a significant number of warm summer days in order for egg hatching. Moreover, box turtles are presently not found in pure coniferous forests and are rare or absent from mixed coniferous/deciduous forest.

References. Galbreath (1938), Milstead (1967), Ray et al. (1968), Holman (1995b).

Galena Mastodont Site, Jo Daviess County, Illinois.
The Galena Mastodont Site, sometimes referred to as the Galena Lead Region Site, is important in that it has yielded not only three species of large, extinct mammals but in that it is one of the northernmost localities for the extinct Jefferson's ground sloth (*Megalonyx jeffersonii*) in the Great Lakes region. Very little stratigraphic data is provided in the literature about the site, but considering the taxonomic identity of the three extinct mammals, it would appear that the Late Wisconsinan is represented. The vertebrates from the site were identified many years ago and include Jefferson's ground sloth (*Megalonyx jeffersonii*), flat-headed peccary (*Platygonus compressus*), and American mastodont (*Mammut americanum*).

References. Allen (1876), Anderson (1905), Baker (1920).

Carthage Tapir Site, Hancock County, Illinois.
The Carthage Tapir Site near Carthage in west-central Illinois is significant because it has yielded the only confirmed record of the Vero tapir (*Tapirus veroensis*) from the Great Lakes region. The Vero tapir is smaller than the Hay's tapir (also known from only one locality in southwestern Indiana in the Great Lakes region), and its distribution is not as well known in North America as the Hay's tapir. The Vero tapir specimen consists of a single tooth collected in the 1860s. Some vertebrate paleontologists, however, might be somewhat skeptical of identifying tapirs to the specific level on the basis of a single tooth.

References. Leidy (1869), Ray and Sanders (1984).

Late Wisconsinan Proboscidean Sites.
Reports on at least forty mastodont and eleven mammoth sites in Illinois have been published, and many more unpublished specimens exist. In fact, in 1996 Bill Wakefield of DeKalb, Illinois, prepared a list of over 188 proboscidean reports from northern Illinois. This listing came from a variety of sources, including newspaper clippings.

The ratio of about four published mastodont sites to every one published mammoth site is similar to the pattern found in Ontario, Michigan, and Ohio. All of the mastodont records represent the American mastodont (*Mammut americanum*) and, with the exception of the Sangamonian Hopwood Farm mastodont, probably represent Late Wisconsinan times. It appears that both the Jefferson mammoth (*Mammuthus jeffersonii*) and the woolly mammoth (*Mammuthus primigenius*) were present in Illinois.

References. Anderson (1905), Crook (1927), Graham and Lundelius (1994).

WISCONSIN

As in Michigan, most records of Wisconsin Pleistocene vertebrates are from the southern part of the state, but all of the state's fossil vertebrate sites appear to represent the Late Wisconsinan age. Wisconsin is unique in the Great Lakes region in that the southwestern part of the state is unglaciated. In fact, the only diverse Pleistocene vertebrate fauna known from Wisconsin comes from a fissure fill in the extreme southwestern corner of the state.

Individual records of large, extinct vertebrates are not rare in Wisconsin, but because of its northern location, there are not as many vertebrate species known as in Ohio, Indiana, and Illinois. A significant aspect of the Pleistocene vertebrate fauna in Wisconsin is the fact that mammoth sites outnumber those of mastodonts, unlike the other Great Lakes provinces and states, where mastodont sites significantly outnumber mammoth sites.

But importantly, there is more lithic evidence that indicates the association of humans and large, extinct mammals in Wisconsin than in any other province or state in the Great Lakes region. For instance, the Chesrow archaeological site near Kenosha in southeastern Wisconsin has produced extensive evidence of an active Paleo-Indian industry for the production of lithic tools, including fluted or basally thin projectile (spear) points, biface tools, gravers, and cutters. Moreover, a heat-treating facility for the production of these tools was found at Chesrow; in southeastern Wisconsin alone, Clovis, Folsom, Gainey, and Clovis/Folsom-type projectile points have been found.

Late Wisconsinan Sites (Rancholabrean)

Moscow Fissure, near Blanchardville, Iowa County, Wisconsin. The Moscow Fissure in unglaciated southwestern Wisconsin is the most important Pleistocene site in the state, not only because it contains the most diverse vertebrate fauna in Wisconsin but also because the fauna probably lived within or almost within sight of the Late Wisconsinan ice sheet. In other words, this site vividly demonstrates that an extensive vertebrate fauna with amphibians, reptiles, and mammals could exist "in the shadow of the glacier."

The fossils vertebrates at the Moscow Site occurred in a horizontal layer in a 5-foot-deep vertical fissure in an outcrop of Ordovician dolomite. Scientists have often argued that snakes and other reptiles found in Pleistocene sites with boreal mammals must have been modern intruders that burrowed into the site. This idea was put to rest, at least at the Moscow Site, by a radiocarbon date of about 17,000 B.P. obtained from about 3 pounds of snake bones from the fossiliferous zone. Sites such as the Moscow Fissure often change ingrained ideas about the Pleistocene.

Lower vertebrates (fishes, amphibians, and reptiles) from the Moscow Fissure Site include: unidentified fish remains, American toad (*Bufo americanus,* perhaps the northern subspecies *Bufo americanus copei*), frog (*Rana* sp.), garter snake or ribbon snake (*Thamnophis* sp.), fox snake (*Elaphe vulpina*), and eastern milk snake (*Lampropeltis triangulum*). It is of considerable interest that the egg-laying fox snake and milk snake were found living so close to the ice.

The small mammal fauna identified from the Moscow Fissure includes: short-tailed shrew (*Blarina brevicauda*), arctic shrew (*Sorex arcticus*), masked shrew (*Sorex cinereus*), pygmy shrew (*Sorex hoyi*), northern water shrew (*Sorex palustris*), thirteen-lined ground squirrel (*Spermophilus tridecemlineatus*), red squirrel (*Tamiasciurus hudsonicus*), northern pocket gopher (*Thomomys talpoides*), deer mouse (*Peromyscus maniculatus*), southern red-backed vole (*Clethrionomys gapperi*), prairie vole (*Microtus ochrogaster*), meadow vole (*Mictotus pennsylvanicus*), yellow-cheeked vole (*Microtus xanthognathus*), heather vole (*Phenacomys intermedius*), collared lemming (*Dicrostonyx torquatus*), northern bog lemming (*Synaptomys borealis*), southern bog lemming (*Synaptomys cooperi*), muskrat (*Ondatra zibethicus*), western jumping mouse (*Zapus princeps*), American porcupine (*Erithizon dorsatum*), and rabbit or hare (Lagomorpha sp.).

Small mammals that invaded southwestern Wisconsin from the north 17,000 years ago include the arctic shrew, northern pocket gopher, heather vole, southern red-backed vole, yellow-cheeked vole, collared lemming, and northern bog lemming. Although the climate during the deposition of the bones must have been colder than at present in southwestern Wisconsin, it must have been warm enough for the eggs of the fox snake and the eastern milk snake to hatch. Presently, neither of these snake species occurs north of the southern border of Lake Superior.

Reference. Foley (1984).

Deerfield Mastodont Site, Dane County, Wisconsin. The mastodont site near Deerfield in southern Wisconsin is well known because it yielded the remains of three mastodonts (*Mammut americanum*) as well as cuttings from the American beaver (*Castor canadensis*). Two partially articulated mastodonts were found buried under 4 feet of peat and 2 feet of clay along the shore of a postglacial lake believed to have held water until at least 9,000 B.P. New radiocarbon dates on these two mastodonts are about 10,910 and about 10,790.

In a second excavation, about 300 feet from where the first two mastodonts were excavated, a third mastodont specimen was found that has been recently radiocarbon dated to about 11,150 B.P. Sediments that contained the three mastodont skeletons, one of which had an almost perfect skull, also contained mollusk and pollen remains that showed that the site was a lake surrounded by a spruce-dominated boreal forest.

References. Dallman (1968), West and Dallman (1980), Overstreet (1998).

Schaefer Site, near Kenosha, Kenosha County, Wisconsin. About 75 percent of a woolly mammoth skeleton (*Mammuthus primigenius*) has been excavated from sediments representing the shore of a Late Wisconsinan glacial lake in Kenosha County in extreme southeastern Wisconsin. Frank Schaefer first discovered the skeleton in 1964 while laying drain tiles on his farm.

Marks found on the early mammoth bones were interpreted as having been made by humans, but later this putative evidence was thought to have been equivocal. When additional bones were recovered, marks on several other mammoth elements, particularly ribs, were thought to be possible human cut marks. Finally, while removing the final element of the skeleton (the pelvis), two human artifacts were discovered immediately below the bone. The artifacts consisted of a flake that originated from a tool sharpened at the site and the broken edge of a general utility tool called a biface. Four radiocarbon dates associated with the Schaefer mammoth ranged from about 12,480 to about 10,960 B.P.

The Schaefer mammoth bone pile occurred as several clusters (fig. 166). Major long bones were separate but occurred near a pile of ribs and articulated vertebrae. The mandible was separated from the skull. All of this indicates that the skeleton was dismembered by humans. The bone site was in a shore or nearshore setting of a low-energy aquatic environment.

References. Joyce (1995), Overstreet et al. (1995), Overstreet (1998).

Hebior Site near Kenosha, Kenosha County, Wisconsin. The Hebior Site, a short distance from the Schaefer

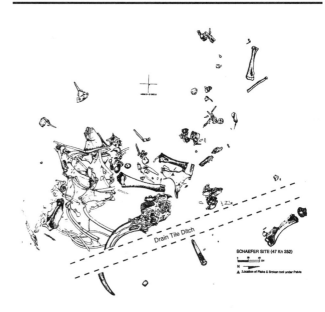

FIGURE 166. Plan view of the Late Wisconsinan woolly mammoth (*Mammuthus primigenius*) at the Schaefer Site, Kenosha County, Wisconsin.

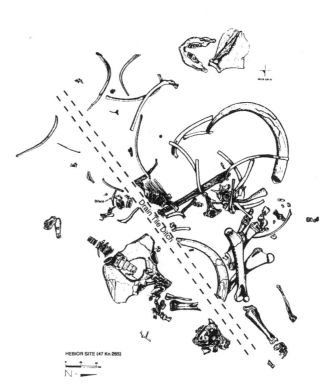

FIGURE 167. Plan view of the Late Wisconsinan woolly mammoth (*Mammuthus primigenius*) at the Hebior Site, Kenosha County, Wisconsin.

Site, yielded 85 to 95 percent of the skeleton of a woolly mammoth (*Mammuthus primigenius*). Two chipped biface tools, a crude dolomite chopper, and a wasted chert flake were discovered among the bones (fig. 167). Two recent radiocarbon dates on the Hebior Site bones are about 12,520 B.P. and about 10,960 B.P.

The organization of the mammoth bone pile at the Hebior Site was similar to that of the Schaefer Site mammoth bone pile. Ribs and vertebrae were separate from the rest of the pile, and the major long bones were separate but near the ribs and vertebrae. Moreover, the mandible had been removed from the skull. This bone mass also appears to have been a product of the dismemberment of the skeleton by humans.

References. Overstreet (1993, 1996, 1998).

Fenske Site, near Kenosha, Kenosha County, Wisconsin. The Fenske Site was discovered in the 1920s, and a complete femur and part of a humerus were originally identified as mastodont (*Mammut americanum*). But presently it is believed that at least the femur, based on length to width ratios, is mammoth rather than mastodont. Putative chop and hack marks have been found on the bones. These marks have been interpreted as being the result of scavenging by humans rather than the dismemberment of a fresh kill. A recent radiocarbon date of about 13,470 B.P. on the Fenske bones is a relatively old one for the Great Lakes region.

Reference. Overstreet (1998).

Mud Lake Site, Bristol Township, southern Kenosha County, Wisconsin. The Mud Lake Site in Bristol Township in southern Kenosha County was discovered in 1936. Proboscidean bones here were identified as woolly mammoth (*Mammuthus primigenius*). The bones consist of twenty-one elements representing the radius, ulna, and manus (front foot). Many of these elements have bone modification marks, some of them being very deep cut marks. It is believed that the Mud Lake bones represent a carcass scavenged by humans rather than a disarticulated kill. A recent radiocarbon date on the Mud Lake Site bones is about 13,440 B.P., also a relatively early date for a proboscidean kill in the Great Lakes region.

Reference. Overstreet (1998).

Interstate Park Bison Kill Site, Polk County, Wisconsin. In the southern part of northwestern Wisconsin, the Interstate Park Site in Polk County, one of the northernmost Pleistocene vertebrate sites in Wisconsin, has produced what appears to be a massive bison (*Bison bison*) kill site. About 300 individual bones attributed to the western bison subspecies *Bison bison occidentalis* were found at a depth of about 3 to 4 feet in a Late Wisconsinan peat bog.

Here, Paleo-Indian artifacts were found in direct association with the bison bones. This is another one of the few sites in midwestern North America where artifacts have been found in direct association with a Pleistocene vertebrate, and it is one of the largest Pleistocene samples of *Bison bison* known.

References. Palmer (1954), West and Dallman (1980).

Summary of Late Wisconsinan Proboscidean Sites. Wisconsin is unique in that it is the only state or province in the Great Lakes region where mammoth sites outnumber those of mastodonts. Mastodont sites significantly outnumber mammoth sites (often by a ratio of four to one) in other Great Lakes states and provinces. About twenty-eight mammoth sites have been identified in Wisconsin, all in the southern half of the state. On the other hand, only twenty mastodont sites have been listed. All of the mastodont records, except for one in Dunn County in west-central Wisconsin, lie well in the southern half of the state.

A very active Paleo-Indian lithic tool-producing industry was present in the Chesrow area of southeastern Wisconsin. Nearby proboscidean sites indicate that humans in southeastern Wisconsin participated in proboscidean hunting, butchery, and scavenging. The northern distribution of mastodonts and mammoths and, for that matter, all authentic records of Pleistocene vertebrates are limited to almost identical latitudes in the southern half of Wisconsin, the southern half of Lower Michigan, and southern Ontario.

References. West and Dallman (1980), Mason (1995), Joyce (1995), Overstreet et al. (1993), Overstreet et al. (1995), Overstreet (1996, 1998).

9

Interpretation of the Fauna

This chapter deals with the interpretation of the Pleistocene vertebrates of the Great Lakes region relative to the ecological structure of vertebrate communities, the two Pleistocene faunal subregions in the region, the vertebrate range adjustments that occurred, and the extinction of the large mammals at the end of the epoch.

PLEISTOCENE VERTEBRATE COMMUNITY STRUCTURE IN THE GREAT LAKES REGION

To put the Great Lakes region Ice Age community structure in perspective, the origin of terrestrial communities dominated by very large land vertebrates is considered here. Such communities first came into being with the diversification and increase in size of dinosaurs in the Jurassic. Dinosaur communities dominated the landscape for about 145 million years, a vast amount of time. At the heart of the dinosaur community were very large herbivores (megaherbivores), in the form of long-necked sauropods that could feed on tall trees. Other megaherbivores were present in the form of upright (bipedal) dinosaurs that traveled in groups and, like present-day birds, minded their nests and

young. Also, there was a host of four-footed armored dinosaurs that fed on plants near the ground.

Preying upon the large herbivores were bipedal dinosaur carnivores. The smaller carnivores attacked the large herbivores in packs, while larger ones stalked the herbivores singly or ate the leftover carcasses.

But many other kinds vertebrates existed somewhat outside of the web of interactions that took place between the various dinosaur groups; fishes, salamanders, frogs, turtles, snakes, birds, and mammals all played a somewhat peripheral role in the Jurassic and Cretaceous dinosaur community structure.

The first true mammals occurred in the Late Triassic, about the same time that the first dinosaurs appeared, but the mammals were very much smaller and not as specifically diverse as the dinosaurs. In fact, the mammals that were contemporaneous with the dinosaurs were mainly furtive creatures of the night. This was adaptive for these furry creatures, for mammals that were active in the daytime would have been eaten by small, carnivorous dinosaurs.

In their own unspectacular way, the Jurassic and Cretaceous mammals evolved into several mammalian groups. Some of these archaic mammalian groups became extinct and are only known by a few fragments of jaws and teeth. Others, however, evolved into the marsupials and placentals, the basal stock for future mammalian evolution.

When dinosaurs became extinct at the end of the Cretaceous, the small mammals of the Early Tertiary rapidly evolved into larger ones that took over many of the ecological niches vacated by the dinosaurs. In fact, herbivore-carnivore-scavenger communities dominated by large mammals evolved quickly in the Paleocene, the first epoch of the Tertiary.

The Paleocene large mammal communities were odd. Large, ungainly, small-brained herbivores were about. Some looked somewhat like elephants or rhino, others looked somewhat like horses or camels, and some even looked like giant guinea pigs. Rounding out these early mammalian terrestrial communities were relatively clumsy and rather small-brained carnivores and scavengers that feasted on the herbivores or their remains.

Later, in the Eocene epoch, modern orders and families of mammals evolved and replaced the earlier versions of herbivores, carnivores, and scavengers. By Miocene times, with the advent of widespread grasslands, the mammalian fauna took on a more modern aspect. Many kinds of hoofed herbivores evolved as the grasslands expanded, and specialized predators evolved along with them. Several species of bone-crushing, doglike animals evolved that cleaned up the leftover carcasses.

Further modernization of mammalian groups took place in the Pliocene, setting the stage for the Pleistocene megaherbivore-dominated mammalian fauna. But in the Pleistocene, the extinction of most of the large mammalian herbivores, and the carnivores and scavengers that were dependant upon them, took place by the end of the epoch, except in Africa where, in a sense, the Pleistocene still exits.

Megaherbivore-Dominated Communities

The late vertebrate paleontologist George Gaylord Simpson correctly pointed out that, during the Pleistocene, Florida had a fauna more like the one that presently occurs in Africa than like the one that presently exists in Florida. He was referring to the diversity of large herbivores and their predators. Unfortunately, most of the animals that Simpson was talking about became extinct by about 10,000 years ago, at the end of the Pleistocene.

In recent years, the ecology of communities in the grasslands and scrublands of Africa has been intensively studied. In some of the managed game preserves, a significant number of large mammalian herbivores exist, which are fed upon by various canids as well as large cats such as lions, cheetahs, and leopards. Hyenas play a dual role as both carnivores and scavengers.

Megaherbivores that weigh more than 2,000 pounds are the most important species in protected game preserves in Africa. They are essential to the stability of the entire community. In modern Africa, the feeding activities of elephants and rhino change wooded savanna to open, shortgrass situations, which are dominated by rapidly regenerating plants that provide food for the many other herbivores in the community. Without megaherbivores, a scrubland grows back that cannot support nearly as many mammalian species as can the rapidly regenerating grasses. Even the dung secreted by the giant herbivores and the habit of creating wallows are important aspects of the African environment.

Furthermore, it has been shown in recent years that the elimination of the megaherbivores, the elephants and rhinos, leads to massive environmental changes and the elimination of the entire community. This situation has been called the domino effect.

The large mammals of the Ice Age in North America existed in megaherbivore-dominated communities. Here, the mastodonts and mammoths that lived all over ice-free North America are thought to have played a role similar to that of the elephants and rhino in modern Africa.

These animals certainly would have been the largest herbivores in the Ice Age community and, like elephants, would have kept the vegetation open, which would allow grasses and other ground plants to regenerate rapidly to the benefit of other Pleistocene herbivores. Therefore, invoking the domino effect, it may not be coincidental that when both mastodonts and mammoths became extinct worldwide at the close of the Pleistocene, other large herbivores followed them into oblivion.

Comments on Great Lakes Pleistocene Megavertebrates and Salt. Questions about Ice Age megaherbivores and their need for salt were first raised in the Great Lakes region by a consideration of the large number of American mastodont and Jefferson mammoth records confined in space and time to Michigan. In fact, all of the more than 260 verified records of these megaherbivores are from the southern half of the Lower Peninsula of the state, where they lived between 12,000 and 10,000 B.P.

The abundance of these megaherbivores is puzzling in light of the scarcity of other vertebrate remains.

Earlier, the late C. W. Hibbard of the University of Michigan suggested that the reason smaller animals were not trapped in Michigan quaking bog burials was because these animals were light enough to run across the top of the mats of vegetation that acted as a trap for the massive megaherbivores (see fig. 13). But we now know that bog sites in Ohio and Indiana do not bear this out. We have seen that the 11,000 B.P. Carter Site bog in west-central Ohio has yielded many small vertebrates and that the 13,000 B.P. Christensen Bog Site in east-central Indiana has yielded a substantial small vertebrate fauna as well as mastodont remains.

Pollen studies indicate that spruce and pine were the dominant elements in the vegetation during the time of mastodonts and mammoths in southern Michigan. Other criteria, such as the scarcity of reptiles during the entire period of occupation of the state by mastodonts and mammoths, indicate a harsh, perhaps even a savagely cold climate.

Why then should so many huge animals, so demanding on biotic resources, be so common in such a severe environment, indeed much more abundant than in any other state or Ontario of the Great Lakes region? The suggestion that was put forward was that the large concentration of mastodonts and mammoths in Michigan might be correlated with the occurrence of widespread salt seeps and shallow saline water that were available in the state at that time.

Modern megaherbivores, the elephants, crave salt and will sometimes travel great distances to reach a salt supply. This is because elephants have high sodium (an ingredient of salt) dietary requirements. The animals depend on a concentrated external salt supply because they must have a proper balance of sodium and potassium to stay alive.

Elephants (and mastodonts and mammoths) eat plants that are rich in potassium but not in sodium. Thus, after a time, these large creatures build up a sodium debt. In fact, this precarious sodium budget is said to be the limiting factor in elephant life in Africa, where the density of elephant populations may be correlated with sodium concentration in water and soil.

In areas of Africa where salt is scarce, elephants will travel to caves to reach salt supplies, treading on rock piles, which they normally refuse to do, and walking along precarious passageways from which they sometimes fall. Moreover, congregations at salt licks are important in the social lives of African elephants, and the salt lick wallows of elephants are familiar landmarks in that continent. Seasonal movements of ele-

phants in Wankie National Park, Zimbabwe, were dictated by the distribution of salt-rich waterholes. When the waterholes dry up, the animals use their tusks to loosen the salt-rich soil, which is then eaten.

Michigan's salt deposits accumulated in shallow, evaporative seas during Silurian and Devonian times in a depressed structure called the Michigan Basin. Subsequent uplift of this area not only exposed the Michigan Basin rocks to millions of years of erosion but also exposed one of the largest commercial salt deposits in the world. About 20 percent of the salt used in the United States is produced in Michigan.

Bromine-rich salt has been vital to the huge chemical industry in the midland area of Michigan. Because surficial salt deposits were abundant in Late Wisconsinan times, paleontologists and geologists at Michigan State University hypothesized that salt might have been an important limiting factor in the distribution of Pleistocene mastodonts and mammoths and that the large concentration of these animals in southern Michigan might be correlated with the existence of widely available salt seeps or shallow saline water (see fig. 149).

As previously discussed, the Mason-Quimby Line marks the northern limits of mastodont and mammoth distribution in Michigan and also most of the Paleo-Indian fluted point finds in Michigan (see fig. 10). All of the historically recorded salt seeps or shallow saline waters in Michigan occur south of or adjacent to the Mason-Quimby Line (see fig. 157). Moreover, many mastodont and mammoth records appear to cluster around specific salt deposits. Noteworthy clusters of both mastodonts and mammoths occur around a single salt deposit in Berrien County in extreme southwestern Michigan and around the large number of salt deposits in the southeastern third of the Lower Peninsula.

If mastodonts and mammoths migrated seasonally into Michigan from other areas to obtain salt, then it seems that it might someday be possible to obtain elemental signatures (chemical evidence) of their salt-procuring activities in Michigan. In some preliminary work at Michigan State University, bromine, strontium, and other chemical elements have been identified in both bones and teeth of Michigan mastodonts and mammoths, but it has not yet been determined whether these elements are from products eaten during the lives of the animals or from changes that occurred in the bones and teeth after death and burial.

Another suggestion that has been offered is that if mastodonts and mammoths seasonally migrated to Michigan for salt, bands of human hunters might have followed the migrating herd to prey upon stragglers or individuals stuck in quaking bogs. There is no scientific evidence, other than the widespread occurrence of Paleo-Indian fluted points in the state, to support this.

Overview of the Pleistocene Terrestrial Vertebrate Fauna

Now that the role of megaherbivores in the Pleistocene terrestrial vertebrate fauna of the Great Lakes region has been defined, we turn to the ecological roles of the other terrestrial Pleistocene vertebrates, namely herbivores, omnivores, and carnivores. Herbivores are those animals that primarily eat plant material. Omnivores are those animals that usually eat both plant and animal material. Carnivores are those animals that primarily feed on other animals.

Herbivores from the terrestrial vertebrate fauna of the Pleistocene of the Great Lakes region include the following turtles, birds, xenarthrans, and ungulates: extinct giant land tortoise (*Hesperotestudo crassicutata*), cf. ruffed grouse (*Bonassa umbellus*), prairie chicken (*Tympanuchus cupido*), extinct Jefferson's ground sloth (*Megalonyx jeffersonii*), extinct complex-toothed horse (*Equus complicatus*), extinct Hay's tapir (*Tapirus haysii*), extinct Vero tapir (*Tapirus veroensis*), extinct camel (*Camelops* sp.), moose (*Alces alces*), extinct Scott's moose (stag moose) (*Cervalces scotti*), wapiti (elk) (*Cervus elaphus*), white-tailed deer (*Odocoileus virginianus*), caribou (*Rangifer tarandus*), American bison (*Bison bison*), extinct giant bison (*Bison latifrons*), tundra muskox (*Ovibos moschatus*), and extinct woodland muskox (*Bootherium bombifrons*).

Other Great Lakes terrestrial Ice Age herbivores include the following rodents, lagomorphs, and proboscideans: southern flying squirrel (*Glaucomys volans*), woodchuck (*Marmota monax*), eastern gray squirrel (*Sciurus carolinensis*), eastern fox squirrel (*Sciurus niger*), thirteen-lined ground squirrel (*Spermophilus tridecemlineatus*), eastern chipmunk (*Tamias striatus*), red squirrel (*Tamiasciurus hudsonicus*), plains pocket gopher (*Geomys bursarius*), northern pocket gopher (*Thomomys talpoides*), eastern woodrat (*Neotoma floridana*), rice rat (*Oryzomys palustris*), southern red-backed vole (*Clethrionomys gapperi*), extinct Cape Deceit vole (*Lasiopodomys deceitensis*), prairie vole (*Microtus ochrogaster*), extinct Hibbard's tundra vole (*Microtus paroperarius*), meadow vole (*Microtus pennsylvanicus*), woodland vole (*Microtus pinetorum*), yellow-cheeked vole (*Microtus xanthognathus*), heather

vole (*Phenacomys intermedius*), Greenland collared lemming (*Dicrostonyx torquatus*), brown lemming (*Lemmus* sp.), northern bog lemming (*Synaptomys borealis*), southern bog lemming (*Synaptomys cooperi*), meadow jumping mouse (*Zapus hudsonius*), western jumping mouse (*Zapus princeps*), common porcupine (*Erithizon dorsatum*), pika (*Ochotona* sp.), snowshoe hare (*Lepus americanus*), eastern cottontail (*Sylvilagus floridanus*), extinct mastodont (*Mammut americanum*), extinct Jefferson mammoth (*Mammuthus jeffersonii*), and extinct woolly mammoth (*Mammuthus primigenius*). One can see that herbivores make up a large part of the Pleistocene terrestrial fauna of the Great Lakes region.

The following Great Lakes Pleistocene terrestrial vertebrates are omnivores: eastern box turtle (*Terrapene carolina*), wild turkey (*Meleagris gallopavo*), raven (*Corvus corax*), extinct beautiful armadillo (*Dasypus bellus*), human (*Homo sapiens*), black bear (*Ursus americanus*), brown or grizzly bear (*Ursus arctos*), raccoon (*Procyon lotor*), extinct long-nosed peccary (*Mylohyus nasutus*), extinct flat-headed peccary (*Platygonus compressus*), extinct Leidy's peccary (*Platygonus vetus*), white-footed mouse (*Peromyscus leucopus*), and deer mouse (*Peromyscus maniculatus*).

Carnivores identified from the terrestrial fauna of the Pleistocene of the Great Lakes region include the following amphibians and reptiles: blue-spotted salamander group (*Ambystoma laterale* complex), cave salamander (*Eurycea lucifuga*), American toad (*Bufo americanus*), Fowler's toad (*Bufo fowleri*), striped chorus frog (*Pseudacris triseriata*), northern leopard frog (*Rana pipiens*), wood frog (*Rana sylvatica*), slender glass lizard (*Ophisaurus attenuatus*), racer (*Coluber constrictor*), fox snake (*Elaphe vulpina*), milk snake (*Lampropeltis triangulum*), smooth green snake (*Opheodrys vernalis*), garter snake (*Thamnophis sirtalis*), and timber rattlesnake (*Crotalus horridus*).

Insectivore and bat species identified from the terrestrial fauna of the Ice Age of the Great Lakes region are all carnivores. These animals include: northern short-tailed shrew (*Blarina brevicauda*), least shrew (*Cryptotis parva*), arctic shrew (*Sorex arcticus*), masked shrew (*Sorex cinereus*), smoky shrew (*Sorex fumeus*), pygmy shrew (*Sorex hoyi*), star-nosed mole (*Condylura cristata*), hairy-tailed mole (*Parascalops breweri*), eastern mole (*Scalopus aquaticus*), gray bat or northern bat (*Myotis grisescens* or *septentrionalis*), small-footed bat (*Myotis leibii*), little brown bat (*Myotis lucifugus*), eastern pipistrelle bat (*Pipistrellus subflavus*), and big-eared bat (*Plecotus* sp.).

The following mammals of the order Carnivora are carnivores that have been identified from the terrestrial fauna of the Pleistocene of the Great Lakes region: extinct dire wolf (*Canis dirus*), coyote (*Canis latrans*), gray fox (*Urocyon cinereoargenteus*), red fox (*Vulpes vulpes*), extinct short-faced bear (*Arctodus simus*), marten (*Martes americana*), fisher (*Martes pennanti*), ermine (*Mustela erminea*), long-tailed weasel (*Mustela frenata*), mink (*Mustela vison*), striped skunk (*Mephitis mephitis*), spotted skunk (*Spilogale putorius*), jaguar (*Panthera onca*), and extinct sabertooth (*Smilodon fatalis*).

Table 4 breaks down the total number and percentages of the herbivores, omnivores, and carnivores of the total terrestrial fauna of the Pleistocene of the Great Lakes region in three ways. When the total terrestrial vertebrate fauna is considered, about 59 percent of the fauna is composed of herbivores and omnivores and about 41 percent of carnivores. When only the small vertebrates (animals smaller than pigs or peccaries) of the terrestrial vertebrate fauna are considered, however, about 51 percent are herbivores and omnivores and about 49 percent are carnivores. On the other hand, when only the large terrestrial vertebrates (animals of pig- or peccary-sized or larger) are considered, about 86 percent of them are herbivores and omnivores and about 14 percent are carnivores. The number of large terrestrial carnivores taken together is

only about 4 percent of the total Great Lakes Pleistocene terrestrial vertebrate fauna.

The data in table 4 generally follows the pyramid of energy concept (see chapter 1) in that herbivores tend to be more numerous than carnivores in terrestrial biotic communities. This is the result of the fact that energy is lost at each trophic (feeding) level. The similarity in numbers between the small herbivore-omnivore and small carnivore groups in the Pleistocene of the Great Lakes region may be explained on the basis that small herbivore species, such as mice and voles, produce exceedingly large numbers of individuals.

In the following section of this chapter, subregions of the Great Lakes region will be compared with one another.

PLEISTOCENE SUBREGIONS

In the first chapter, I discussed the fact that two Pleistocene faunal subregions exist in the Great Lakes region: a northern one (Subregion I) and a southern one (Subregion II) (see fig. 9). Pleistocene faunal differences between Subregion I and Subregion II mainly reflect: the existence of some cold-adapted terrestrial species in Subregion I that may not have been individually able to have invaded Subregion II during cold climates; the existence of cold-tolerant marine fishes, seals, and whales that entered Ontario in the Late Wisconsinan by way of the encroaching Champlain Sea; the existence of more warm-adapted species in Subregion II, especially those species of Sangamonian age; and the existence of many more caves in Subregion II, which trapped a wider variety of species than the kettle and shallow basin sites of the north.

Important vertebrate species that are present in Subregion I but absent in Subregion II include:

Lake Trout - *Salvelinus namaycush*
Capelin - *Mallotus villosus*
Rainbow Smelt - *Osmorus mordax*
Atlantic Tomcod - *Microgadus tomcod*
Threespine Stickleback - *Gasterosteus aculeatus*
Hook-eared Sculpin - *Artediellus uncinatus*
Lumpfish - *Cyclopterus lumpus*
Brown or Grizzly Bear - *Ursus arctos*
Marten - *Martes americana*
Ermine - *Mustela erminea*
Bearded Seal - *Erignathus barbatus*

TABLE 4. Summary of Terrestrial Vertebrate Herbivores, Omnivores, and Carnivores in the Pleistocene of the Great Lakes Region. (Small vertebrates are defined as being smaller than a pig or peccary; large vertebrates are defined as being pig- or peccary-sized or larger.)

Number of Vertebrate Species		% of Total Unit
Total Number of All Herbivores	– 48	46.60%
Total Number of All Omnivores	– 13	12.62%
Total Number of All Carnivores	– 42	40.78%
	= 103	
Total Number of Small Herbivores	– 30	40.54%
Total Number of Small Omnivores	– 6	8.12%
Total Number of Small Carnivores	– 38	51.35%
	= 74	
Total Number of Large Herbivores	– 18	62.07%
Total Number of Large Omnivores	– 7	24.14%
Total Number of Large Carnivores	– 4	13.79%
	= 29	
Total Vertebrate Species	– 103	
Total Number of Large Carnivores	– 4	3.88%

Harp Seal - *Phoca groenlandica*
Ringed Seal - *Phoca hispida*
White Whale - *Delphinapterus leucas*
Humpback Whale - *Megaptera novaeangliae*
Bowhead Whale - *Balaena mysticetus*
Moose - *Alces alces*
Northern Pocket Gopher - *Thomomys talpoides*
Brown Lemming - *Lemmus* sp.
Collared Lemming - *Dicrostonyx torquatus*
Common Porcupine - *Erithizon dorsatum*

Important vertebrate species that are present in Subregion II but absent in Subregion I include:

Cave Salamander - *Eurycea lucifuga*
Common Musk Turtle - *Sternotherus odoratus*
Cooter - *Pseudemys* sp.
Map Turtle - *Graptemys* sp.
Extinct Giant Land Tortoise – *Hesperotestudo crassiscutata*
Slender Glass Lizard - *Ophisaurus attenuatus*
Plain-bellied Water Snake - *Nerodia erythrogaster*
Timber Rattlesnake - *Crotalus horridus*
Extinct Beautiful Armadillo - *Dasypus bellus*
Big-eared Bat - *Plecotus* sp.
Dire Wolf - *Canis dirus*
Jaguar - *Panthera onca*
Extinct Sabertooth - *Smilodon fatalis*
Extinct Long-nosed Peccary - *Mylohyus nasutus*
Extinct Leidy's Peccary- *Platygonus vetus*
Extinct American Camel - *Camelops* sp.
Rice Rat - *Oryzomys palustris*

Table 5 indicates several possible adaptive reasons for the presence or absence of some of the fauna of both subregions. These animals include cold-adapted species present in Subregion I that may have been unable to invade and survive in more southern areas during glacial advances in the Pleistocene and warm-adapted species in Subregion II that expanded northward during the Sangamonian warm period or during the glacial retreat in the Late Wisconsinan.

Since most Pleistocene vertebrate faunas in the Great Lakes region represent the Late Wisconsinan, it would seem instructive to compare the community structure of the Late Wisconsinan terrestrial vertebrate fauna of Subregion I with that of Subregion II. Thus the following paragraphs analyze the Late Wisconsinan terrestrial vertebrate herbivores, omnivores, and carnivores of the two subregions.

TABLE 5. Possible Adaptive Reasons for Some Pleistocene Faunal Differences in Subregion I and Subregion II of the Great Lakes Region.

Subregion I Species Absent from Subregion II

Cold-adapted Champlain Sea Species	Cold-adapted Terrestrial Species
Lake Trout	Brown or Grizzly Bear
Capelin	Marten
Rainbow Smelt	Ermine
Atlantic Tomcod	
Threespine Stickleback	Moose
Hook-eared Sculpin	Northern Pocket Gopher
Lumpfish	Brown Lemming
Bearded Seal	Collared Lemming
Harp Seal	Common Porcupine
Ringed Seal	
White Whale	
Humpback Whale	
Bowhead Whale	

Subregion II Species Absent from Subregion I

Sangamonian Warm-adapted Species	Late Wisconsinan Warm-adapted Species
Giant land Tortoise	Cave Salamander
Jaguar	Common Musk Turtle
Sabertooth	Cooter Turtle
	Slender Glass Lizard
	Plain-bellied Water Snake
	Timber Rattlesnake
	Beautiful Armadillo
	Rice Rat

Subregion I

Herbivores identified from the terrestrial fauna of the Late Wisconsinan of Subregion I include the following birds, xenarthrans, and artiodactyls: cf. ruffed grouse (*Bonasa umbellus*), prairie chicken (*Tympanuchus cupido*), extinct Jefferson's ground sloth (*Megalonyx jeffersonii*), moose (*Alces alces*), extinct Scott's moose (stag moose) (*Cervalces scotti*), wapiti (elk) (*Cervus elaphus*), white- tailed deer (*Odocoileus virginianus*), caribou (*Rangifer tarandus*), American bison (*Bison bison*), tundra muskox (*Ovibos moschatus*), and extinct woodland muskox (*Bootherium bombifrons*). Among the large herbivores here are species (moose, caribou, and tundra muskox) that represent the Far North.

Herbivores identified from the terrestrial vertebrate

fauna of the Late Wisconsinan of Subregion I include the following rodents, lagomorphs, and proboscideans: woodchuck (*Marmota monax*), tree squirrel (*Sciurus* sp.), thirteen-lined ground squirrel (*Spermophilus tridecemlineatus*), eastern chipmunk (*Tamias striatus*), red squirrel (*Tamiasciurus hudsonicus*), northern pocket gopher (*Thomomys talpoides*), woodrat (*Neotoma* sp.), southern red-backed vole (*Clethrionomys gapperi*), prairie vole (*Microtus ochrogaster*), meadow vole (*Microtus pennsylvanicus*), yellow-cheeked vole (*Microtus xanthognathus*), heather vole (*Phenacomys intermedius*), collared lemming (*Dicrostonyx torquatus*), northern bog lemming (*Synaptomys borealis*), southern bog lemming (*Synaptomys cooperi*), western jumping mouse (*Zapus princeps*), common porcupine (*Erithizon dorsatum*), hare or jackrabbit (*Lepus* sp.), extinct American mastodont (*Mammut americanum*), extinct Jefferson mammoth (*Mammuthus jeffersonii*), and extinct woolly mammoth (*Mammuthus primigenius*). Many of the far northern rodents (northern pocket gopher, southern red-backed vole, yellow-cheeked vole, heather vole, collared lemming, and northern bog lemming) are a significant part of the small herbivore fauna of Subregion I.

Omnivores identified from the terrestrial vertebrate fauna of the Late Wisconsinan of Subregion I include: eastern box turtle (*Terrapene carolina*), wild turkey (*Meleagris gallopavo*), raven (*Corvus corax*), human (*Homo sapiens*), black bear (*Ursus americanus*), brown or grizzly bear (*Ursus arctos*), raccoon (*Procyon lotor*), extinct flat-headed peccary (*Platygonus compressus*), and deer mouse (*Peromyscus maniculatus*). A significant northern omnivore is the brown or grizzly bear (*Ursus arctos*).

Carnivores identified from the terrestrial vertebrate fauna of the Late Wisconsinan of Subregion I include the following amphibians, reptiles, insectivores, and bats: blue-spotted salamander complex (*Ambystoma laterale* complex), American toad (*Bufo americanus*), Fowler's toad (*Bufo fowleri*), striped chorus frog (*Pseudacris triseriata*), northern leopard frog (*Rana pipiens*), wood frog (*Rana sylvatica*), racer (*Coluber constrictor*), fox snake (*Elaphe vulpina*), milk snake (*Lampropeltis triangulum*), smooth green snake (*Opheodrys vernalis*), common garter snake (*Thamnophis sirtalis*), northern short-tailed shrew (*Blarina brevicauda*), arctic shrew (*Sorex arcticus*), masked shrew (*Sorex cinereus*), smoky shrew (*Sorex fumeus*), pygmy shrew (*Sorex hoyi*), and northern bat (*Myotis septentrionalis*).

True carnivores of the order Carnivora found in the terrestrial vertebrate fauna of the Late Wisconsinan of Subregion I include: red fox (*Vulpes vulpes*), extinct short-faced bear (*Arctodus simus*), marten (*Martes americana*), fisher (*Martes pennanti*), ermine (*Mustela erminea*), long-tailed weasel (*Mustela frenata*), mink (*Mustela vison*), and striped skunk (*Mephitis mephitis*). The presence of only one large terrestrial carnivore, the short-faced bear, in Subregion I is puzzling.

Subregion II

Herbivores of the terrestrial vertebrate fauna of the Late Wisconsinan of Subregion II include the following xenarthrans, perissodactyls, and artiodactyls: extinct Jefferson's ground sloth (*Megalonyx jeffersonii*), extinct Hay's tapir (*Tapirus haysii*), extinct long-nosed peccary (*Mylohyus nasutus*), extinct flat-headed peccary (*Platygonus compressus*), extinct Scott's moose (stag moose) (*Cervalces scotti*), white-tailed deer (*Odocoileus virginianus*), American bison (*Bison bison*), and extinct woodland muskox (*Bootherium bombifrons*). The lack of far northern artiodactyls in this segment of the late Wisconsinan herbivores of Subregion II is significant.

Herbivores of the terrestrial vertebrate fauna of the Late Wisconsinan of Subregion II include the following rodents, lagomorphs, and proboscideans: southern flying squirrel (*Glaucomys volans*), woodchuck (*Marmota monax*), eastern gray squirrel (*Sciurus carolinensis*), eastern fox squirrel (*Sciurus niger*), thirteen-lined ground squirrel (*Spermophilus tridecemlineatus*), eastern chipmunk (*Tamias striatus*), red squirrel (*Tamiastriatus hudsonicus*), plains pocket gopher (*Geomys bursarius*), eastern wood rat (*Neotoma floridana*), rice rat (*Oryzomys palustris*), southern red-backed vole (*Clethrionomys gapperi*), prairie vole (*Microtus ochrogaster*), meadow vole (*Microtus pennsylvanicus*), woodland vole (*Microtus pinetorum*), yellow-cheeked vole (*Microtus xanthognathus*), heather vole (*Phenacomys intermedius*), northern bog lemming (*Synaptomys borealis*), southern bog lemming (*Synaptomys cooperi*), meadow jumping mouse (*Zapus hudsonius*), snowshoe hare (*Lepus americanus*), hare or jackrabbit (*Lepus* sp.), eastern cottontail (*Sylvilagus floridanus*), extinct American mastodont (*Mammut americanum*), extinct Jefferson mammoth (*Mammuthus jeffersonii*), and extinct woolly mammoth (*Mammuthus primigenius*). The presence of several invaders from the north, the small northern herbivores (southern red-backed vole, yellow-cheeked vole, heather vole, and northern bog lemming), and the wooly mammoth in Subregion II is noteworthy.

Omnivores of the terrestrial vertebrate fauna of the Late Wisconsinan of Subregion II include: extinct beautiful armadillo (*Dasypus bellus*), human (*Homo sapiens*), black bear (*Ursus americanus*), raccoon (*Procyon lotor*), flat-headed peccary (*Platygonus compressus*), long-nosed peccary (*Mylohyus nasutus*), white-footed mouse (*Peromyscus leucopus*), and deer mouse (*Peromyscus maniculatus*). Omnivores are essentially similar to what they were in the Late Wisconsinan terrestrial fauna of Subregion I.

Carnivores of the terrestrial vertebrate fauna of the Late Wisconsinan of Subregion II include the following amphibians, reptiles, insectivores, and bats: mole salamander (*Ambystoma* sp.), cave salamander (*Eurycea lucifuga*), toad (*Bufo* sp.), northern leopard frog (*Rana pipiens*), wood frog (*Rana sylvatica*), slender glass lizard (*Ophisaurus attenuatus*), smooth green snake (*Opheodrys vernalis*), garter or ribbon snake (*Thamnophis* sp.), timber rattlesnake (*Crotalus horridus*), northern short-tailed shrew (*Blarina brevicauda*), least shrew (*Cryptotis parva*), arctic shrew (*Sorex arcticus*), masked shrew (*Sorex cinereus*), smoky shrew (*Sorex fumeus*), pygmy shrew (*Sorex hoyi*), star-nosed mole (*Condylura cristata*), hairy-tailed mole (*Parascalops breweri*), eastern mole (*Scalopus aquaticus*), gray bat or northern bat (*Myotis grisescens* or *septentrionalis*), small-footed bat (*Myotis leibii*), northern bat (*Myotis septentrionalis*), eastern pipistrelle (*Pipistrellus subflavus*), and big-eared bat (*Plecotus* sp.).

The large number of small predators in Subregion II compared to Subregion I obviously reflects the fact that cave faunas are much more abundant in Subregion II. Cave salamanders often live in or near caves; small insectivores are usually derived in the form of pellets regurgitated by owls; and bats roost in caves.

True carnivores of the order Carnivora present in the Late Wisconsinan terrestrial fauna of Subregion II include: extinct dire wolf (*Canis dirus*), coyote (*Canis latrans*), gray fox (*Urocyon cinereoargenteus*), fisher (*Martes pennanti*), long-tailed weasel (*Mustela frenata*), mink (*Mustela vison*), and spotted skunk (*Spilogale putorius*).

Table 6 breaks down the total number and percentages of herbivores, omnivores, and carnivores of the Late Wisconsinan fauna of Subregion I. In general, the total fauna reflects the pyramid of energy concept in that herbivores and omnivores outnumber carnivores in the fauna. About 63 percent of the total

fauna is composed of herbivores and omnivores, and about 37 percent of the fauna is composed of carnivores. In the entire Pleistocene Great Lakes region, vertebrate fauna is about 59 percent herbivore/omnivore and 41 percent carnivore (see table 4).

TABLE 6. Summary of Terrestrial Vertebrate Herbivores, Omnivores, and Carnivores in the Late Wisconsinan of Subregion I of the Great Lakes Region. (Small vertebrates are defined as being smaller than a pig or peccary; large vertebrates are defined as being pig- or peccary-sized or larger.)

Number of Vertebrate Species		% of Total Unit
Total Number of All Herbivores	− 32	47.76%
Total Number of All Omnivores	− 10	14.93%
Total Number of All Carnivores	− 25	37.31%
	= 67	
Total Number of Small Herbivores	− 20	40.82%
Total Number of Small Omnivores	− 5	10.20%
Total Number of Small Carnivores	− 24	48.98%
	= 49	
Total Number of Large Herbivores	− 12	66.67%
Total Number of Large Omnivores	− 5	27.78%
Total Number of Large Carnivores	− 1	5.56%
	= 18	
Total Vertebrate Species	− 67	
Total Number of Large Carnivores	− 1	1.49%

When only the small vertebrates (animals smaller than pigs) or peccaries are considered, however, about 50 percent are herbivores and omnivores and about 50 percent are carnivores. In the entire Pleistocene Great Lakes region, terrestrial small vertebrate fauna is about 49 percent herbivore/omnivore and 51 percent carnivore (see table 4). This reflects the high reproductive rate of small herbivores. On the other hand, when only the large vertebrates (animals pig- or peccary-sized or larger) are considered, the large herbivores and omnivores outnumber the carnivores by about 94 percent to about 6 percent. This reflects the fact that only one large carnivore, the short-faced bear, has been identified from the Late Wisconsinan of Subregion I, unless one wishes to count the undetermined canids that have been recognized on the basis of fragmentary remains from a few sites. This contrasts somewhat with entire Pleistocene vertebrate fauna of the Great Lakes region, where the large herbivores and omnivores comprise about 86 percent of the fauna and the large carnivores comprise about 14 percent (see table 4).

Table 7 breaks down the total number and percentages of herbivores, omnivores, and carnivores of the Late Wisconsinan fauna of Subregion II. In general, the total fauna reflects the pyramid of energy concept in that herbivores and omnivores comprise about 58 percent of the total fauna, and carnivores comprise about 42 percent. In the entire Great Lakes region, Pleistocene vertebrate fauna is about 59 percent herbivore/omnivore and 41 percent carnivore (see table 4). In the Late Wisconsinan fauna of Subregion I, this percentage is about 63 percent herbivore/omnivore and about 37 percent carnivores (see table 6).

When only the small vertebrates (animals smaller than pigs or peccaries) are considered in the Late Wisconsinan of Subregion II, however, about 46 percent are herbivores and omnivores and about 54 percent are carnivores. These figures are remarkably consistent with the total vertebrate Pleistocene fauna in the Great Lakes region (see table 4) as well as with the Late Wisconsinan vertebrate fauna of Subregion I (see table 6). This once more reflects the high reproductive rate of small herbivores. When only the large vertebrates (animals pig- or peccary-sized or larger) are considered, the herbivore/omnivore complex in Subregion II is about 94 percent and the carnivores about 6 percent. This corresponds fairly closely with the same percentage for the entire Great Lakes region Pleistocene terrestrial vertebrate fauna, where the large herbivore/omnivore complex comprises about 96 percent the fauna and the carnivores about 4 percent (see table 4).

Comments about Subregions

In light of all of the variables one faces in the interpretation of fossil faunas, comparisons of species and composition of faunas in the two subregions is of interest but should not be not taken as absolute information. Let us take, for instance, the dearth of large carnivores in both subregions.

Carnivores consist mainly of large canids and cats, both of which are very intelligent. Would it be more parsimonious to suggest that there was a dearth of carnivores in the Late Wisconsinan of the Great Lakes region or to suggest that the wily carnivores were adept at avoiding quaking bogs in the north or falling into caves in the south?

TABLE 7. Summary of Terrestrial Vertebrate Herbivores, Omnivores, and Carnivores in the Late Wisconsinan of Subregion II of the Great Lakes Region. (Small vertebrates are defined as being smaller than a pig or peccary; large vertebrates are defined as being pig- or peccary-sized or larger.)

Number of Vertebrate Species		% of Total Unit
Total Number of All Herbivores	− 32	46.38%
Total Number of All Omnivores	− 8	11.59%
Total Number of All Carnivores	− 29	42.03%
	= 69	
Total Number of Small Herbivores	− 21	40.38%
Total Number of Small Omnivores	− 3	5.77%
Total Number of Small Carnivores	− 28	53.85%
	= 52	
Total Number of Large Herbivores	− 11	64.71%
Total Number of Large Omnivores	− 5	29.41%
Total Number of Large Carnivores	− 1	5.88%
	= 17	
Total Vertebrate Species	− 69	
Total Number of Large Carnivores	− 1	1.45%

On the other hand, the relatively abundant fossil artiodactyls in the Pleistocene of the Great Lakes region might reflect the near-sightedness or lack of quick wits of these large herbivores. In fact, in cave regions in modern Florida, many cattle fall into natural sinkhole traps. This happens so often that most Florida ranchers circle the entrances of such pitfalls with barbed-wire fences.

The utter randomness associated with accumulation and burial of fossils is also an important consideration. In other words, just because we find a species of rodent, shrew, or bat in a cave in southern Indiana in Subregion II and not in Sheriden Pit in northwest Ohio in Subregion I does not necessarily mean that the species did not occur in the northern region. Such occurrences may have to do with the prey preference of the owls roosting in the cave or the relative abundance of specific rodents, bats, and shrews at the particular season represented.

Continued paleontological studies not only will add to the species lists in both regions but also will add more ecological information about each region as the pieces of the puzzle slowly accumulate.

VERTEBRATE RANGE ADJUSTMENTS AND EXTINCTIONS IN THE PLEISTOCENE

Two primary events occurred in vertebrate populations during the Ice Age: range adjustments due to the obliteration of habitats by advances of the ice sheet and associated climatic changes, and widespread extinctions. We shall consider patterns of vertebrate range adjustments here and deal with the problem of extinctions in the next section of this chapter.

The main vertebrate range adjustments in the Pleistocene of the Great Lakes region were: southward movements of northern species during glacial advances and their reoccupation of ranges during glacial retreats; invasion of saltwater fishes and marine mammals into the St. Lawrence River Valley as the Champlain Sea penetrated the interior of Ontario and elimination of these species as the sea retreated; and northward movement of several southern species during Sangamonian times and their retreat during the onset of the Wisconsinan glaciation.

Many vertebrate species south of the glaciated regions tended to stay in place during the Pleistocene. This was not generally recognized by early scientists. The idea put forth up to the late 1950s (and still taught in some classrooms) was that advancing and retreating ice altered the climate in such a regular way that communities were uniformly displaced southward during glacial advances and uniformly advanced northward during glacial retreats. In other words, uniform bands of tundra, coniferous forest, and deciduous forest, with their animals included, were thought to have moved uniformly southward during ice sheet advances and uniformly northward during ice sheet retreats (see fig. 8). This classic stripe hypothesis was a vast oversimplification.

Tenaciously invoking the stripe hypothesis, many paleontologists believed that climates in the southern United States during glacial advances were much cooler in the Pleistocene than they actually were. For instance, some scientists in the early 1950s believed that the Pleistocene climate in northern Florida was sometimes similar to that of Virginia—hardly a climate that would support the fauna that we know continuously existed in Florida during the Pleistocene. Moreover, other scientists believed cold climates in the southwestern United States forced many reptiles into Mexican "refugia"—a concept that has been recently negated by the Pleistocene fossil record of reptiles.

The late 1950s marked the beginning of the end of the stripe hypothesis, for after the study of scores of North American Pleistocene vertebrate faunas, it became apparent that many of these faunas were coexisting mixtures of northern and southern species. These communities became known as disharmonius communities, as no such communities are presently known in North America.

To deal with the concept of disharmonious communities, the idea of colder communities far south of the glacial boundaries had to be abandoned. The only parsimonious explanation for the occurrence of northern and southern extralimital species existing together in the same communities was that cooler summers allowed for the presence of the northern species and that warmer winters allowed for the presence of the southern ones. This idea has currently been accepted by the majority of vertebrate paleontologists in North America and has become known as the Pleistocene climatic equability model. In other studies that grew out of the contemplation of disharmonious communities and the Pleistocene climatic equability hypothesis, it was sensibly pointed out that each species of plant or animal adjusts in its own individual way to glacial and interglacial events and that it was naive to assume that entire biomes would march in place from north to south and back again.

In many areas south of the maximum penetration of the ice sheets, including the southern Great Lakes region, we have seen that certain northern species adjusted to the situations in the newly invaded land and that other northern species obviously could not make these adjustments. This leads to a mosaic community composed of the original fauna plus the surviving extralimital invaders. This reasoning has lead to the so-called plaid hypothesis.

In summary, we now know that many Pleistocene vertebrate communities south of the glaciated areas were mosaic ones with extralimital invading species superimposed on the original fauna. It is believed that these faunas existed in a climate that was more equable than at present. We shall examine these concepts further in the light of the distribution of Great Lakes region Ice Age vertebrates.

Vertebrate Range Adjustments in Subregion I

The northern portion of the Great Lakes region, essentially the area north of the Mason-Quimby Line (see fig. 10) and north of similar latitudes in Ontario and Wisconsin, has, up to this point, yielded no authentic records of Pleistocene vertebrate fossils. Why is there a lack of Pleistocene vertebrate fossils in this part of the Great Lakes region?

Some have maintained that the lack of Pleistocene mastodonts and mammoths in the northern part of the Great Lakes area relates to the lack of plowed fields that would expose vertebrate fossils. This idea has some merit, but there are many other excavations in the north—such as those associated with housing developments, road construction, lake-access channels, pipeline ditches, and borrow pits—and enough plowed fields that one would expect that at least a few Pleistocene vertebrates would have turned up.

Alternatively, it is proposed here that the lack of Pleistocene vertebrate fossils in the northern part of the Great Lakes region is due to the fact that, during most of the Pleistocene, the area north of the Mason-Quimby Line and its equivalents was either covered with ice or was a proglacial area where few stable biotic communities were able to develop. This is not to say that populations of small vertebrates did not occur in this area from time to time or that the occasional mastodont or mammoth did not wander north of the Mason-Quimby Line or its equivalents. But it is suggested that vertebrate populations in this area were so low or so temporary that there was little chance that a significant number of vertebrate fossils could accumulate. Moreover, small lakes and basins were probably in the early stages of ecological succession during most of the Pleistocene in this area; thus there would have been relatively few quaking bogs that could have acted as traps for Pleistocene vertebrates.

South of the Mason-Quimby Line and its equivalents in Ontario and Wisconsin, stable biotic communities had developed (pollen and plant fossil evidence), salt licks and shallow saline water were available to attract mastodonts and mammoths in Michigan (geological evidence), and many mature shallow kettles and basins had become covered with mats of aquatic vegetation, making them potential traps for vertebrate animals.

Latitudinal climatic differences in Subregion I. Most sites in Subregion I represent the time period from about 12,000 to 10,000 years ago, in other words, the last gasp of the Pleistocene after the final re-treat of the ice sheet in the Late Wisconsinan. During this interval, the pollen, plant, and vertebrate fossil records suggest that southern Ontario and southern Michigan had a cold, proglacial climate. The vertebrate fossil record, however, suggests that during this period of time the climate was actually milder or at least more equable in northern Ohio and central Indiana and Illinois than at present.

Evidence for a cold climate in southern Ontario and Michigan includes the fact that Late Wisconsinan vertebrate species consist only of mastodonts and mammoths, a few other large herbivorous mammals, and a few small vertebrates, including only one unquestionable reptile. However, in northern Ohio and central Indiana and Illinois, not only are mastodonts, mammoths, and other large herbivorous mammals well known, but several diverse vertebrate faunas and a moderately large number of reptiles have been reported.

Reptiles, especially egg-laying species, appear to be very important indicators of a moderate climate, as these species need a certain minimum number of warm days in the summer for their eggs to hatch. The only unquestionable record of a Late Wisconsinan reptile in Michigan is that of a painted turtle (*Chrysemys picta*) found in association with a mastodont in southeastern Michigan (a softshell turtle may have been associated with the same deposit). The painted turtle has the most northern distribution of any turtle in modern North America, occurring north to the Great Slave Lake in Canada. The painted turtle is resistant to damage from the freezing of its tissues, and its hatchlings often overwinter in nests above the frost line. It is no wonder that it is considered to be the most cold-tolerant turtle in North America.

In the Late Wisconsinan of northwest and west-central Ohio, central Indiana, and central Illinois, the following turtle species have been recorded: snapping turtle (*Chelydra serpentina*), painted turtle (*Chrysemys picta*), Blanding's turtle (*Emydoidea blandingii*), eastern box turtle (*Terrapene carolina*), slider turtle (*Trachemys scripta*), and softshell turtle (*Apalone spinifera*). At least three species of turtles have been found at the Sheriden Pit Cave Site in northwest Ohio, along with several species of snakes. At the Clear Creek Site in central Illinois, three species of turtles have been found.

On the other hand, evidence from the 17,000 B.P. Late Pleistocene Moscow Fissure in southwestern Wisconsin offered some surprising results. This site yielded a large vertebrate fauna, with three species of snakes (including the fox snake [*Elaphe vulpina*] and the milk snake [*Lampropeltis triangulum*], both egg-laying species) from an area so close to the glacier that the ice mass possibly could have been seen on the northern horizon by any humans who might have been around. The fact that these reptiles could exist so near the glacier is somewhat of a dilemma that has not yet been explained. There certainly is much more to be learned about the distribution of reptiles in the Late Wisconsinan of the northern part of Subregion I.

The late vertebrate paleontologist C. W. Hibbard suggested that the reason that Late Wisconsinan vertebrates, other than mastodonts and mammoths, were practically unknown in the kettle bog and shallow depression sites in the Late Wisconsinan of southern Lower Michigan was that the smaller vertebrates could run over the top of quaking bogs without falling through and thus would not have been preserved as fossils. But this does not explain the fact that the Carter Bog Site in west-central Ohio and the Christensen Bog Site in central Indiana provided rather large vertebrate faunas, including many small vertebrate species such as turtles. There is also much to be learned before the lack of small vertebrate species in the Late Wisconsinan of Ontario and Michigan can be completely explained.

Southward movement of extralimital northern species into Subregion I. Several species of mammals that presently occur far north of Subregion I invaded the area a number of times in the Pleistocene.

Representing pre-Illinoian times in the Middle Pleistocene (Rancholabrean II) between about 830,000 and 730,000 B.P., the County Line Locality in Hancock County in west-central Illinois yielded two extinct boreal voles, the only extinct small rodents known in the Pleistocene of the Great Lakes region. These voles, the Cape Deceit vole (*Lasiopodomys deceitensis*) and Hibbard's tundra vole (*Microtus paroperarius*), occurred in association with two shrews, a rabbit, and a deer mouse that were identified only to the generic level.

Representing Sangamonian interglacial times, the Don Valley Brickyard Site near Toronto has yielded the remains of a barren ground (tundra) muskox (*Ovibos moschatus*, listed as *Ovibos* sp. by authors cited here). Presently, natural populations of muskoxen are restricted in their distribution to far northern Canada and the sea islands to the north, as well as to coastal areas in Greenland. The muskox is associated with a fauna that contained the woodchuck (*Marmota monax*) and white-tailed deer (*Odocoileus* sp.), mammals that are presently typical of much more southern latitudes than the muskox.

In Early Wisconsinan times, sites of the Pottery Road Formation on the Scarborough Bluffs on the north shore of Lake Ontario east of Toronto and in Toronto have yielded the remains of a grizzly bear (*Ursus arctos*) and the barren ground (tundra) muskox (*Ovibos moschatus*, listed as *Obvibos* sp. by authors cited here). The presence of muskoxen indicates cold conditions for the times, but the presence of the white-tailed deer (*Odocoileus virginianus*) indicates a lack of deep snow or a savagely cold climate for the area.

A Middle Wisconsinan site, the Woodbridge Borrow-pit in York County, southern Ontario, has yielded the remains of two animals of the Far North, the woolly mammoth (*Mammuthus primigenius*) and the grizzly bear (*Ursus arctos*). The occurrence of these two mammals together has been taken as an indication that the climate was cold in southern Ontario during this portion of the Middle Wisconsinan.

The pre-Late Wisconsinan Mill Creek Site near Port Huron in southeastern Michigan has yielded, among other vertebrate remains, four rodents that presently live far to the north of the area. The lemming (*Dicrostonyx* sp.) presently occurs in Alaska, in northern Canada and the Arctic Sea islands to the north, as well as in coastal areas of Greenland. The brown lemming (*Lemmus* sp.) presently occurs in northern Canada, Alaska, and arctic islands. The yellow-cheeked vole (*Microtus xanthognathus*) occurs in Alaska and northern Canada at present. The modern range of the northern bog lemming (*Synaptomys borealis*) is from Alaska and Canada to the extreme northwestern tip of the United States.

Extralimital northern mammals also occur in the terrestrial fauna of the Late Wisconsinan of Subregion I. The Glacial Lake Iroquois sites near Toronto, Ontario, yielded the remains of the barren ground (tundra) muskox (*Ovibos moschatus*, listed as *Ovibos* sp. by authors cited here). If remains previously identified as those of woolly mammoth (*Mammuthus primigenius*) from these sites actually represent that species, then one might expect that the Toronto region of southern Ontario remained rather cold in the Late Wisconsinan. But it has been questioned whether woolly mammoth or Jefferson mammoth (*Mammuthus jeffersonii*) or both species actually occurred at the Lake Iroquois sites.

The Sheriden Pit Cave in Wyandot County in northwestern Ohio has produced the largest vertebrate fauna in Subregion I and one of the largest vertebrate faunas in the entire Great Lakes region. This fauna is about 11,700 years old and has provided more information than any other site in the subregion. Adding to the importance of the site is the fact that an antler spear point found in stratigraphic contest with the vertebrate fossils shows that humans were present during the time of the deposition of the bones.

At the Sheriden Pit Cave Site, several extralimital northern mammals occur in a fauna of more than sixty vertebrate species. The northern species consist of insectivores, small carnivores, an artiodactyl, and several rodents, including: pygmy shrew (*Sorex hoyi*), ermine (*Mustela erminea*), American marten (*Martes americana*), fisher (*Martes pennanti*), caribou (*Rangifer tarandus*), yellow-cheeked vole (*Microtus xanthognathus*), heather vole (*Phenacomys intermedius*), northern bog lemming (*Synaptomys borealis*), and porcupine (*Erethizon dorsatum*).

Thus the vertebrate assemblage at the Sheriden Pit Cave Site perfectly fits the concept of the mosaic Pleistocene fauna discussed earlier, as the fishes, amphibians, reptiles, a bird, and many of the mammalian species are those that presently occur in northwest Ohio. This association strongly suggests that a somewhat equable climate occurred in northwest Ohio about 11,700 years ago, where not too far to the north in southern Ontario and Michigan a cold, proglacial climate apparently existed. The human that made the antler spear point certainly existed in an area where he or she was surrounded by a fauna with no modern analog.

The Carter Bog Site in west-central Ohio, where vertebrate fossils accumulated from about 12,000 to about 10,000 B.P., has also produced a large vertebrate fauna of fishes, turtles, birds, and mammals. Many of these animals currently live in the area. Nevertheless, a fisher (*Mustela pennanti*), a northern extralimital species, is present in the fauna.

The Christensen Bog Site near Greenfield in central Indiana yielded a fauna of amphibians, turtles, a bird, and mammals. The extant vertebrates of the fauna lived in central Indiana during historical times, but a caribou (*Rangifer tarandus*) represents an extralimital northern species. The northern extralimital caribou was also found at the Kolarik Mastodont Site in northwestern Indiana in association with the deer mouse (*Peromyscus maniculatus*) and the meadow vole (*Microtus pennsylvanicus*), species that currently occur in the area.

The Moscow Fissure Site in extreme southwestern Wisconsin was deposited about 17,000 B.P. very near the edge of the continental ice sheet. In fact, if humans had been around the vicinity at the time, they possibly could have seen the glacier looming up as an ice mountain in the northeastern horizon. The Moscow Fissure yielded a very numerous small mammal fauna, several of which represent extralimital northern species, including: arctic shrew (*Sorex arcticus*), northern pocket gopher (*Thomomys talpoides*), southern red-backed vole (*Clethrionomys gapperi*), yellow-cheeked vole (*Microtus xanthognathus*), heather vole (*Phenacomys intermedius*), collared lemming (*Dicrostonyx torquatus*), and northern bog lemming (*Synaptomys borealis*). These northern species occurred with mammals that presently may be found in the area, as well as with two egg-laying snakes, the fox snake (*Elaphe vulpina*) and the milk snake (*Lampropeltis triangulum*). This is yet another example of a mosaic fauna in Subregion I but is of special interest because of the proximity of the site to the ice margin.

In summary, there is considerable evidence that several extralimital northern species entered Subregion I from Illinoian through Late Wisconsinan times. These northern animals were probably forced south by the advances of the continental ice sheet and coexisted with vertebrate species that are presently found in the fossil area. These kinds of faunas with no modern analogs are called mosaic faunas, and it has been suggested that these faunas existed in climates in the region that were more equable than they are presently.

Recolonization of formerly glaciated areas. We have seen that extralimital northern species driven southward by the advancing ice sheet coexisted with more southern forms in Subregion I throughout much of the Pleistocene. The other side of the coin involves the fact that several times vertebrate species reoccupied formerly glaciated areas during times of glacial retreat.

Recently, a model was developed to suggest patterns of amphibian and reptile reoccupation of Late Wisconsinan postglacial Michigan based on the interpretation of geological data, paleobotanical and paleovertebrate assemblages, and ecological tolerances of the modern herpetofauna (amphibian and reptile fauna). Three categories of invading species were proposed: primary invaders, whose ecological tolerances included tundralike conditions; secondary invaders adapted for coniferous associations; and tertiary invaders, rather narrowly adapted to exist in broadleaf forest areas. Actually, the model may be applied to birds and mammals as well as to amphibians and reptiles.

Here we shall consider the Late Wisconsinan vertebrate reoccupation of Michigan based on the fine pollen record that outlines the redevelopment of vegetational communities in the state during the last glacial retreat.

The 14,800 to 12,500 B.P. Interval Tundralike Conditions (Primary Invaders)

Pollen records from southern Lower Michigan indicate that at the beginning of the final glacial retreat in Michigan, which began about 14,800 B.P., the vegetation was tundra and that marshes and muskegs characterized the lowlands. More land for plants and animals became available by 13,000 B.P., as it appears that about half of Lower Michigan became deglaciated. At this time, nearly all of the drained landscape supported tundralike vegetation or scattered stands of pioneer trees such as juniper, aspen, ash, and spruce, as well as shrubs such as willow, silverberry, and crowberry.

It is possible that Michigan was a virtual "herpetological desert" during many of these years. Only one Michigan herpetological species, the American toad (*Bufo americanus*) (see fig. 40), has been recorded from Michigan during this interval.

Without a fossil record to substantiate its occurrence, it has been suggested that the wood frog (*Rana sylvatica*) was possibly the first herpetological invader of the area after the last gasp of the continental ice sheet in the Wisconsinan. This frog presently penetrates tundra areas in North America, and it has recently been demonstrated that the wood frog in Ottawa has the ability to freeze solid every winter during its hibernation period and to thaw out successfully each spring to resume its breeding activities. A mating pair of wood frogs was frozen solid together in a laboratory in Ottawa and, when thawed out, continued the mating process.

Unfortunately, we do not have a mammalian Pleistocene fossil record for Michigan for this time period, but we can certainly suggest that tundra-tolerant species presently living in Michigan—such as the masked shrew (*Sorex cinereus*), black bear (*Ursus americanus*), mink (*Mustela vison*), northern river otter (*Lutra canadensis*), coyote (*Canis latrans*), gray wolf (*Canis lupus*), red squirrel (*Tamiasciurus hudsonicus*), American beaver (*Castor canadensis*), meadow vole (*Microtus pennsylvanicus*), muskrat (*Ondatra zibethicus*), common porcupine (*Erethizon dorsatum*), snowshoe hare (*Lepus americanus*), and moose (*Alces alces*)—were likely primary invaders.

The 12,500 to 10,000 B.P. Interval Coniferous Forests (Secondary Invaders)

The pollen record of the 12,500 to 10,000 B.P. interval shows that a boreal forest dominated by spruce trees had developed in southern Michigan and that there were areas of open woodland and parkland. At least fourteen species of vertebrates, including six extinct mammals, are known from the Pleistocene of southern Lower Michigan during this time period.

The only accepted amphibian records of this time period are based on a frog scapula and an ilium from the Shelton Mastodont Site in southeastern Michigan that spanned the time between from about 12,300 to about 11,700 B.P. The scapula and the ilium were identified, respectively, as green frog (*Rana clamitans*) and bullfrog (*Rana catesbeiana*), but some doubt has been expressed about the identification of the bullfrog. Both of these animals may presently be found in coniferous forest areas in Michigan.

Modern Michigan amphibian species unrecorded from the Pleistocene but presently found within coniferous forest regions were probable secondary invaders. These are: blue-spotted salamander (*Ambystoma laterale*), spotted salamander (*Ambystoma maculatum*), tiger salamander (*Ambystoma tigrinum*), four-toed salamander (*Hemidactylium scutatum*), mudpuppy (*Necturus maculosus*), eastern newt (*Notophthalmus viridescens*), eastern red-backed salamander (*Plethodon cinereus*), gray treefrog (*Hyla versicolor*), spring peeper (*Pseudacris crucifer*), striped chorus frog (*Pseudacris triseriata*), pickerel frog (*Rana palustris*), and mink frog (*Rana septentrionalis*).

The only unquestioned reptile fossils of this interval are two individuals of the painted turtle (*Chrysemys picta*) found in association with a mastodont in southeastern Michigan (a softshelled turtle was possibly associated with this fauna). The painted turtle presently has the most northern distribution and is probably the most cold-tolerant turtle in North America. Late-hatching baby painted turtles characteristically freeze solid in their nests in Michigan. When spring comes, they thaw out and find their way to water.

Other reptile species that presently occur within coniferous forest areas in Michigan but that are not recorded in the Pleistocene of the state include: snapping turtle (*Chelydra serpentina*), wood turtle (*Clemmys insculpta*), Blanding's turtle (*Emydoidea blandingii*), five-lined skink (*Eumeces fasciatus*), ring-necked snake

(*Diadophis punctatus*), western fox snake subspecies (*Elaphe vulpina*), northern water snake (*Nerodia sipedon*), smooth green snake (*Opheodrys vernalis*), red-bellied snake (*Storeria occipitomaculata*), and common garter snake (*Thamnophis sirtalis*).

Mammals that are known to have recolonized southern Lower Michigan during this interval based on their occurrence in the Michigan fossil record include: extinct giant beaver (*Castoroides ohioensis*), meadow vole (*Microtus pennsylvanicus*), muskrat (*Ondatra zibethicus*), black bear (*Ursus americanus*), extinct flat-headed peccary (*Platygonus compressus*), wapiti (elk) (*Cervus elaphus*), extinct Scott's moose (stag moose) (*Cervalces scotti*), white-tailed deer (*Odocoileus virginianus*), caribou (*Rangifer tarandus*), extinct woodland muskox (*Bootherium bombifrons*), American mastodont (*Mammut americanum*), and Jefferson mammoth (*Mammuthus jeffersonii*).

Although they are not represented in the fossil record, most of the remainder of the modern Michigan mammalian fauna probably invaded southern Lower Michigan during this time period, possibly near the end of the interval. These species probably include such familiar mammals as the eastern mole (*Scalopus aquaticus*), northern short-tailed shrew (*Blarina brevicauda*), common raccoon (*Procyon lotor*), American badger (*Taxidea taxus*), striped skunk (*Mephitis mephitis*), eastern Gray fox (*Urocyon cinereoargenteus*), bobcat (*Lynx rufus*), woodchuck (*Marmota monax*), gray squirrel (*Sciurus carolinensis*), eastern fox squirrel (*Sciurus niger*), whitefooted mouse (*Peromyscus leucopus*), and eastern cottontail (*Sylvilagus floridanus*).

About 10,000 B.P. Mixed Coniferous/Broadleaf Forests (Secondary Invaders)

It appears that by about 10,000 B.P. most of the modern mammalian fauna would have reinvaded Michigan, for by this time the vegetation of southern Lower Michigan had become more diverse, with mixed coniferous/broadleaf forests of white and red pine, yellow and paper birch, aspen, oak, red and white elm, and ironwood/blue beech. This is similar to the vegetation that presently occurs in the northern half of Lower Michigan. Although additional mammalian secondary invaders might have been minimal, a few species of amphibians and quite a few reptiles that presently reach the northern limits of their distribution in this habitat in Michigan might have first been able to establish themselves in Michigan by about 10,000 B.P.

These animals would have included: northern cricket frog (*Acris crepitans*), Fowler's toad (*Bufo fowleri*), Cope's gray treefrog (*Hyla chrysoscelis*), spotted turtle (*Clemmys guttata*), common map turtle (*Graptemys geographica*), common box turtle (*Terrapene carolina*), spiny softshell turtle (*Apalone spinifera*), racer snake (*Coluber constrictor*), eastern fox snake subspecies (*Elaphe vulpina*), eastern hog-nosed snake (*Heterodon platirhinos*), milk snake (*Lampropeltis triangulum*), queen snake (*Regina septemvittata*), massasauga rattlesnake (*Sistrurus catenatus*), Dekay's brown snake (*Storeria dekayi*), Butler's garter snake (*Thamnophis butleri*), and eastern ribbon snake (*Thamnophis sauritus*).

The 9,900 to 9,000 B.P. Interval Broadleaf Forest (Tertiary Invaders)

More amphibian and reptile reoccupations as well as the last few mammalian ones probably occurred in Michigan by about 9,900 B.P., as glacial ice had left the state except for the northern edge of the Upper Peninsula. A vegetation of mixed-hardwood forest dominated by birch, ash, ironwood/blue beech, elm, and oak, with lesser amounts of hickory, walnut, butternut, basswoods and smaller amounts of white pine, existed in southern Lower Michigan.

All of the amphibians and reptiles considered to have been probable tertiary invaders are confined to southern Lower Michigan at present, where most of them exist in isolated or peripheral habitats. These animals include: marbled salamander (*Ambystoma opacum*), small-mouthed salamander (*Ambystoma texanum*), lesser siren (*Siren intermedia*), common musk turtle (*Sternotherus odoratus*), red-eared slider turtle (*Trachemys scripta*), six-lined racerunner (*Cnemidophorus sexlineatus*), Kirtland's water snake (*Clonophis kirtlandii*), and plain-bellied water snake (*Nerodia erythrogaster*).

Among the mammals, tertiary invaders were probably the least shrew (*Cryptotis parva*), Indiana bat (*Myotis sodalis*), evening bat (*Nycticeius humeralis*), and prairie vole (*Microtus ochrogaster*). Except for the prairie vole, which has extended its range northward in modern times, all forms are restricted to very southern portions of Lower Michigan today.

Invasion of marine vertebrates into the interior of Ontario. When the Late Wisconsinan ice sheet retreated north of the St. Lawrence Valley, the Atlantic Ocean flooded the low areas in the valley and produced an inland marine body of water called the Champlain Sea. At its greatest extent at about 12,000

B.P., the Champlain Sea covered over 20,000 square miles in Quebec and Ontario (see fig. 154). Based on evidence from marine mollusk shells, it has been suggested that the early Champlain Sea environment was similar to that of the present-day Gulf of St. Lawrence.

Champlain Sea marine vertebrates that invaded Ontario between about 11,400 and 10,300 B.P. include: lake trout (*Salvelinus namaycush*), capelin (*Mallotus villosus*), rainbow smelt (*Osmorus mordax*), Atlantic tomcod (*Microgadus tomcod*), three-spine stickleback (*Gasterosteus aculeatus*), hook-eared sculpin (*Artediellus uncinatus*), lumpfish (*Cyclopterus lumpus*), bearded seal (*Erignathus barbatus*), harp seal (*Phoca groenlandica*), ringed seal (*Phoca hispida*), white whale (*Delphinapterus leucas*), humpback whale (*Megaptera novaeangliae*), and bowhead whale (*Balaena mysticetus*). Several of the marine mammals are presently associated with fast ice.

As the continental glacier continued to retreat northward between 12,000 and 10,000 B.P., the climate became warmer and the Champlain Sea started to contract and get less salty. Finally, the sea reached a freshwater lake stage called Lampsilis Lake, and the saltwater fishes and marine mammals left the area.

Vertebrate Range Adjustments in Subregion II

The main vertebrate range adjustments in Subregion II involve the southward movement of extralimital northern species into the region during the Pleistocene. Moreover, a few species typical of more southern regions moved into Subregion II in Sangamonian times.

Southward movement of northern species into Subregion II. At the Middle Wisconsinan Alton Mammoth Site near the Ohio River in southern Indiana, the southern red-backed vole (*Clethrionomys gapperi*) and the heather vole (*Phenacomys intermedius*) are extralimital northern forms that occur with the meadow vole (*Microtus pennsylvanicus*), which presently occurs in the area. The nearest the southern red-backed vole gets to the site at present is in central Lower Michigan and southeastern Wisconsin. The heather vole ranges south at present only to the northern borders of Lake Superior and Lake Huron.

The Middle to Late Wisconsinan stratum of the Megenity Peccary Cave Site in Crawford County in southern Indiana has produced a number of extralimital northern mammalian species. These include the arctic shrew (*Sorex arcticus*), southern red-backed vole (*Clethrionomys gapperi*), yellow-cheeked vole (*Micro-*

tus xanthognathus), and northern bog lemming (*Synaptomys borealis*). At present, the nearest the arctic shrew gets to the site is in southeastern Wisconsin; the southern limits of the yellow-cheeked vole are in central Manitoba, Saskatchewan, and Alberta; and the southern limits of the northern bog lemming are in extreme northern Minnesota. The Megenity Peccary Cave fauna is a truly mosaic Pleistocene fauna, as the northern extralimital species occur with three shrews, four carnivores, and two mice that occur presently or at least were historically in the area.

The Late Wisconsinan King Leo Pit Cave in Harrison County, also in southern Indiana, has produced a mosaic fauna. Extralimital northern invaders include the fisher (*Martes pennanti*), southern red-backed vole (*Clethrionomys gapperi*), and heather vole (*Phenacomys intermedius*). The nearest the fisher presently occurs to the site is in central Wisconsin (although the fisher has recently been successfully reintroduced into northern Michigan). At least eight insectivores and bats and seven rodents identified from the King Leo Pit Cave fauna presently occur in Harrison County, Indiana.

Still another Late Wisconsinan site in Indiana, the Prairie Creek D Site in southwest Indiana, has yielded a mosaic Pleistocene vertebrate fauna. Extralimital northern forms include Blanding's turtle (*Emydoidea blandingii*), star-nosed mole (*Condylura cristata*), fisher (*Martes pennanti*), red squirrel (*Tamiasciurus hudsonicus*), southern red-backed vole (*Clethrionomys gapperi*), and yellow-cheeked vole (*Microtus xanthognathus*). At present, Blanding's turtle mainly occupies northern Indiana. The star-nosed mole now occurs nearest the site in northeastern Indiana and the red squirrel in central Indiana. But the rest of the amphibians and reptiles as well as five insectivores, two carnivores, and six rodents presently occur in the Prairie Creek area.

In summary, several southern Indiana sites ranging from Middle Wisconsinan through Late Wisconsinan times consisted of mosaic faunas composed of large, extinct mammals; extralimital northern forms; and those that just stayed put.

Sangamonian sites with invading southern species. It has been suggested that the Sangamonian climate in southern Indiana and Illinois was much warmer and drier than it is today. This is borne out by evidence of the identification of species typical of more southern areas from three Sangamonian sites.

In the Sangamonian Harrodsburg Crevice Site near Bloomington in south-central Indiana, the jaguar

(*Panthera onca*) and the sabertooth (*Smilodon fatalis*), forms typical of more southern areas, occur along with the grassland forms, the plains pocket gopher (*Geomys bursarius*) and the prairie vole (*Microtus ochrogaster*).

Moreover, giant land tortoise remains (*Hesperotestudo* sp.) were found in a Sangamonian stratum in the Megenity Peccary Cave in Crawford County in southern Indiana. Giant land tortoises of the genus *Hesperotestudo* are important climatic indicators in that they suggest temperatures consistently above freezing during the winter. Also, giant land tortoises are considered to prefer a dry climate with ample grasslands.

The Hopwood Farm Sangamonian site near Fillmore in south-central Illinois also yielded the remains of a giant land tortoise (*Hesperotestudo crassiscutata*), indicating a very warm, dry climate in the Sangamonian of that state.

Other invading southern species. Several sites in southern Indiana and Ohio without radiocarbon dates but probably representing postglacial Wisconsinan times have produced species typical of more southern areas. These include records of horses, tapirs, and camels.

Summary of Vertebrate Range Adjustments in the Great Lakes Region

We have seen that vertebrates in the glaciated parts of the Great Lakes region were displaced southward by the advancing ice sheet but were able to reoccupy the areas from which they were displaced when the ice retreated. This must have happened several times. Moreover, we know that, in Sangamonian times, there was movement of typically southern animals such as the giant land tortoise into southern Illinois, and that giant land tortoises, jaguars, and sabertooths moved into southern Indiana. Horses, tapirs, and camels reached the southern part of Subregion II several times in the Late Pleistocene. Together, these animals indicate a much warmer, drier climate than presently occurs in these areas.

At the same time, there is compelling evidence that a large number of vertebrate species south of the glaciated areas merely stayed put. These include all of the fishes and amphibians, most of the reptiles, and many mammalian species identified from the Pleistocene of the region. This negates the formerly held idea that advancing and retreating ice sheets altered the North American climate in such a way that communities were uniformly displaced southward during glacial advances and uniformly advanced northward during glacial retreats.

Finally, vertebrate fossils in Ontario tell the story of the Champlain Sea, which filled in the low areas of the St. Lawrence River Valley in the Late Wisconsinan and brought with it marine fishes, seals, and whales. These saltwater fishes and marine mammals departed when the sea turned into a freshwater lake and finally disappeared.

THE GREAT PLEISTOCENE EXTINCTION

Several devastating episodes of extinction have greatly changed the world from time to time, but none have affected humankind as much as the one that occurred during the Pleistocene Ice Age. Coupled with the mass extinction of most of the large herbivorous mammals, human expansion toward the end of the Pleistocene sparked the beginning of events that led to the world as we know it today. In our modern, human-dominated world, edible domestic animals such as cattle, sheep, goats, and pigs, as well as beasts of burden such as llamas, horses, mules, oxen, and elephants, have taken the place of the large, extinct herbivores. Moreover, grass cereals such as corn, wheat, oats, barely, and rice have largely replaced natural grasslands.

The Pleistocene extinction occurred throughout most of the world, although it affected some parts of the world more than others. For example, North America took much more of a "hit" than Africa. The Ice Age extinction mainly affected larger mammals and to a lesser extent birds. This extinction knocked out the majority of large herbivorous mammals, most of the large carnivores that preyed upon them, and even the mammalian and avian scavengers that cleaned up the carcasses.

For instance, at the famous late Pleistocene Rancho La Brea tar pits in Los Angeles, the large mammalian herbivores and the carnivorous dire wolves, sabertooths, and American lions that fed upon them became extinct. Extremely large condors that fed on the carcasses of the large herbivores, as well as several kinds of dung beetles that utilized the droppings of herbivores, also died out.

On the other hand, most of the smaller mammals in North America, at least those that existed in the Late Wisconsinan, were spared. Fishes and amphibians mainly survived the Ice Age unscathed, and reptiles, other than giant tortoises and some island species, were also largely spared.

Several striking kinds of large vertebrates became extinct in North America near the end of the Ice Age. Giant land tortoises ranged from Pennsylvania and south-central Illinois to Florida and west to Texas and possibly as far as California. Giant condors occurred in the western states. Among several sloth species, the giant ground sloth stood over 15 feet high and ranged far and wide in North America. The armadillo-like glyptodonts, with shells over 6 feet long, occurred mainly in southern areas. Smaller armadillos were also important south of the glaciated regions. Packs of dire wolves roamed throughout much of North America, and short-faced bears were top predators that existed in many areas.

An American lion that was larger and had longer legs than the African lion existed in the western United States. The sabertooth was thought to be an important predator on young mammoths and mastodonts, and the jaguar ranged far north of its present range.

Various species of extinct horses, some of them thought to be striped like zebras, occurred mainly in areas south of the glacial margins. Several kinds of extinct tapirs extended far north of the present range of tapirs in the Western Hemisphere. Moreover, several kinds of extinct peccaries were important omnivores, and camels also existed. One very large rodent, the capybara, and one giant rodent, the giant beaver, were important denizens of Pleistocene wetlands. Finally, the American mastodont and several kinds of mammoths ranged far and wide in North America.

The Pleistocene vertebrate fauna of the world was the product of millions of years of evolution during the Tertiary period. The most important terrestrial communities of the Tertiary were similar to the dinosaur communities of the Jurassic and Cretaceous in many ways. The most important animals were the megaherbivores that ate massive quantities of vegetation. The large amount of dung that the megaherbivores returned to the environment was an important fertilizer.

As pointed out in a previous section, modern African elephants and rhino are megaherbivores, and when these animals are lost to the community, a domino effect leads to the disruption and possible extinction of much of the community. In North America, the mastodonts and mammoths were the megaherbivores of the Ice Age and the most important vertebrates in biotic communities from Michigan to Florida and west to California. The extinction of all mastodonts and mammoths in North America at the end of the Pleistocene may have paved the way for the extinction of the other Ice Age vertebrates.

Many magazine articles, scientific papers, books, and international meetings of scientists have dealt partially or entirely with the "why" of the Pleistocene extinction. Over the years, several hypotheses have been formulated to explain this extinction. These hypotheses usually fit into one or the other of two broad explanations: overkill by early humans was responsible for the extinction or environmental changes were the cause. Here I will include some of the more interesting hypotheses as a prelude to a discussion of Pleistocene extinction in the Great Lakes region.

The Pleistocene overkill hypothesis. My spouse, an archaeologist, once said to me with tongue in cheek, "The trouble with a pet hypothesis is that you have to feed it," and I remember thinking that it would certainly be a mistake to let a worthy pet starve to death. The Pleistocene overkill hypothesis has been championed for decades by Paul Martin of the University of Arizona and, although challenged many times, remains one of the two viable broad explanations for the Pleistocene mammalian extinction

In this hypothesis, it is pointed out that Pleistocene vertebrates survived several periods of climatic change during the Pleistocene, yet the major extinction came at the end of the epoch. Martin's hypothesis suggests that the extinction follows the chronology of the spread of humans and their development as hunters of big game.

He reasons that when humans arrived in North America by the way of the land bridge that connected Siberia and Alaska, they came upon herds of animals that were quite vulnerable to human predation because they had never before seen or been confronted by humans. Moreover, the invading humans had an unfair advantage in that their hunting skills had been previously sharpened by hunting large mammals that were wary of them. Thus humans are believed to have spread rapidly through North America to the tip of South America, killing off the large herbivorous mammals as they went, in what has been termed a blitzkrieg.

Opponents to this hypothesis point out that human kill sites are known from only a relatively few places and that human populations were far too small to have caused such a mass extinction. One must admit, however, that it is a most intriguing hypothesis, for otherwise it is hard to explain why so few small vertebrates became extinct. If one looks at the reptiles of the Pleistocene, for instance, the only significant extinction that took place was that of the giant land tortoises and some of their smaller relatives, and these reptiles would have been very vulnerable to human predation. Actually, a huge Pleistocene tortoise that had been putatively cooked over a fire and had either been killed by a wooden spear or manipulated over the fire by the spear was recently found in a sinkhole in southwestern Florida.

The Pleistocene climatic equability model. The Pleistocene climatic equability model is built on the concept that the climate in North America south of the glaciers must have been more equable than at present in that it had cooler summers and milder winters. The parsimonious assumption is that cool summers explain the presence of the extralimital northern animals in North American Pleistocene sites and that mild winters explain the presence of southern ones. It is further reasoned that our modern climate, with colder winters and hotter summers, must have originated about 10,000 years ago and that this somehow had such a negative affect on the large mammalian herbivores that it initiated a mass extinction. A second hypothesis associated with the Pleistocene climatic equability hypothesis tries to explain how the changing climates could affect the megaherbivore community. This has been called the out-of-step mating hypothesis.

The out-of-step mating hypothesis. The out-of-step mating hypothesis assumes that the Ice Age equable climate changed to the modern one of extremes about 10,000 years ago. It matches the extinction of many large mammals with the onset of severe modern climates in the following manner. Many large ungulates (large herbivorous mammals) have gestation periods that are precisely in step with existing climatic patterns. In other words, breeding in the fall is timed so that the young are born in the spring when new grasses and shrubs are available.

Let us suppose that we have an equable climate, say in northeast Nebraska, and that in-step mating allows the young to be born in early March. Since plenty of moisture would be available at this time, ample grasses and low herbs would be available. But what if the modern climate of northeastern Nebraska arrived suddenly? The young would be born when there was a covering of snow and ice on the ground and probably would die. Moreover, severe feeding stress would occur

in both parents, already weakened by reproductive stress. It seems unlikely that genetic changes could occur fast enough for these mammals to adapt to the new climate.

Drought hypothesis. Still another hypothesis suggests that a general drought and lowering of the water table at the end of the Pleistocene might have been at least partially responsible for the great extinction. This hypothesis takes into account the fact that small animals are capable of existing and reproducing in much smaller bits of land than are the large ones and that when conditions such a major drought eliminate large parts of a particular habitat, the small animals are likely to survive. In other words, it takes many acres to support a single megaherbivore but only a few square feet to support a mouse or a shrew. A lowered water table that would have dried up the saline water supply for mastodonts and mammoths in the Late Wisconsinan of southern Lower Michigan could have had a disastrous effect on migrating herds of animals searching for salt to maintain the critical sodium-potassium balance in their bodies.

Coevolutionary disequilibrium model. Coevolution is the concept that the evolution of one group of organisms affects the evolution of another group. Thus it has been suggested that when the communities are in a coevolutionary equilibrium, they evolve together in a harmonious way and that relatively few extinctions occur. But when these communities begin to evolve in a disharmonious way (for instance, the structure of the plant community may have changed to the detriment of the mastodonts and mammoths at the end of the Pleistocene), the disequilibrium might cause a massive extinction.

Vertebrate Extinction Percentages in the Pleistocene of the Great Lakes Region

As far as I am aware, regional accounts of vertebrate Pleistocene extinction percentages, such as the ones that follow, have not been done before in North America. Thus in the future, it will be interesting to see how extinction percentages in other North American regions compare with the Great Lakes region. It becomes strikingly apparent that in the Great Lakes region, extinction devastated the large mammals and hardly affected the other Ice Age vertebrates (table 8).

Extinction in fishes. Pleistocene fish species from the Great Lakes region reflect the marine environment of the

Champlain Sea in the Late Wisconsinan of Ontario as well as species from freshwater situations. Of the at least twenty-nine species of fishes that have been recorded, none are extinct. This is not an unusual situation in other regions in North America in the Pleistocene.

Extinction in amphibians. Pleistocene amphibian species from the Great Lakes region reflect mainly marsh, damp woodland, and cave habitats. Only nine species were recorded, and the low number is attributed to the fragile bones of amphibians, lack of screening of sediments in many sites, and the fact that amphibians at some Great Lakes sites have not yet been identified Of the at least nine species of amphibians that have been recorded, none are extinct. This is

also not an unusual situation in other regions in the Pleistocene of North America.

Extinction in reptiles. Pleistocene reptiles species from the Great Lakes region reflect a large number of reptilian habitats from dry grasslands to damp woodlands, marshes, ponds, and lakes. Of the twenty reptile species recorded, only one, the giant land tortoise (*Hesperotestudo crassicutata*) from a Sangamonian site in south-central Illinois, and *Hesperotestudo* sp. from a probable Sangamonian layer in a cave in southern Indiana represent an extinct form. This is the general situation in most of continental North America, where the most common extinct reptiles are the giant land tortoises and some of their smaller relatives.

Extinction in birds. Only nine species of birds, none of which is extinct, have been recorded from the Great Lakes region. The small number of birds recovered probably reflects the fragility of their bones and the fact that arboreal species are often absent from fossil sites. A few extinct genera and a number of extinct species of Pleistocene birds have been described from Pleistocene sites in California and Florida, but the validity of some of these fossil species has recently been questioned.

Extinction in large nonmarine mammals. Of the thirty large land mammals that have been recorded from the Pleistocene of the Great lakes region, nineteen are extinct (63.33 percent). This extinction percentage is striking and speaks to the fact that the Pleistocene extinction mainly affected the large mammals.

Extinction in small land mammals. Sixty small land mammals were identified from a very wide variety of habitats in the Great Lakes region, and although some of them were forced into major range adjustments in the Pleistocene, only two (3.33 percent) became extinct. This is in remarkable contrast to the 63.33 percent extinction rate of large nonmarine mammals of the Great Lakes region.

Extinction in large marine mammals. Six marine mammals, three seals and three whales, have been recorded from the Late Wisconsinan Champlain Sea of Ontario, and none of them are extinct species. The fact that large marine mammals, including the gigantic whales, survived the Pleistocene extinction episode unscathed is a fascinating subject that obviously reflects on the causes of the extinction of the large nonmarine land mammals and that has thus far not been addressed in great detail.

TABLE 8. Vertebrate Extinction in the Pleistocene of the Great Lakes Region.

Number of all Vertebrate Species	– 163
Number of Extinct Vertebrate Species	– 22
% of Extinct Vertebrate Species	– 13.50%
Number of Fish Species	– 29
Number of Extinct Fish Species	– 0
% of Extinct Fish Species	– 0.00%
Number of Amphibian Species	– 9
Number of Extinct Amphibian Species	– 0
% of Extinct Amphibian Species	– 0.00%
Number of Reptile Species	– 20
Number of Extinct Reptile Species	– 1
% of Extinct Reptile Species	– 5.00%
Number of Bird Species	– 9
Number of Extinct Bird Species	– 0
% of Extinct Bird Species	– 0.00%
Number of Large Land Mammal Species	– 30
Number of Extinct Large Land Mammal Species	– 19
% of Extinct Large Land Mammal Species	– 63.33%
Number of Small Land Mammal Species Recorded	– 60
Number of Extinct Small Land Mammal Species	– 2
% of Extinct Small Land Mammal Species	– 3.33%
Number of Marine Mammal Species Recorded	– 6
Number of Extinct Marine Mammal Species	– 0
% of Extinct Marine Mammal Species	– 0.0%

The Why of the Extinction of the Large Land Mammals with Reference to the Great Lakes Region

This section will deal with the author's view of the why of the extinction of the large land mammals in the Great Lakes region with reference to several topics that are especially relevant to the area.

Ectotherms versus endotherms. Ectotherms ("cold-blooded" animals) are generally unable to produce enough internal heat to maintain a constant body temperature and, aside from behavioral adaptations (such as basking when they are cold or retreating beneath the ground when they begin to overheat), are more or less at the mercy of the environment. Fishes, amphibians and reptiles are ectothermic vertebrates. Endotherms ("warm-blooded" animals) are able to maintain a more or less constantly warm body temperature by internal means. Birds and mammals are endothermic vertebrates. Several suggestions have been put forth why ectothermic amphibians and reptiles might have had advantages over the endotherms in times of Ice Age stress.

Modern amphibians and reptiles in the Great Lakes region are well adapted to survive the area's long, cold winters by the process of hibernation (more precisely called brumation). Thus, during the most stressful time of the year, when most mammals and nonmigratory birds are struggling to find food or to escape predators, amphibians and reptiles are below the ground, awaiting the return of more favorable temperatures.

Some Great Lakes amphibians and reptiles have the remarkable ability to freeze solid without damage to their body tissues. These animals become frozen in the winter under a few inches of soil and thaw out again in the spring to resume their lives. This is accomplished by converting substances in the liver to other substances that enter individual cells to keep them from breaking apart during the freezing process. Neither birds nor mammals are able to freeze solid without dying.

Great Lakes species that are able to freeze and thaw successfully on an annual basis include the gray tree frog (*Hyla versicolor*), chorus frog (*Pseudacris triseriata*), wood frog (*Rana sylvatica*), painted turtle (*Chrysemys picta*), and box turtle (*Terrapene carolina*). Of great interest here is that although relatively few amphibians and reptiles have been recorded from the Pleistocene of the Great Lakes region, all of the above species, with the exception of the gray tree frog, have

been recorded from the Pleistocene of the area. One might imagine amphibians and reptiles snuggled comfortably in their places of hibernation or frozen in "deep-freeze sites" while the large mammals are dying of starvation in the savage proglacial winters.

Another very likely possibility is that few North American amphibians and reptiles were directly dependant on the large endothermic mammals that became extinct during the Pleistocene, either as predators, scavengers, or commensals (one of a pair of interacting organisms gains benefit from a common food supply without harming the other). At the late Pleistocene Rancho La Brea Site in Los Angeles, the scavenging or commensal birds and dung beetles that were dependant upon the large herbivorous mammals became extinct along with the mammals, but all the amphibian and reptile species survived into modern times, with the possible exception of a single species of tortoise. On the other hand, in Australia, the giant monitor lizard (*Megalania*) and the large constricting snake (*Wonambi*) were top predators that became extinct in the Pleistocene because many of the large marsupial herbivores that they preyed upon died out. Almost all of the amphibian and reptile species in the Great Lakes region in the Pleistocene and at present are small predators that would not be as affected by the extinction of large land mammals.

Parenting in birds and mammals is a part of the reproductive stress syndrome, and it would have been a special drain on the energy budget of these animals during Pleistocene climatic and vegetational changes. Parenting is lacking in all of the Great Lakes region amphibians and reptiles and is a stress that they would not have had to contend with during the Pleistocene.

Finally, the generally small Great Lakes region Pleistocene amphibians and reptiles would have been much less a sustaining food resource for Paleo-Indians than the larger herbivorous mammals. In other words, why should small frogs be hunted when several families could be fed for weeks on a single mammoth kill?

Small mammals versus large mammals. We have seen that 63.33 percent of the thirty large mammals identified from the Great Lakes region became extinct by the end of the Pleistocene, whereas only 3.33 percent of the sixty-one small mammals identified from the area did not survive into modern times.

Small mammalian predators (for example, shrews, bats, weasels, and foxes) probably survived because they were not dependant upon large herbivores for their food supply, instead eating insects, rodents, small birds, and rabbits. The surplus of vegetation left over after the extinction of the large mammalian herbivores might have greatly benefited the small herbivorous rodents, which in turn would have enhanced the lives of small mammalian predators.

Another reason for the survival of the small mammals obviously deals with the fact that they do not need as much space to live in. Therefore the fragmentation and shrinkage of habitats that must have taken place in the Pleistocene would have affected the large herbivores, especially the megaherbivores, much more than the small species.

Finally, everyone knows about the reproductive potential of small animals such as mice and rabbits compared to the large herbivores, which often have only a single offspring per year. A mouse may have a new litter every few weeks, but a mastodont or mammoth probably had about a nine-month gestation period and probably produced one offspring at a time. Differences in reproductive potential have favored small vertebrates in times of environmental change down through the ages and most likely did so during the stresses of the Pleistocene.

Salt and Megaherbivores. The suggestion that the widespread occurrence of salt licks and shallow saline water attracted mastodonts and mammoths to southern Lower Michigan in great numbers in the Pleistocene has previously been put forth. It has also been suggested that mastodonts and mammoths might have engaged in yearly migrations to Michigan from other parts of the Great Lakes region to obtain salt, which was an essential element in their diets. One of the hypotheses as to the extinction of the large mammals in the Pleistocene suggests that drought conditions caused a lowering of the water table, which created great stress in the larger mammals. In the Pleistocene of Michigan, a lowering of the water table would have eliminated a large area of available surficial saline water. It would seem that this in itself might have caused a local or perhaps even a regional extinction in these vulnerable beasts.

Out-of-step mating hypothesis. The out-of-step mating hypothesis suggests that the sudden replacement of Pleistocene equable climates by modern ones, with hot summers and cold winters, would devastate large herbivores programmed to bear their young in early spring. In other words, the birthing process would likely occur in cold weather with snow on the ground rather than in warm spring weather with new green grass available for feeding.

All indications are that the northern parts of Subregion I had cold climates throughout the Pleistocene. Thus the out-of-step mating hypothesis would not necessarily apply to southern Ontario, southern Lower Michigan, and southern Wisconsin. On the other hand, the mosaic vertebrate faunas in Ohio and Indiana suggest more equable climates existed in these areas. It is quite possible that the large Pleistocene herbivores there might have been highly affected by an abrupt change to cold weather during the birthing season.

Coevolutionary disequilibrium in the Pleistocene of the Great Lakes region. If mastodonts coevolved with specific vegetational associations (that is a certain mixture of plants required in the their diet), any changes in the coevolutionary balance might be harmful to mastodonts by increasing competition with other species.

It has recently been shown on the basis of pollen records from basins in Ohio and Indiana that the period between 14,000 and 9,000 B.P. was characterized by major vegetational changes. Late glacial warming led to the replacement of spruce forests and woodlands by mixed coniferous and deciduous forest by around 14,000 B.P. Vegetation stabilized until 11,000 B.P., but between 11,000 and 10,000 B.P. a series of quick, intense changes led to the development of complex vegetation patterns unlike any that may be presently found in the area. By 10,000 B.P., modern oak-dominated forests were established. Perhaps these changes caused a disruption of the coevolutionary equilibrium of plants and large herbivorous mammals in the Great Lakes region.

Considering the general existence of mosaic communities in the Pleistocene south of the glaciated areas, let us say that megaherbivores coevolved with and were dependant upon a variety of plants that occurred in these mosaic communities. But the abrupt change to the modern climate at the end of the Pleistocene caused a change to sharply divided climatic zones, each zone with a decreased variety of plants. Let us say that in Subregion II, the megaherbivores' choice of plants dwindled from ten to five species. Now the megaherbivores would have to compete with other herbivores for the limited plant resources.

Mammoths, for instance, seemed to have been able to exist in mosaic woodland areas in the Great Lakes region because patches of grasses existed there. But if these patches were replaced by pure stands of timber at the end of the Pleistocene when discrete vegetational zones were established, mammoths would have had to look for grassland elsewhere and obviously would be forced into intense competition with other mammoths.

There are modern analogies for the coevolutionary dependence of one species on another, which is one reason why biologists worry about the loss of biological diversity in the world. The box turtle, a species that occurred in the Pleistocene and presently occurs in the southern part of the Great Lakes region, feeds on mayapple fruit and is said to be essential for the propagation of this plant because the seed coat is digested away as it passes through the digestive tract of the turtle. The mayapple seed cannot germinate enveloped in its seed coat; thus the box turtle, a veritable "Johnny mayapple seed," spreads viable mayapple seeds as it wanders about in the woodlands.

Interactions with humans. If the Pleistocene overkill hypothesis is true, it would seem that we should find evidence of the blitzkrieg in the Great Lakes region, especially in Michigan where so many mastodonts and mammoths occurred. There are putative reports of mastodont butchery in Michigan, but other than some well-documented records in southeastern Wisconsin, there is little direct evidence of the interactions between Pleistocene humans and vertebrates in the Great Lakes region.

Humans lived in Europe and Asia about half a million years ago, but they did not reach North America until the Late Wisconsinan when the last ice sheet was retreating. People are thought to have reached the northern part of North America by way of a complete or partly complete land bridge that connected Siberia and Alaska. They then moved down through the unglaciated region between the Laurentide Ice Sheet in the East and the Cordilleran Ice Sheet in the West through the United States, across the Central American land bridge, and finally all the way to the southern tip of South America.

One should not get the idea that these were primitive "cave men." If these people wore modern clothes, they would not look any different from people today with similar genetic backgrounds. As these people entered the Great Lakes region near the end of the Ice Age, they left some of their spear points and other tools behind them, probably by accident, as these artifacts were very essential to their existence. The scattered spear points are direct evidence that Paleo-Indian big-game hunters lived in the Great Lakes region, and that is why *Homo sapiens* is included in the faunal list in the appendix section. Nevertheless, the relatively few

Paleo-Indian occupation sites known are thought to have been only briefly inhabited.

Paleo-Indian spear points are called fluted projectile points and are quite easily distinguished from the small, triangular points (and other points notched at the base) left behind by people who lived in the region much later (see fig. 11). As one might imagine, the fluted points are much less common than the later points. Fluted points lack notches at the base and are shaped somewhat like elongated leaves. They have a groove down the middle of one or both sides called a flute, which acted like the groove on a bayonet and promoted bleeding. The fluted point was inserted into a notch at the end of the shaft of the spear and bound there tightly by twine made from animal hide.

Some have suggested that there were few bands of Paleo-Indian hunters in the Great Lakes region, but others point out that they may have been rather numerous, based on the number of fluted point finds in the region. There are more than a hundred of these points recorded in Michigan alone, and many private collectors have as many as thirty fluted points in their collections.

Other Paleo-Indian tools include stone hide-scrapers, stone knives, and small, pointed stone tools called gravers, which are used to make holes in hides and other objects. Moreover, there are symmetrically shaped rocks that people suggest were used by Paleo-Indians to smash or grind up objects, as well as cores that were in the process of being shaped into spear points when they were lost or discarded.

Great Lakes fluted points occurred at the same time that mastodonts and mammoths were present in the area in large numbers. Some of the Michigan points are called Gainey points and are very similar to the Clovis points of the plains and Southwest made by people about 11,500 to 11,000 years ago. Fluted points in other states of the Great Lakes region are generally similar to Gainey and Clovis points, and Clovis, Gainey, and Folsom points have been recovered in southeastern Wisconsin. Still other fluted points discovered in the Great Lakes region are considered to be about 2,000 years younger than the Clovis-like points and are thought to have been used for hunting other kinds of vertebrates.

Clovis points are frequently discovered near mammoth skeletons in the plains and southwestern United States. One Arizona mammoth died with eight Clovis points in its body, indicating that several people took part in the slaying. The Clovis-point people tended to make their camps on ridges and terraces near streams where large mammals came to eat or drink.

Researchers have experimented with Clovis points by thrusting spears armed with these points into dead African elephants. The results showed that the spears could have easily have caused fatal wounds. It has been hypothesized that Clovis-point people hunted mammoths in pairs and that one person would attract the animal's attention while the other hunter either hurled the spear or thrust it directly into the animal.

The western Clovis-point people, as well as people in the Great Lakes region with similar spear points, are believed to have been key players in the overkill hypothesis, the suggestion being that the large herbivores did not fear these people who had sharpened their wits as well as their spear points for thousands of years on more wary species in Europe and Asia. In other words, why should a mastodont or mammoth fear a relatively small, stick-bearing mammal?

The evidence that mastodonts and mammoths were significantly impacted by Paleo-Indians in the Great Lakes region is inconclusive at present. There are no sites in Ontario, Michigan, Indiana, or Illinois as far as I know, where mastodonts or mammoths were directly associated with any Paleo-Indian fluted points or flint tools, although such sites exist in southeastern Wisconsin.

Some paleontologists, however, claim that stone tools and spear points were not expendable items in the Great Lakes region, which is not known for its abundance of flint and chert, rock types used to make stone artifacts. This might explain the lack of these objects in association with mastodont and mammoth finds in most of the region.

But other kinds of evidence that humans killed, butchered, and ate mastodonts in Michigan and Ohio has been presented. This evidence includes marks made on mastodont bones believed to have been produced when meat was removed, and other marks have been found on the ends of mastodont bones that have been interpreted as having been made when the bones were disarticulated (taken apart) during the process of butchering.

Other mastodont bones have been interpreted as having been blackened by fire during the cooking process, and it has been suggested that certain disarticulated mastodont bones were used as probes to assist in stripping off the hide. A bone in the tongue skeleton called the stylohyoid bone is shaped like a probe, and some have proposed that polished mastodont stylohyoid bones were used in stripping hides from the carcass.

These suggestions have been backed up by the study of marks on bones under the scanning electron microscope, which appear to show that alterations made during the butchering process differ from other alterations such as shovel marks or other marks made by humans digging up mastodont skeletons.

Moreover, certain mastodont bones look like they were broken up for the purpose of marrow extraction, and other important parts of the skeleton are missing, as if they had been carried away. Recently, material has been found that has been interpreted as evidence that mastodont meat was cached in mastodont intestines under the ice and that the caches were identified by stone markers.

But, relative to the Pleistocene overkill hypothesis, is this enough evidence to indicate that Paleo-Indians alone wiped out the mastodonts and mammoths of the Great Lakes region? Were a relatively few bands of Paleo-Indians, armed with simple spears, able to kill off the literally thousands of mastodonts and mammoths that roamed the Great Lakes region? This is a very important question because the last gasp of all mastodonts and mammoths occurred about 10,000 years ago, and the demise of these key megaherbivores may have triggered the final extinction of the large terrestrial mammals at the end of the Pleistocene.

The "everything went wrong at once" analogy, a derivative of the "multiple working hypothesis." Thomas C. Chamberlain, an eminent late-nineteenth-century student of Pleistocene events, cautioned that one must often employ a multiple working hypothesis to explain important geological events. It seems that, as humans, we want to find a single explanation for each important event. What caused the dinosaurs to become extinct? What causes continental ice sheets to form? Why did so many large terrestrial mammals become extinct worldwide at the end of the Pleistocene? Obviously, many hypotheses have been put forth to explain each of these events, and each "pet hypothesis" tends to be "fed" by its originator. But what if everything went wrong at once?

Two cars approach each other in opposite lanes at night; the driver of one car falls asleep momentarily and swerves into the wrong lane, but the other driver is alert and efficiently dodges the errant vehicle. Or two cars approach each other in opposite lanes at night; the driver of one car falls asleep momentarily and swerves into the other lane. The other driver has been drinking alcohol so his or her reaction time is slowed and the cars graze one another. Both drivers are shaken but survive. Or two cars approach each other at night in opposite lanes. Both cars are packed with people. The driver of one car falls asleep momentarily and swerves into the wrong lane. The driver of the other car is drunk, the car's brakes are faulty, and the road is slick with ice. A head-on collision based on "everything going wrong at once" results in a major tragedy and everyone is killed.

Let us say that everything went wrong at once at the end of the Pleistocene. The climate changes from an equable one to a nonequable one, mating and birthing in large herbivores become out of step with the new climate, mosaic communities give way rapidly to less diverse communities, large herbivores are thrown into intense competition with one another, and salt supplies for salt-dependant megaherbivores diminish because of lower water tables. Humans and other mammals emigrating from Eurasia to North American bring new diseases to which the New World mammals lack immunity. Finally, for some unexplained reason, bands of experienced, intelligent hunters "lose it" and kill every large mammal in sight as they move from Alaska to the tip of South America.

Perhaps the most one can say about the extinction of large mammals at the end of the Pleistocene at this point in time is that a complex of environmental and human-related impacts appear to have coincided at the end of the Pleistocene in a way that was devastating to the ecologically dominant, large mammalian members of terrestrial communities.

10

The Holocene and the Aftermath of the Ice Age

Most of the world's scientists now agree that the epoch we call the Pleistocene ended about 10,000 years ago. The epoch that followed the Pleistocene, the one we are living in now, is called the Holocene. The event that marked the end of the Pleistocene was the extinction of a large number of families, genera, and species of large mammals. Based on fossil evidence, it has been demonstrated that equable climates in many parts of the world were replaced by the zonal climates that are familiar to us today and that in these zones, mosaic communities were replaced by much less diverse communities.

In North America, although the ranges of many large, wild mammals have been greatly restricted by human big-game hunting and environmental disturbances, relatively few extinctions of large mammals have been recorded in the Holocene. Moreover, there have been fewer range adjustments of smaller mammals in the Holocene than in the Pleistocene, and the range adjustments that have occurred have usually involved much shorter distances.

One of the main climatic events in the Holocene of North America was a warm spell, or hypsithermal period, that occurred about midway in the epoch. Unfortunately, in the Great Lakes region, we have few large vertebrate faunas from the mid-Holocene warm spell. One of these faunas (the Harper Site fauna) was found near Lansing in south-central Lower Michigan (fig. 168) and yielded a diverse assemblage of turtles, with species present that are not currently found in the area. This has been interpreted as an indication of climatic warming in southern Michigan about 5,800 years ago.

Obviously, the most important event of the Holocene has been the spread of humans throughout the world, along with a sequence of modifications of their original hunting-gathering way of life. As small bands of hunter-gatherers, Pleistocene humans were part of the natural world. In the Holocene, however, humans became agriculturalists, then industrialists, and finally technologists, all to the detriment of natural communities. Fields of grain have replaced the natural grasslands and woodlands of the Pleistocene, and large domestic herbivores have replaced the extinct Pleistocene ones.

In other words, in the Pleistocene, the energy from the sun was captured by a myriad of photosynthetic plants and then was widely redistributed to a tremendously diverse fauna. Now, much of the energy from the sun flows into a few human crops and is narrowly distributed to humans and their domestic animals.

Biodiversity, which suffered a tremendous blow at the end of the Pleistocene and took another hit when monocultural agriculture replaced natural plant communities, continues to diminish in an almost out-of-control fashion, as natural communities are replaced helter-skelter by artificial human habitations.

All geological epochs must come to an end, and the Holocene will inevitably be replaced by another unit of geological time at some point in the future. Hopefully, the new epoch will be marked by natural geological processes rather than by a human-induced catastrophic extinction.

FIGURE 168. Some vertebrates from the mid-Holocene Harper Site near Lansing, Michigan. A wapiti (elk) (*Cervus elaphus*) is in the background. A painted turtle (*Chrysemys picta*) is basking in the right foreground. A common musk turtle (*Sternotherus odoratus*) basks near the water in the left foreground.

Appendix 1

Subregional and Age Distribution of Great Lakes Region Pleistocene Vertebrates

KEY TO SYMBOLS: *= extinct, YES= present in region, NO= absent in region, PLW= Pre-Late Wisconsinan, LW= Late Wisconsinan, PL= Late Pleistocene undesignated, PU= Pleistocene undesignated.

	SUBREGION I	SUBREGION II		SUBREGION I	SUBREGION II
FISHES: (Pisces)			MINNOWS		
			(Cyprinidae)	YES	NO
MUDMINNOWS			Central Stoneroller		
(Umbridae)	YES	NO	(*Campostoma anomalum*)	LW	NO
Central Mudminnow			Silver Chub		
(*Umbra limi*)	LW	NO	(*Macrhybopsis storeriana*)	LW	NO
			Hornyhead Chub		
PIKES			(*Nocomis biguttatus*)	LW	NO
(Esocidae)	YES	NO	Golden Shiner		
Northern Pike			(*Notemigonus crysoleucas*)	LW	NO
(*Esox lucius*)	PLW, LW	NO	Shiner		
Muskellunge			(cf. *Notropis* sp.)	PLW	NO
(*Esox masquinongy*)	LW	NO	Creek Chub		
Pike or Muskellunge			(*Semotilus atromaculatus*)	LW	NO
(*Esox* sp.)	PLW	NO			
			SUCKERS		
SALMON GROUP			(Catostomidae)	YES	NO
(Salmonidae)	YES	NO	White Sucker		
Lake Trout			(*Catostomus commersoni*)	LW	NO
(*Salvelinus namaycush*)	LW	NO	Shorthead Redhorse		
"trout" sp.	PLW	NO	(*Moxostoma*		
"white fish" sp.	PLW	NO	*macrolepidotum*)	LW	NO
SMELTS			NORTH AMERICAN FW CATFISHES		
(Osmeridae)	YES	NO	(Ictaluridae)	YES	NO
Capelin			Brown Bullhead		
(*Mallotus villosus*)	LW	NO	(*Ameiurus nebulosus*)	LW	NO
Rainbow Smelt			Channel Catfish		
(*Osmerus mordax*)	LW	NO	(*Ictalurus punctatus*)	PLW, LW	NO
CODS			STICKLEBACKS		
(Gadidae)	YES	NO	(Gasterosteidae)	YES	NO
Atlantic Tomcod			Threespine Stickleback		
(*Microgadus tomcod*)	LW	NO	(*Gasterosteus aculeatus*)	LW	NO
Burbot			Ninespine Stickleback		
(*Lota lota*)	PLW	NO	(*Pungitius* sp.)	PLW	NO

	SUBREGION I	SUBREGION II		SUBREGION I	SUBREGION II
SUNFISHES AND BASSES			**TOADS**		
(Centrarchidae)	YES	NO	(Bufonidae)	YES	YES
Smallmouth Bass			American Toad		
(*Micropterus dolomieu*)	LW	NO	(*Bufo americanus*)	LW	NO
Largemouth Bass			Fowler's Toad		
(*Micropterus salmoides*)	LW	NO	(*Bufo fowleri*)	LW	NO
Crappie			Toad		
(*Pomoxis* sp.)	LW	NO	(*Bufo* sp.)	NO	LW
PERCHES			**TREEFROGS**		
(*Percidae*)	YES	NO	(Hylidae)	YES	NO
Yellow Perch			Striped Chorus Frog		
(*Perca flavescens*)	PLW, LW	NO	(*Pseudacris triseriata*)	LW	NO
DRUMS AND CROAKERS			**TRUE FROGS**		
(Sciaenidae)	YES	NO	(Ranidae)	YES	YES
Freshwater Drum			Bullfrog		
(*Aplodinotus grunniens*)	PLW	NO	(*Rana catesbeiana*)	LW	LW
			Green Frog		
SCULPIN			(*Rana clamitans*)	LW	LW
(Cottidae)	YES	NO	Northern Leopard Frog		
Hook-eared Sculpin			(*Rana pipiens*) and/or complex	LW	LW
(*Artediellus uncinatus*)	LW	NO	Wood Frog		
Cottidae indet	PLW	NO	(*Rana sylvatica*)	LW	LW
LUMPFISH AND SNAILFISHES			**REPTILES: (Reptilia)**		
(Cyclopteridae)	YES	NO			
Lumpfish			**MUD AND MUSK TURTLES**		
(*Cyclopterus lumpus*)	LW	NO	(Kinosternidae)	NO	YES
			Common Musk Turtle		
AMPHIBIANS: (Amphibia)			(*Sternotherus odoratus*)	NO	LW
MOLE SALAMANDERS			**SNAPPING TURTLES**		
(Ambystomatidae)	YES	YES	(Chelydridae)	YES	YES
Blue-spotted Salamander complex			Snapping Turtle		
(*Ambystoma laterale*) complex	LW	NO	(*Chelydra serpentina*)	LW	LW
Mole Salamander					
(*Ambystoma* sp.)	NO	LW	**NEW WORLD POND TURTLES**		
			(Emydidae)	YES	YES
			Painted Turtle		
LUNGLESS SALAMANDERS			(*Chrysemys picta*)	LW	LW
(Plethodontidae)	NO	YES	Blanding's Turtle		
Cave Salamander			(*Emydoidea blandingii*)	PLW, LW	LW
(*Eurycea lucifuga*)	NO	LW	Map Turtle		
			(*Graptemys* sp.)	NO	LW
			Cooter		
			(*Pseudemys* sp.)	NO	LW

	SUBREGION I	SUBREGION II		SUBREGION I	SUBREGION II
Eastern Box Turtle			Queen Snake		
(*Terrapene carolina*)	LW	NO	(*Regina septemvittata*)	LW	NO
Box Turtle			Common Garter Snake		
(*Terrapene* sp.)	NO	PLW	(*Thamnophis sirtalis*)	LW	NO
Slider			Garter or Ribbon Snake		
(*Trachemys scripta*)	LW	LW	(*Thamnophis* sp.)	LW	LW
TORTOISES					
(Testudinidae)	NO	YES	VIPERS		
*Giant Land Tortoise			(Viperidae)	NO	YES
(**Hesperotestudo*			Timber Rattlesnake		
crassiscutata*)	NO	PLW	(*Crotalus horridus*)	NO	LW
(**Hesperotestudo* sp.)	NO	PLW			
			BIRDS: (Aves)		
SOFTSHELL TURTLES					
(Trionychidae)	YES	YES	DUCKS, GEESE, and SWANS		
Spiny Softshell Turtle			(Anatidae)	YES	NO
(*Apalone spinifer*)	LW	LW	Northern Shoveler		
Softshell Turtle			(*Anas clypeata*)	LW	NO
(*Apalone* sp.)	LW	NO	Mallard		
			(*Anas platyrhynchos*)	LW	NO
LIZARDS:			Lesser Scaup		
(Sauria)			(*Aythya affinis*)	LW	NO
			Ring-necked Duck		
ANGUID LIZARDS			(*Aythya collaris*)	LW	NO
(Anguidae)	NO	YES	Canada Goose		
Slender Glass Lizard			(*Branta canadensis*)	LW	NO
(*Ophisaurus attenuatus*)	NO	LW	TURKEYS		
			(Meleagridae)	LW	NO
SNAKES: (Serpentes)			Wild Turkey		
			(*Meleagris gallopavo*)	LW	NO
COLUBRID SNAKES					
(Colubridae)	YES	YES	GROUSE		
Racer			(Tetraonidae)	YES	NO
(*Coluber constrictor*)	LW	NO	cf. Ruffed Grouse		
Fox Snake			(*Bonasa umbellus*)	LW	NO
(*Elaphe vulpina*)	LW	NO	Prairie Chicken		
Milk Snake			(*Tympanuchus cupido*)	LW	NO
(*Lampropeltis triangulum*)	LW	NO			
Plain-bellied Water Snake			CROWS AND JAYS		
(*Nerodia erythrogaster*)	NO	LW	(Corvidae)	YES	NO
Northern Water Snake			Raven		
(*Nerodia sipedon*)	LW	NO	(*Corvus corax*)	LW	NO
Smooth Green Snake					
(*Opheodrys vernalis*)	LW	LW			

	SUBREGION I	SUBREGION II
MAMMALS (Given in Groups Below):		
(Xenarthrans, Insectivores, Bats)		
ARMARDILLOS		
(Dasypodidae)	NO	YES
*Beautiful Armadillo		
(*Dasypus bellus)	NO	LW
*GROUND SLOTHS		
(*Megalonychidae)	YES	YES
*Jefferson's Ground Sloth		
(*Megalonyx jeffersonii)	LW	LW
SHREWS		
(Soricidae)	YES	YES
Northern Short-tailed Shrew		
(Blarina brevicauda)	LW	LW
Least Shrew		
(Cryptotis parva)	NO	LW
Arctic Shrew		
(Sorex arcticus)	LW	LW
Masked Shrew		
(Sorex cinereus)	LW	LW
Smoky Shrew		
(Sorex fumeus)	LW	LW
Pygmy Shrew		
(Sorex hoyi)	PLW, LW	LW
Water Shrew		
(Sorex palustris)	LW	NO
MOLES		
(Talpidae)	NO	YES
Star-nosed Mole		
(Condylura cristata)	NO	LW
Hairy-tailed Mole		
(Parascalops breweri)	NO	LW
Eastern Mole		
(Scalopus aquaticus)	NO	LW
COMMON BATS		
(Vespertilionidae)	YES	YES
Small-footed Bat		
(Myotis leibii)	NO	LW
Little Brown Bat		
(Myotis lucifugus)	LW	NO
Northern Bat		
(Myotis septentrionalis)	LW	LW
Mouse-eared bat species but not one of the above		
Myotis sp.	NO	LW

	SUBREGION I	SUBREGION II
Eastern Pipistrelle		
(Pipistrellus subflavus)	NO	LW
Big-Eared Bat		
(Plecotus sp.)	NO	LW
MAMMALS:		
(Primates, Carnivores, Cetaceans)		
HOMONIDS		
(Hominidae)	YES	YES
Humans		
(Homo sapiens)	LW	LW
DOGS AND RELATIVES		
(Canidae)	YES	YES
*Dire Wolf		
(*Canis dirus)	NO	PLW
Coyote		
(Canis latrans)	NO	LW
Gray Fox		
(Urocyon cinereoargenteus)	NO	LW
Red Fox		
(Vulpes vulpes)	LW	NO
BEARS		
(Ursidae)	YES	YES
*Short-faced Bear		
(*Arctodus simus)	LW	NO
Black Bear		
(Ursus americanus)	LW	LW
Brown and Grizzly Bears		
(Ursus arctos)	PLW, LW	NO
RACCOONS AND RELATIVES		
(Procyonidae)	YES	YES
Raccoon		
(Procyon lotor)	LW	LW
MUSTELIDS		
(Mustelidae)	YES	YES
Marten		
(Martes americana)	LW	NO
Fisher		
(Martes pennanti)	LW	LW
Ermine		
(Mustela erminea)	PLW, LW	NO
Long-tailed Weasel		
(Mustela frenata)	LW	LW

	SUBREGION I	SUBREGION II		SUBREGION I	SUBREGION II
Mink			TAPIRS		
(*Mustela vison*)	LW	LW	(Tapiridae)	YES	YES
Striped Skunk			*Hay's Tapir		
(*Mephitis mephitis*)	LW	NO	(*Tapirus haysii*)	NO	LW
Spotted Skunk			*Vero Tapir		
(*Spilogale putorius*)	NO	PLW, LW	(*Tapirus veroensis*)	LP	NO
River Otter			PECCARIES		
(*Lutra canadensis*)	LW	LW	(Tayassuidae)	YES	YES
CATS			*Long-nosed Peccary		
(Felidae)	NO	YES	(*Mylohyus nasutus*)	NO	LP, LW
Jaguar			*Flat-headed Peccary		
(*Panthera onca*)	NO	PLW	(*Platygonus compressus*)	LP, LW	LW
*Sabertooth			*Leidy's Peccary		
(*Smilodon fatalis*)	NO	PLW	(*Platygonus vetus*)	NO	PLW
TRUE SEALS			CAMELS AND LLAMAS		
(Phocidae)	YES	NO	(Camelidae)	NO	YES
Bearded Seal			*American Camel		
(*Erignathus barbatus*)	LW	NO	(*Camelops* sp.)	NO	LP
Harp Seal			DEER AND RELATIVES		
(*Phoca groenlandica*)	LW	NO	(Cervidae)	YES	YES
Ringed Seal			Moose		
(*Phoca hispida*)	LW	NO	(*Alces alces*)	LW	NO
WHITE WHALES AND NARWHALS			*Scott's Moose (Stag Moose)		
(Monodontidae)	YES	NO	(*Cervalces scotti*)	PLW, LW	LW
White Whale			Wapiti (Elk)		
(*Delphinapterus leucas*)	LW	NO	(*Cervus elaphus*)	LW	NO
RORQUALS			White-tailed Deer		
(Balaenopteridae)	YES	NO	(*Odocoileus virginianus*)	LP, LW	LW
Humpback Whale			Caribou		
(*Megaptera novaeangliae*)	LW	NO	(*Rangifer tarandus*)	PU, LW	NO
BOWHEAD AND RIGHT WHALES			BOVIDS		
(Balaenidae)	YES	NO	(Bovidae)	YES	YES
Bowhead Whale			American Bison		
(*Balaena mysticetus*)	LW	NO	(*Bison bison*)	LW	LW
MAMMALS:			*Giant Bison		
(Perissodactyls, Artiodactyls)			(*Bison latifrons*)	PLW	PLW?
HORSES			Barren Ground (Tundra) Muskox		
(Equidae)	YES	YES	(*Ovibos moschatus*)	PL, PLW, LW	PU
*Complex-toothed Horse			*Helmeted		
(*Equus complicatus*)	LP	PLW	(*Bootherium bombifrons*)	LW	LW

	SUBREGION I	SUBREGION II
MAMMALS:		
(Rodents, Lagomorphs, Proboscideans)		
SQUIRRELS		
(Sciuridae)	YES	YES
Southern Flying Squirrel		
(*Glaucomys volans*)	NO	LW
Woodchuck		
(*Marmota monax*)	PLW, LW	LW
Eastern Gray Squirrel		
(*Sciurus carolinensis*)	NO	LW
Fox Squirrel		
(*Sciurus niger*)	NO	LW
Tree Squirrel		
(*Sciurus* sp.)	LW	LW
Thirteen-lined Ground Squirrel		
(*Spermophilus tridecemlineatus*)	LW	LW
Eastern Chipmunk		
(*Tamias striatus*)	LW	LW
Red Squirrel		
(*Tamiasciurus hudsonicus*)	LW	LW
POCKET GOPHERS		
(Geomyidae)	YES	YES
Plains Pocket Gopher		
(*Geomys bursarius*)	NO	LW
Northern Pocket Gopher		
(*Thomomys talpoides*)	LW	NO
BEAVERS		
(Castoridae)	YES	YES
American Beaver		
(*Castor canadensis*)	LW	NO
*Giant Beaver		
(*Castoroides ohioensis*)	LP, PLW, LW	LW
RATS, MICE, AND RELATIVES		
(Muridae)	YES	YES
Eastern Woodrat		
(*Neotoma floridana*)	NO	PLW, LW
Woodrat		
(*Neotoma* sp.)	LW	NO
Rice rat		
(*Oryzomys palustris*)	NO	LW
White-footed Mouse		
(*Peromyscus leucopus*)	NO	LW
Deer Mouse		
(*Peromyscus maniculatus*)	LW	LW

	SUBREGION I	SUBREGION II
White-footed or Deer Mouse		
(*Peromyscus* sp.)	PLW, LW	LW
Southern Red-backed Vole		
(*Clethrionomys gapperi*)	LW	LW
Red-backed Vole		
(*Clethrionomys* sp.)	PLW	NO
*Cape Deceit Vole		
(*Lasiopodemys deceitensis*)	NO	PLW
Prairie Vole		
(*Microtus ochrogaster*)	LW	LW
*Hibbard's Tundra Vole		
(*Microtus paroperarius*)	NO	PLW
Meadow Vole		
(*Microtus pennsylvanicus*)	PLW, LW	LW
Woodland Vole		
(*Microtus pinetorum*)	NO	LW
Yellow-cheeked Vole		
(*Microtus xanthognathus*)	PLW, LW	LW
Heather Vole		
(*Phenacomys intermedius*)	LW	LW
Collared Lemming		
(*Dicrostonyx torquatus*)	LW	NO
Lemming		
(*Dicrostonyx* sp.)	PLW	NO
Brown Lemming		
(*Lemmus* sp.)	PLW	NO
Northern Bog Lemming		
(*Synaptomys borealis*)	PLW, LW	LW
Southern Bog Lemming		
(*Synaptomys cooperi*)	LW	LW
Muskrat		
(*Ondatra zibethicus*)	PLW, LW	LW
JUMPING MICE AND RELATIVES		
(Zapodidae)	YES	YES
Meadow Jumping Mouse		
(*Zapus hudsonius*)	NO	LW
Western Jumping Mouse		
(*Zapus princeps*)	LW	NO
Jumping Mouse		
(*Zapus* sp.)	PLW	NO
NEW WORLD PORCUPINES		
(Erithizontidae)	LW	NO
Common Porcupine		
(*Erithizon dorsatum*)	LW	NO

	SUBREGION I	SUBREGION II
PIKAS		
(Ochotonidae)	YES	NO
Pika		
(*Ochotona* sp.)	LW	NO
HARES, JACKRABBITS, AND RABBITS		
(Leporidae)	YES	YES
Snowshoe Hare		
(*Lepus americanus*)	LW	LW
Hare or Jackrabbit		
(*Lepus* sp.)	LW	LW
Eastern Cottontail		
(*Sylvilagus floridanus*)	LW	LW

	SUBREGION I	SUBREGION II
*MASTODONTS		
(*Mammutidae)	YES	YES
*American Mastodont		
(*Mammut americanum*)	LW	PLW, LW
MAMMOTHS AND ELEPHANTS		
(Elephantidae)	YES	YES
*Jefferson Mammoth		
(*Mammuthus jeffersonii*)	PLW, LW	LW
*Woolly Mammoth		
(*Mammuthus primigenius*)	LP, PLW, LW	LW

References

This bibliography contains references for both the general and specific topics in the individual chapters of the book. The reference section for chapters 7 and 8 are so extensive they are subdivided into general, provincial, and state subsections.

PREFACE AND INTRODUCTION

Bates, R. L., and J. A. Jackson, eds. 1976. *Dictionary of geological terms*..3rd ed. Garden City, N.Y.: Anchor Press.

Behrensmeyer, A. K., and A. P. Hill. 1980. *Fossils in the making: Vertebrate taphonomy and paleoecology*. Chicago: University of Chicago Press.

Brewer, R. 1988. *The science of ecology*. Philadelphia: Saunders College Publishing.

Dodson, E. O., and P. Dodson. 1985. *Evolution, process and product*. 3rd ed. Belmont, Calif.: Wadsworth Publishing.

Eicher, D. L. 1976. *Geologic time*.. 2nd ed. Englewood Cliffs, N.J.: Prentice-Hall.

Eicher, D. L., A. L. McAlester, and M. L. Rottman. 1980. *History of the earth's crust*. Englewood Cliffs, N.J.: Prentice-Hall.

Eldredge, N. 1985. *Unfinished synthesis, biological hierarchies and modern evolutionary thought*. New York: Oxford University Press.

Futuyma, D. J., and M. Slatkin. 1983. *Coevolution*. Sunderland, Mass.: Sinauer Associates.

Hecker, R. F. 1965. *Introduction to paleoecology*. New York: Elsevier Press.

Hildebrand, M. 1974. *Analysis of vertebrate structure*. New York: John Wiley and Sons.

Holman, J. A. 1994. *Vertebrate life of the past*. 5th rev. .With excerpts from *Dinosaurs* by Spencer Lucas. Dubuque, Iowa: W. C. Brown.

———. 1995. *Ancient life of the Great Lakes Basin*. Ann Arbor: University of Michigan Press.

Imbrie, J., and N. D. Newell. 1964. *Approaches to paleoecology*. New York: John Wiley and Sons.

Kukla, G. 1998. The Eeemian, local sequences, global perspectives. Abstracts, SEQS Conference, Kerkgrade, The Netherlands, 1–11 September 1998.

Leet, L. D., S. Judson, and M. E. Kauffman. 1978. *Physical geology*. 5th ed. Englewood Cliffs, N.J.: Prentice Hall.

Lincoln, R. J., G. A. Boxshall, and P. F Clark. 1982. *A dictionary of ecology, evolution, and systematics*. Cambridge: Cambridge University Press.

Lyman, R. L. 1994. *Vertebrate taphonomy*. Cambridge: Cambridge University Press.

McKinney, M. L. 1993. *Evolution of life: Processes, patterns, and prospects*. Englewood Cliffs, N.J.: Prentice-Hall.

MacStalker, A. 1977. The megablocks, or giant erratics of the Canadian prairies. Abstracts of the tenth INQUA Congress, August 1977, Birmingham, England, 32.

Odum, E. P. 1983. *Basic ecology*. Philadelphia: Saunders College Publishing.

Pough, F. H., J. B. Heiser, and W. N. McFarland. 1996. *Vertebrate life*. 4th ed. Upper Saddle River, N.J.: Prentice-Hall.

Shelton, S. Y., ed. 1995. Amateur paleontologists in the news. *Outreach* 1:7–8.

Simpson, G. G. 1953. *The major features of evolution*. New York: Columbia University Press.

Wallace, R. A. 1987. *Biology, the world of life*. 4th ed. Glenview, Ill.: Scott, Foresman.

CHAPTER 2: THE PLEISTOCENE ICE AGE

Bell, C. J. 2000. Biochronology of North American microtine rodents. In *Quaternary geochronology, methods and applications*, edited by J. S. Noller, J. M. Sowers, and W. R. Lettis, 379–405. Washington, D.C.: American Geophysical Union.

Bowen, D. Q. 1978. *Quaternary geology: A stratigraphic framework for multidisciplinary work*. London: Pergamon Press.

Bradley, R. S. 1985. *Quaternary paleoclimatology*. Boston: Allen and Unwin.

Chamberlain, T. C. 1895. The classification of American glacial deposits. *Journal of Geology* 3:270–77.

Flint, R. F. 1971. *Glacial and Quaternary geology*. New York: John Wiley and Sons.

Geikie, J. 1874. *The great Ice Age*. London: W. Isbiter.

Gribbin, J., ed. 1978. *Climatic change*. Cambridge: Cambridge University Press.

Harland, W. B., R. L. Armstrong, A. V. Cox, L. E. Craig, A. G. Smith, and D. G. Smith. 1989. *A geologic time scale*. Cambridge: Cambridge University Press.

Holman, J. A. 1995. *Pleistocene amphibians and reptiles in North America*. New York: Oxford University Press.

John, B. S. 1977. *The ice age, past and present*. London: Collins.

Kurtén, B. 1972. *The ice age*. London: Ruperet Hart-Davis.

Lamb, H. H. 1977. *Climate, present, past and future*. London: Methuen.

Lundelius, E. L., T. Downs, E. H. Lindsay, H. A. Semken, R. J. Zakrzewski, C. S. Churcher, C. R. Harington, G. E.

Schultz, and S. D. Webb. 1987. The North American Quaternary sequence. In *Cenozoic mammals of North America, geochronology and biostratigraphy*, edited by M. O. Woodburne, 211–35. Berkeley and Los Angeles: University of California Press.

Matsch, C. L. 1976. *North America and the great Ice Age*. New York: McGraw Hill.

Mead, J. I., and D. J. Meltzer. 1984. North American Late Quaternary extinctions and the radiocarbon record. In *Quaternary extinctions—a prehistoric revolution*, edited by P. S. Martin and R. G. Klein, 440–50. Tucson: University of Arizona Press.

Mickelson, D. M., L. Clayton, D. S. Fullerton, and H. W. Borns Jr. 1983. The Late Wisconsin glacial record of the Laurentide Ice Sheet in the United States. In *Late Quaternary environments of the United States*. Vol. 1, *The Late Pleistocene*, edited by H. E. Wright Jr., 3–37. Minneapolis: University of Minnesota Press.

Morner, N. A. 1973. The Plum Point Interstadial: Age, climate, and subdivision. *Canadian Journal of Earth Sciences* 8:1423–31.

Repenning, C. A. 1987. Biochronology of the Microtine rodents of the United States. In *Cenozoic mammals of North America, geochronology and biostratigraphy*, edited by M. O. Woodburne, 236–68. Berkeley and Los Angeles: University of California Press.

Sutcliffe, A. J. 1985. *On the track of Ice Age mammals*. London: British Museum (Natural History).

West, R. G. 1968. *Pleistocene geology and biology*. London: Longmans, Green.

Woodburne, M. O., 1987. Mammal ages, stages, and zones. In *Cenozoic mammals of North America, geochronology and biostratigraphy*, edited by M. O. Woodburne, 18–23. Berkeley and Los Angeles: University of California Press.

———., ed. 1987. *Cenozoic mammals of North America, geochronology and biostratigraphy*. Berkeley and Los Angeles: University of California Press.

Wright, H. E., Jr., and D. G. Frey, eds. 1965. *The Quaternary of the United States*. Princeton, N.J.: Princeton University Press.

Wright, W. B. 1936. *The Quaternary ice age*. 2nd ed. London: Macmillan.

Zeuner, F. E. 1959. *The Pleistocene period*. London: Hutchinson.

CHAPTER 3: THE PLEISTOCENE IN THE GREAT LAKES REGION

Anderson, S. T. 1954. A late-glacial pollen diagram from southern Michigan, U.S.A. *Denmarks Geologiske Undersogelse* 49:1–16.

Artist, R. C. 1936. Stratigraphy and preliminary pollen analysis of a Lake County, Illinois, bog. *Butler University Botanical Studies* 3:191–98.

Attig, J. W., L. Clayton, and D. M. Mickelson. 1985. Correlation of Late Wisconsinan glacial phases in the western Great Lakes area. *Geological Society of America Bulletin* 96:1585–93.

Baker, R. G. 1970. A radiocarbon dated pollen chronology for Wisconsin: Disterhaft Farm bog revisited. *Geological Society of America Abstracts* 2:488.

Barnett, J. 1937. A pollen study of Cranberry Pond near Emporia, Madison County, Indiana. *Butler University Botanical Studies* 4:55–64.

Berti, A. A. 1975. Paleobotany of Wisconsinan interstadials, eastern Great Lakes region, North America. *Quaternary Research* 5:591–619.

Cleland, C. E., M. B. Holman, and J. A. Holman. 1998. The Mason-Quimby Line revisited. In From the northern tier: Papers in honor of Ronald J. Mason, edited by C. E. Cleland and R. A. Birmingham. *Wisconsin Archeologist* 79:8–27.

Coleman, A. P. 1933. The Pleistocene of the Toronto region (including the Toronto interglacial formation). *Ontario Department of Mines Annual Report* 41:1–55.

Dorr, J. A., and D. F. Eschman. 1970. *Geology of Michigan*. Ann Arbor: University of Michigan Press.

Dorwin, J. T. 1966. Fluted points and Late Pleistocene geochronology of Indiana. *Indiana Historical Society, Prehistory Research Series* 4:145–88.

Dreimanis, A. 1964a. Lake Warren and the Two Creeks Interval. *Journal of Geology* 72:247–50.

———. 1964b. Notes on the Pleistocene time-scale in Canada. In *Geochronology in Canada*, edited by F. Osborne, 139–56. Toronto: University of Toronto Press.

Dreimanis, A., J. Terasmae, and G. D. McKenzie. 1966. The Port Talbot interstade of the Wisconsinan glaciation. *Canadian Journal of Earth Sciences* 3:305–25.

Farrand, W. R., and D. F. Eschman. 1974. Glaciation of the southern peninsula of Michigan. *Michigan Academician* 7:31–56.

Flint, R. F. 1971. *Glacial and Quaternary geology*. New York: John Wiley and Sons.

Frye, J. C., H. B. Willman, and R. P. Black. 1965. Outline of glacial geology in Illinois and Wisconsin. In *The Quaternary of the United States*, edited by H. E. Wright Jr. and D. G. Frey, 43–61. Princeton, N.J.: Princeton University Press.

Goldthwait, R. P. 1979. Ice over Ohio. In *Ohio's natural heritage*, edited by M. B. Lafferty, 32–47. Columbus: Ohio Academy of Science.

Goldthwait, R. P., G. W. White, and J. L. Forsyth. 1979. *Glacial map of Ohio*. Washington, D.C.: United States Geological Survey.

Halsey, J. R., and M. D. Stafford. 1999. Retrieving Michigan's buried past: The archaeology of the Great Lakes State. *Cranbrook Institute of Science Bulletin* 64:1–477.

Holman, J. A. 1991. New records of Michigan Pleistocene vertebrates with comments on the Mason-Quimby Line. *Michigan Academician* 27:409–24.

———. 1995. *Ancient life of the Great Lakes Basin.* Ann Arbor: University of Michigan Press.

Hough, J. L. 1958. *Geology of the Great Lakes.* Urbana: University of Illinois Press.

Kapp, R. O. 1977. Late Pleistocene postglacial plant communities of the Great Lakes region. In *Geobotany*, edited by R. C. Romans, 1–27. New York: Plenum.

———. 1999. Michigan Late Pliocene, Holocene, and presetlement vegetation. In Retrieving Michigan's buried past: The archaeology of the Great Lakes State, edited by J. R. Halsey and M. D. Stafford. *Cranbrook Institute of Science Bulletin* 64:31–38

Kapp, R. O., S. G. Beld, and J. A. Holman. 1990. Paleontological resources in Michigan: An overview. In *Cultural and paleontological effects of siting a low-level radioactive waste storage facility in Michigan*, edited by W. Stoffle, 14–34. Ann Arbor: Institute of Social Research Publication.

Karrow, P. F. 1969. Stratigraphic studies in the Toronto Pleistocene. *Proceedings of the Geological Association of Canada* 20:4–16.

Kelly, R. W., and W. R. Farrand. 1967. The glacial lakes around Michigan. *Michigan Geological Survey Bulletin* 4:1–23.

Larsen, C. E. 1999. A century of Great Lakes levels research: Finished or just beginning? In Retrieving Michigan's buried past: The archaeology of the Great Lakes State, edited by J. R. Halsey and M. D. Stafford. *Cranbrook Institute of Science Bulletin* 64:1–30.

Leverett, F., and F. B. Taylor. 1915. The Pleistocene of Indiana and Michigan and the history of the Great Lakes. *United States Geological Survey Monograph* 53:1–529.

Lyell, C. 1830. *Principles of geology.* London: J. Murray.

Paull, R. K., and R. A. Paull. 1977. *Geology of Wisconsin and Upper Michigan.* Dubuque, Iowa: Kendall-Hunt.

Terasmae, J. 1960. A palynological study of the Pleistocene interglacial beds at Toronto. *Bulletin of the Geological Survey of Canada* 56:23–41.

Terasmae. J., P. F. Karrow, and A. Dreimanis. 1972. Quaternary stratigraphy and geomorphology of the eastern Great Lakes region of southern Ontario. *Twenty-fourth International Geological Congress, Excursion A42 Guidebook*, 26–32.

Wayne, W. J., and J. H. Zumberge. 1965. Pleistocene geology of Indiana and Michigan. In *The Quaternary of the United States*, edited by H. E. Wright Jr. and D. G. Frey, 63–84. Princeton, N.J.: Princeton University Press.

CHAPTERS 4, 5, AND 6: FINDING, COLLECTING, AND DATING VERTEBRATE FOSSILS

Bretz, J. H., and S. E. Harris Jr. 1961. *Caves of Illinois.* Urbana: Illinois Geological Survey.

Burleigh, R., ed. 1980. Progress in scientific dating methods. *British Museum Occasional Paper* 21:1–96.

Hibbard, C. W. 1949. Techniques of collecting microvertebrate fossils. *Contributions from the Museum of Paleontology, the University of Michigan* 8:7–19.

Holman, J. A. 1975. Michigan's fossil vertebrates. *Publications of the Museum, Michigan State University, Educational Bulletin* 2:1–54.

———. 1985. Herpetofauna of Ladds Quarry. *National Geographic Research* 1:423–36.

———. 1994. *Vertebrate life of the past.* 5th rev. With excerpts from dinosaurs by Spencer Lucas. Dubuque, Iowa: W. C. Brown.

———. 1995a. *Ancient life of the Great Lakes Basin.* Ann Arbor: University of Michigan Press.

———. 1995b. *Pleistocene amphibians and reptiles in North America.* New York: Oxford University Press.

Holman, J. A., D. C. Fisher, and R. O. Kapp. 1986. Recent discoveries of fossil vertebrates in the Lower Peninsula of Michigan. *Michigan Academician* 18:431–63.

Holman, J. A., and F. Grady. 1987. Herpetofauna of New Trout Cave. *National Geographic Research* 3:305–17.

La Plante, L. 1977. *The weekend fossil hunter.* New York: Drake.

MacFall, R. P., and J. Wollin. 1983. *Fossils for amateurs.* 2nd ed. New York: Van Nostrand Reinhold.

Mohr, C. M., and T. L. Poulson. 1966. *The life of the cave.* New York: McGraw-Hill.

Moore, P. D., and J. A. Webb. 1978. *An illustrated guide to pollen analysis.* London: Hodder and Stroughton.

Murray, M. 1967. *Hunting for fossils.* New York: Macmillan.

Oakley, K. P. 1969. *Frameworks for dating fossil man.* 3rd ed. London: Weidenfield and Nicholson.

Olsen, S. J. 1972. Osteology for the archaeologist: The American mastodon and the woolly mammoth. *Papers of the Peabody Museum of Archaeology and Ethnology, Harvard University* 56:1–47.

Parker, S., and R. L. Berner. 1990. *The practical paleontologist.* New York: Simon and Schuster.

Powell, R. L. 1961. *Caves of Indiana.* Indianapolis: Department of Conservation.

Raup, R. L. Cmn. Paleontological collecting. Washington, D.C.: National Academy Press.

Richards, R. L. 1982. Hunting Indiana bears. *Outdoor Indiana* 47:16–18.

———. 1983. Getting down to the bear bones. *Outdoor Indiana* 48:32–34.

———. 1984. It's the pits. *Outdoor Indiana* 49:25–27.

———. 1988. Cave graves. *Outdoor Indiana* 53:4–7.

Skeels, M. A. 1962. The mastodons and mammoths of Michigan. *Papers of the Michigan Academy of Science, Arts, and Letters* 47:103–33.

Sutcliffe, A. J. 1985. *On the track of ice age mammals*. London: British Museum (Natural History).

West, R. W. 1989. State regulation of geological, paleontological, and archaeological collecting. *Curator* 32: 281–319.

———. 1991. State regulation of geological, paleontological, and archaeological collecting. *Curator*. 34:199–209.

CHAPTERS 7 AND 8: A BESTIARY OF GREAT LAKES ICE AGE VERTEBRATES AND MAJOR VERTEBRATE LOCALITIES IN THE GREAT LAKES REGION

General References

Anderson, E. 1984. Who's who in the Pleistocene: A mammalian bestiary. In *Quaternary extinctions—a prehistoric revolution*, edited by P. S. Martin and R. G. Klein, 40–89. Tucson: University of Arizona Press.

Auffenberg, W. 1957. A note on an unusually complete specimen of *Dasypus bellus* (Simpson) from Florida. *Quarterly Journal of the Florida Academy of Sciences* 20:233–37.

Auffenberg, W., and W. W. Milstead. 1965. Reptiles in the Quaternary of North America. In *The Quaternary of the United States*, edited by H. E. Wright Jr. and D. Frey, 557–67. Princeton, N.J.: Princeton University Press.

Baker, F. C. 1920. The life of the Pleistocene or glacial period. *University of Illinois Bulletin* 17:1–476.

Banks, R. C., R. W. McDiarmid, and A. L. Gardner. 1987. Checklist of vertebrates of the United States, the U.S. territories, and Canada. *United States Department of the Interior Fish and Wildlife Service Resource Publication* 166:1–79.

Barbour, E. H., and H. J. Cook. 1914. A new Saber-toothed cat from Nebraska. *Nebraska Geological Survey* 4: 235–38, 1 pl.

Bell, C. J. 2000. Synopsis of terrestrial and non-marine aquatic fossil groups. In *Quaternary geochronology, methods and applications*, edited by J. S. Noller, J. M. Sowers, and W. R. Lettis, 407–11. Washington, D.C.: American Geophysical Union.

Berry, E. W. 1929. *Paleontology*. New York: McGraw-Hill.

Brodkorb, P. 1964. Catalogue of fossil birds part 2 (Anseriformes through Galliformes). *Bulletin of the Florida State Museum Biological Sciences* 8:195–335.

Burt, W. H., and R. P. Grossenheider. 1998. *A field guide to the mammals*. Boston: Houghton Mifflin.

Cleland, C. E. 1966. The prehistoric animals ecology and ethnozoology of the upper Great Lakes region. *Anthropological Papers, Museum of Anthropology, University of Michigan* 29:1–294.

Conant, R., and J. T. Collins. 1998. *A field guide to reptiles and amphibians of eastern and central North America*. Boston: Houghton Mifflin.

Fagan, B. M. 1991. *Ancient North America: The archaeology of a continent*. London: Thames and Hutton.

Graham, R. W. 1991. Variability in the size of North American black bears (*Ursus americanus*) with the description of a fossil black bear from Bill Neff Cave, Virginia. In Beamers, bobwhites, and blue-points: Tributes to the career of Paul W. Parmalee, edited by J. R. Purdue, W. E. Klippel, and B. W. Styles. *Illinois State Museum Scientific Papers* 23:237–50.

Graham, R. W. 1992. Late Pleistocene faunal changes as a guide to understanding effects of greenhouse warming on the mammalian fauna of North America. In *Global warming and biological diversity*, edited by R. L. Peters and T. E. Lovejoy, 76–87. New Haven, Conn.: Yale University Press.

Graham, R. W., and E. L. Lundelius Jr. 1994. Faunmap: A database documenting Late Quaternary distributions of mammal species of the United States. 2 vols. *Illinois State Museum Scientific Papers* 25:1–690.

Guilday, J. E. 1971. the distributional history of the biota of the Southern Appalachians. Part III.: Vertebrates. *Virginia Polytechnic Institute and State University, Blacksburg, Virginia Research Division Monograph* 4:233–62.

Guilday, J. E., P. W. Parmalee, and H. W. Hamilton. 1977. The Clark's Cave Bone Deposit and the Late Pleistocene Paleoecology of the Central Appalachian Mountains of Virginia. *Bulletin of Carnegie Museum of Natural History* 2:1–88

Harington, C. R. 1996. Giant beaver. *Beringian Research Notes* 6:1–4.

———. 1999. Ancient caribou. *Beringian Research Notes* 12:1–4.

Hay, O. P. 1912. The Pleistocene period and its vertebrata. *Indiana Department of Geology and Natural Resources 36th Annual Report* 1911:541–782.

———. 1923. The Pleistocene of North America and its vertebrated animals from the states east of the Mississippi River and from the Canadian provinces east of longitude 95 degrees. *Carnegie Institution of Washington Publication* 322A:1–499.

Haynes, G. 1991. *Mammoths, mastodonts, and elephants*. Cambridge: Cambridge University Press.

Hibbard, C. W. 1940a. The occurrence of *Cervalces scotti*

Lydekker in Kansas. *Transactions of the Kansas Academy of Sciences* 43:411–15.

———. 1940b. A new Pleistocene fauna from Meade County, Kansas. *Transactions of the Kansas Academy of Sciences* 43: 417–25.

———. 1950. Mammals of the Rexroad Formation from Fox Canyon, Kansas. *Contributions from the Museum of Paleontology University of Michigan* 8: 113–92.

Hibbard, C. W., and O. Mooser. 1963. A porcupine from the Pleistocene of Aguascalientes, Mexico. *Contributions from the Museum of Paleontology The University of Michigan* 18:245–50.

Hibbard, C. W., D. E. Ray, D. E. Savage, D. W. Taylor, and J. E. Guilday. 1965. Quaternary mammals of North America. In *The Quaternary of the United States*, edited by H. E. Wright Jr. and D. G. Frey, 509–25. Princeton, N.J.: Princeton University Press.

Holman, J. A. 1976. Snakes and stratigraphy. *Michigan Academician* 8:387–96.

———. 1981. A review of North American Pleistocene snakes. *Publications of the Museum, Michigan State University, Paleontological Series* 1:263–306.

———. 1992. Late Quaternary herpetofaunas of the central Great Lakes region, U.S.A.: Zoogeographical and paleoecological implications. *Quaternary Science Reviews* 11:345–51.

———. 1995a. *Ancient life of the Great Lakes Basin.* Ann Arbor: University of Michigan Press.

———. 1995b. *Pleistocene amphibians and reptiles in North America.* New York: Oxford University Press.

———. 2000. *Fossil snakes of North America; Origin, evolution, distribution, paleoecology.* Bloomington: Indiana University Press.

Kurta, A. 1995. *Mammals of the Great Lakes region.* Ann Arbor: University of Michigan Press.

Kurtén, B., and E. Anderson. 1980. *Pleistocene mammals of North America.* New York: Columbia University Press.

Lincoln, R. J., G. A. Boxshall, and P. F. Clark. 1982. *A dictionary of ecology, evolution, and systematics.* Cambridge: Cambridge University Press.

Lundelius, E. L., Jr., R. W. Graham, E. Anderson, J. Guilday, J. A. Holman, D. W. Steadman, and S. D. Webb. 1983. Terrestrial vertebrate faunas. In *Late Quaternary environments of the United States.* Vol.1, *The Late Pleistocene*, edited by H. E. Wright Jr., 311–53. Minneapolis: University of Minnesota Press.

Maglio, V. J. 1973. Origin and evolution of the Elephantidae. *Transactions of the American Philosophical Society, New Series* 63(3):1–149.

Martin, L. D., and A. M. Neuner. 1978. The end of the Pleistocene in North America. *Transaction of the Nebraska Academy of Sciences* 6: 117–26.

Martin, L. D., B. M. Gilbert, and S. A. Chomko. 1979. *Di-*

crostonyx (Rodentia) from the Late Pleistocene of Wyoming. *Journal of Mammology* 60:193–95.

Martin, P. S., and J. E. Guilday. 1967. A bestiary for Pleistocene biologists. In *Pleistocene extinctions: The search for a cause*, edited by P. S. Martin and H. E. Wright Jr., 1–62. New Haven, Conn.: Yale University Press.

Martin, R. A. 1968. Late Pleistocene distribution of *Microtus pennsylvanicus*. *Journal of Mammology* 49: 265–71.

———. 1989. Early Pleistocene zapodid rodents from the Java Local Fauna of north-central South Dakota. *Journal of Vertebrate Paleontology* 9: 101–9.

———. 1991. Evolutionary relationships and biogeography of Late Pleistocene prairie voles from the eastern United States. In *Beamers, bobwhites, and blue-points: Tributes to the career of Paul W. Parmalee*, edited by J. R. Purdue and W. W. Klippel, 227–35. Springfield: Illinois State Museum.

Miller, R. R. 1965. Quaternary freshwater fishes of North America. In *The Quaternary of the United States*, edited by H. E. Wright Jr. and D. G. Frey, 569–81. Princeton, N.J.: Princeton University Press.

Nelson, J.S. 1984. *Fishes of the world*, 2d Edition. New York: John Wiley and Sons.

Page, L. M., and B. M. Burr. 1991. *A field guide to freshwater fishes: North America north of Mexico.* Boston: Houghton Mifflin.

Peterson, R.T., and V.M. Peterson. 1998. *A field guide to the birds: A completely new guide to all of the birds of Eastern and Central North America.* Boston: Houghton Mifflin.

Preston, R. E. 1979. Late Pleistocene cold-blooded vertebrate faunas from the mid-continental United States, I. Reptilia; Testudines, Crocodilia. *University of Michigan Museum of Paleontology, Papers in Paleontology* 19:1–53.

Reynolds, S. H. 1913. *The vertebrate skeleton; second edition.* Cambridge: Cambridge University Press.

Robbins, C. R., G. C. Ray, and J. Douglass. 1986. *A field guide to Atlantic Coast fishes of North America.* Boston: Houghton Mifflin.

Schultz, C. B., and J. M. Hillerud. 1997. The antiquity of *Bison latifrons* (Harlan) in the Great Plains of North America. *Transactions of the Nebraska Academy of Sciences* 4: 103–16.

Schultz, C. B., L. D. Martin, and M. R. Schultz. 1985. A Pleistocene jaguar from north-central Nebraska. *Transactions of the Nebraska Academy of Sciences* 13:93–98.

Scott, W.B., and E.J. Crossman. 1973. Freshwater fishes of Canada. *Fisheries Research Board of Canada Bulletin* 184:1–966.

Selander, R. K. 1965. Avian speciation in the Quaternary. In *The Quaternary of the United States*, edited by H. E. Wright Jr. and D. G. Frey, 527–42. Princeton, N.J.: Princeton University Press.

Semken, H. A., and C. D. Griggs. 1965. The long-nosed peccary, *Mylohyus nasutus*, from McPherson County, Kansas. *Papers of the Michigan Academy of Science, Arts, and Letters* 50:267–74.

Slaughter, B. H. 1963. Some observations concerning the genus Smilodon, with special reference to *Smilodon fatalis*. *Texas Journal of Science* 15:68–81.

———. 1966. The Moore Pit Local Fauna; Pleistocene of Texas. *Journal of Paleontology* 40: 78–91.

Slaughter, B. H., and W. L. McClure. 1965. The Sims Bayou Local Fauna: Pleistocene of Houston, Texas. *Texas Journal of Sciences* 17:404–17.

Wetmore, A. 1956. A check-list of the fossil and prehistoric birds of North America and the West Indies. *Smithsonian Miscellaneous Collections* 131(5):1–105.

Woodburne, M. O., ed. 1987. *Cenozoic mammals of North America: Geochronology and biostratigraphy*. Los Angeles: University of California Press.

Ontario

Ami, H. M. 1892. Additional notes on the geology and palaeontology of Ottawa and its environs. *Ottawa Naturalist* 6:73–78.

———. 1897. Contribution to the palaeontology of the post-Pliocene deposits of the Ottawa Valley. *Ottawa Naturalist* 11:20–26.

———. 1898. The mastodon in western Ontario. *Science* 7:80.

Bensley, B. A. 1913. A *Cervalces* antler from the Toronto interglacial. *University of Toronto Studies* 8:1–3.

———. 1923. A muskox skull from the Iroquois Beach deposits at Toronto: *Ovibos proximus*, sp. nov. *University of Toronto Studies, Biology Series* 23:1–11.

Churcher, C. S. 1968. Mammoth from the Middle Wisconsin of Woodbridge, Ontario. *Canadian Journal of Zoology* 46:219–21.

Churcher, C. S., and R. R. Dods. 1979. *Ochotona* and other vertebrates of possible Illinoian age from Kelso Cave, Halton County, Ontario. *Canadian Journal of Earth Sciences* 16:1613–20.

Churcher, C. S., and P. F. Karrow. 1977. Late Pleistocene muskox (*Ovibos*) from the Early Wisconsin at Scarborough Bluffs, Ontario, Canada. *Canadian Journal of Earth Sciences* 14:326–31.

Churcher, C. S., and A. V. Morgan. 1976. A grizzly bear from the Middle Wisconsin of Woodbridge, Ontario. *Canadian Journal of Earth Sciences* 13:341–47.

Churcher, C. S., and R. L. Peterson. 1982. Chronologic and environmental implications of a new genus of fossil deer from Late Wisconsin deposits at Toronto, Canada. *Quaternary Research* 18:184–95.

Churcher, C. S., J. J. Pilny, and A. V. Morgan. 1990. Late Pleistocene vertebrate, plant, and insect remains from the Innerkip Site, southwestern Ontario. *Geographie physique et Quaternaire* 44:299–308.

Coleman, A. P. 1894. Interglacial fossils from the Don Valley. *American Geologist* 13:85–89.

———. 1895. Glacial and interglacial deposits near Toronto. *Journal of Geology* 3:622–45.

———. 1899. The Iroquois Beach. *Transactions of the Canadian Institute* 6:29–44.

———. 1933. The Pleistocene of the Toronto region. *Ontario Department of Mines, Annual Report* 41:1–69.

Crossman, E. J., and C. R. Harington. 1970. Pleistocene pike, *Esox lucius*, and *Esox* sp. from the Yukon Territory and Ontario. *Canadian Journal of Earth Sciences* 7:1130–38.

Deller, D. B., and C. J. Ellis. 1984. Crowfield: A preliminary report on a probable Paleo-Indian cremation in southwestern Ontario. *Archaeology of Eastern North America* 12:41–71.

———. 1988. Early Paleo-Indian complexes in southwestern Ontario. In Late Pleistocene and Early Holocene paleoecology and archeology of the eastern Great Lakes region, edited by R. S. Laub, N. G. Miller, and D. W. Steadman. *Bulletin of the Buffalo Society of Natural Sciences* 33:251–63.

Dreimanis, A. 1964. Notes on the Pleistocene time scale in Canada. In *Geochronology of Canada*, edited by F. Osborne, 139–56. Toronto: University of Toronto Press.

———. 1967. Mastodons, their geologic age and extinction in Ontario. *Canadian Journal of Earth Sciences* 4:663–75.

Eyles, N., and B. M. Clark. 1988. Last interglacial sediments of the Don Valley Brickyard, Toronto, Canada, and their paleoenvironmental significance. *Canadian Journal of Earth Sciences* 25:1102–22.

Eyles, C. H., and E. Eyles. 1983. Sedimentation in a large lake: A reinterpretation of the Late Pleistocene stratigraphy at Scarborough Bluffs, Ontario, Canada. *Geology* 11:146–52.

Eyles, N., and N. E. Williams. 1992. The sedimentary and biological record of the last interglacial-glacial transition at Toronto, Canada. *Geological Society of America Special Paper* 270:119–37.

Gerrad, C. 1971. Ontario fluted point survey. *Ontario Archaeology* 16:3–18.

Goldring, W. 1922. The Champlain Sea. *New York State Museum Bulletin* 239–40:153–87.

Harington, C. R. 1971. The Champlain Sea and its vertebrate fauna. Part I. The history and environments of the Champlain Sea. *Trail and Landscape* 5:137–41.

———. 1972. The Champlain Sea and its vertebrate fauna. Part II. Vertebrates of the Champlain Sea. *Trail and Landscape* 6:33–39.

———. 1977. Marine mammals in the Champlain Sea and the Great Lakes. In Amerinds and their paleoenvironments in northeastern North America, edited by W. S.

Newman and B. Salwen. *Annals of the New York Academy of Sciences* 288:508–37.

———. 1978. Vertebrate faunas of Canada and Alaska and their suggested chronological sequence. *Syllogeus* 15: 1–105.

———. 1981. Whales and seals of the Champlain Sea. *Trails and Landscape* 15:32–47.

———. 1988. Marine mammals of the Champlain Sea, and the problem of whales in Michigan. In The Late Quaternary development of the Champlain Sea basin, edited by N. R. Gadd. *Geological Association of Canada Special Paper* 35:225–40.

———. 1989. Ice-Age fossils and vanished vertebrates. *Legacy, the Natural History of Ontario* 1989:156–64.

———. 1990. Vertebrates of the last Interglacial in Canada: a review, with new data. *Géographie physique et Quaternaire* 44:375–87.

Jackson, L. J. 1978. Late Wisconsin environments and Paleo-Indian occupation of southern Ontario, Canada. *Quaternary Research* 19:388–99.

———. 1988. Fossil cervids and fluted point hunters: A review for southern Ontario. *Ontario Archaeology* 48:27–41.

———. 1989. Late Pleistocene caribou from northern Ontario. *Current Research in the Pleistocene* 6:72–74.

Jackson, L. J., and H. McKillop. 1991. Approaches to Palaeo-Indian economy: An Ontario and Great Lakes perspective. *Midcontinental Journal of Archaeology* 16:34–68.

Karrow, P. F. 1961. The Champlain Sea and its sediments. *Royal Society of Canada Special Publication* 3:97–108.

———. 1967. Pleistocene geology of the Scarborough area. *Ontario Department of Mines, Geological Report* 46:1–108.

———. 1969. Stratigraphic studies in the Toronto Pleistocene. *Proceedings of the Geological Association of Canada* 20:4–16.

Karrow, P. F., C. S. Churcher, and A. V. Morgan. [1979]. 1980. *Quaternary paleontology of the Toronto region.* Chart published by the Department of Earth Sciences, University of Waterloo, Ontario.

Kurtén, B., and E. Anderson. 1980. *Pleistocene mammals of North America.* New York: Columbia University Press.

Leidy, J. 1856. Note on the remains of a species of seal from the post-Pliocene deposit of the Ottawa River. *Proceedings of the Philadelphia Academy of Natural Science* 8:90–91.

———. 1857. Notice of the remains of a species of seal from the post-Pliocene deposit of the Ottawa River. *Canadian Naturalist and Geologist* 1:238–39.

Mason, R. L. 1960. Early man and the age of the Champlain Sea. *Journal of Geology* 68:366–76.

McAllister, D. E., S. L. Cumbaa, and C. R. Harington. 1981. Pleistocene fishes (*Coregonus, Osmerus, Microgadus,*

Gasterosteus) from Green Creek, Ontario, Canada. *Canadian Journal of Earth Sciences* 18:1356–64.

McAndrews, J. H., and L. J. Jackson. 1988. Age and environment of Late Pleistocene mastodont and mammoth in southern Ontario. In Late Pleistocene and Early Holocene paleoecology and archeology of the eastern Great Lakes region, edited by R. S. Laub, N. G. Miller, and D. W. Steadman. *Bulletin of the Buffalo Society of Natural Sciences* 33:161–72.

Mead, J. I., and F. Grady. 1996. *Ochotona* (Lagomorpha) from Late Quaternary Cave deposits in Eastern North America. *Quaternary Research* 45:93–101.

Mott, R. J. 1968. A radiocarbon-dated marine algal bed of the Champlain Sea episode near Ottawa, Ontario. *Canadian Journal of Earth Sciences* 5:319–24.

Naldrett, D. L. 1988. Seal (*Phoca* sp.) from Champlain Sea deposits near Ottawa, Canada. *Canadian Journal of Earth Sciences* 25:787–90.

Peterson, R. L. 1965. A well-preserved grizzly bear skull, recovered from a late glacial deposit near Lake Simcoe, Ontario. *Nature* 208:1233–34.

Pilny, J. J., and A. V. Morgan. 1987. Paleontology and paleoecology of a possible Sangamonian site near Innerkip, Ontario. *Quaternary Research* 28:157–74.

Robins, C. R., G. C. Ray, and J. Douglass. 1986. *A field guide to Atlantic Coast fishes of North America.* Boston: Houghton Mifflin.

Roosa, W. B. 1977. Great Lakes Paleoindian: The Parkhill Site, Ontario. *Annals of the New York Academy of Science* 288:349–54.

Russell, L. S. 1948. Post-glacial occurrence of mastodon remains in southwestern Ontario. *Transactions of the Royal Canadian Institute* 27:57–64.

Sternberg, C. M. 1930. New records of mastodons and mammoths in Canada. *Canadian Field Naturalist* 54:59–65.

———. 1951. White whale and other Pleistocene fossils from the Ottawa Valley. *National Museum of Canada Bulletin* 123:259–61.

———. 1963. Additional records of mastodons and mammoths in Canada. *National Museum of Canada, Natural History Papers* 19:1–11.

Terasmae, J. 1960. A palynological study of the Pleistocene interglacial beds at Toronto, Ontario. *Geological Survey of Canada Bulletin* 56:23–41.

Wagner, F. J. E. 1970. Faunas of the Pleistocene Champlain Sea. *Geological Survey of Canada Bulletin* 181:1–104.

———. 1984. Fossils of Ontario Part 2: Macroinvertebrates and vertebrates of the Champlain Sea with a listing of nonmarine species. *Royal Ontario Museum Life Sciences Miscellaneous Publication*: 1–64.

Whiteaves, J. F. 1907. Notes on the skeleton of a white whale or beluga, recently discovered in the Pleistocene deposits at Pakenham, Ontario. *Ottawa Naturalist* 20:214–16.

Williams, N. E., J. A. Westgate, D. D. Williams, A. Morgan, and

A. V. Morgan. 1981. Invertebrate fossils (Insecta: Trichoptera, Diptera, Coleoptera) from the Pleistocene Scarborough Formation at Toronto, and their paleoenvironmental significance. *Quaternary Research* 16:146–66.

Michigan

Abraczinskas, L. M. 1993. Pleistocene proboscidean sites in Michigan: New records and an update on published sites. *Michigan Academician* 25:443–90.

Barondess, M. M. 1996. Backhoes, bulldozers, and behemoths. *Michigan History* (January–February):23–33.

Bearss, R. E., and R. O. Kapp. 1987. Vegetation associated with the Heisler Mastodon Site, Calhoun County, Michigan. *Michigan Academician* 19:133–40.

Benninghoff, W. S., D. F. Eschman, and H. J. Scherzer. 1977. An inter-ice florule from Mill Creek, St. Clair County, Michigan. *Ecological Society of America Bulletin* 58:54.

Benninghoff, W. S., and C. W. Hibbard. 1961. Fossil pollen associated with a late-glacial woodland musk ox in Michigan. *Papers of the Michigan Academy of Science, Arts, and Letters* 46:155–59.

Burt, W. H. 1942. A caribou antler from the Lower Peninsula of Michigan. *Journal of Mammalogy* 23:214.

Case, E. C. 1915. On a nearly complete skull of *Symbos cavifrons* (Leidy) from Michigan. *Occasional Papers of the Museum of Zoology, University of Michigan, Scientific Papers of the University of Michigan* 14:1–3.

Case, E. C., I. D. Scott, B. M. Badenoch, and T. E. White. 1935. Discovery of *Elaphas primigenius americanus* in the bed of Glacial Lake Mogodore, in Cass County, Michigan. *Papers of the Michigan Academy of Science, Arts, and Letters* 20:449–54.

Case, E. C., and G. M. Stanley. 1935. The Bloomfield Hills mastodon. *Cranbrook Institute of Science Bulletin* 4:1–8.

Cleland, C. E. 1965. Barren ground caribou (*Rangifer arcticus*) from an early man site in southeastern Michigan. *American Antiquity* 30:350–51.

Cleland, C. E., M. B. Holman, and J. A. Holman. 1998. The Mason-Quimby Line revisited. In From the northern tier: Papers in honor of Ronald J. Mason, edited by C. E. Cleland and R. A. Birmingham. *Wisconsin Archeologist* 79:8–27.

DeFauw, S. L., and J. Shoshani. 1991. *Rana clamitans* and *Rana catesbeiana* from the Late Pleistocene of Michigan. *Journal of Herpetology* 25:95–99.

Dorr, J. A., and D. F. Eschman. 1970. *Geology of Michigan*. Ann Arbor: University of Michigan Press.

Dorr, V., N. Goebel, J. Haslock, K. Lehto, J. Shoshani, P. Sujdak, M. A. Vaerten, F. Zoch, and P. Zoch. 1982. *A guide to the Groleau–White Lake mastodon*. Dearborn: Oakland University College Press.

Eshelman, R. E. 1974. Black bear from Quaternary deposits in Michigan. *Michigan Academician* 6:291–98.

Eshelman, R., E. Evenson, and C. Hibbard. 1972. The peccary, *Platygonus compressus*, from beneath Late Wisconsinan till, Washtenaw County, Michigan. *Michigan Academician* 5:243–56.

Fisher, D. C. 1984a. Mastodon butchery by North American Paleo-Indians. *Nature* 308:271–72.

———.1984b. Taphonomic analysis of Late Pleistocene mastodon occurrences: Evidence of butchery by North American Paleo-Indians. *Paleobiology* 10:338–57.

———. 1990. Age, sex, and season of death of the Grandville mastodont. In Tribute to Richard E. Flanders, part 1, edited by T. J. Martin and C. E. Cleland. *Michigan Archaeologist* 36:141–60.

Fitting, J. E. 1966. Part I: The Holcombe Site. *Anthropological Papers, Museum of Anthropology, University of Michigan* 27:1–81.[3]

———. 1975. *The archeology of Michigan: A guide to the prehistory of the Great Lakes region*. Bloomfield Hills, Mich.: Cranbrook Institute of Science.

Frankforter, W. D. 1966. Some recent discoveries of Late Pleistocene fossils in western Michigan. *Papers of the Michigan Academy of Science, Arts, and Letters* 51:209–20.

———. 1991. On the trail of the mighty mastodon. *Museum (Public Museum of Grand Rapids)* 1991:1–43.

Garland, E. B., and J. W. Cogswell. 1985. The Powers Mastodont Site, Van Buren County, Michigan. *Michigan Archaeologist* 31:3–39.

Gilbert, S. 1981. The Jolman Mastodon Site. *Coffinberry Bulletin of the Michigan Archaeological Society* 28:32.

Green, A. R. 1967. Paleo-Indian and mammoth were contemporaneous. *Michigan Archaeologist* 13:1–10.

Halsey, J. R., and M. D. Stafford, eds. 1999. Retrieving Michigan's buried past: The archaeology of the Great Lakes State. *Cranbrook Institute of Science Bulletin* 64:1–478.

Handley, C. O., Jr. 1953. Marine mammals in Michigan Pleistocene beaches. *Journal of Mammalogy* 34:252–53.

Harington, C. R. 1988. Marine mammals in the Champlain Sea, and the problem of whales in Michigan. In The Late Quaternary development of the Champlain Sea basin, edited by N. R. Gadd. *Geological Association of Canada Special Paper* 35:225–40.

Hatt, R. T. 1963. The mastodon of Pontiac. *Cranbrook Institute of Science Newsletter* 32:62–64.

———.1965a. Fossil proboscidea at Cranbrook. *Cranbrook Institute of Science Newsletter* 35:24–25.

———. 1965b. The littlest mastodon. *Cranbrook Institute of Science Newsletter* 35:20–23.

Hay, O. P. 1923. The Pleistocene of North America and its vertebrated animals from the states east of the Mississippi River and from the Canadian provinces east of longitude 35 degrees. *Carnegie Institution of Washington Publication* 322A:1–499.

Held, E. R., and R. O. Kapp. 1969. Pollen analysis at the Thaller Mastodont Site, Gratiot County, Michigan. *Michigan Botanist* 8:3–10.

Hibbard, C. W. 1951. Animal life in Michigan during the Ice Age. *Michigan Alumnus Quarterly Review* 57:200–208.

———. 1952. Remains of barren ground caribou in Pleistocene deposits of Michigan. *Papers of the Michigan Academy of Science, Arts, and Letters* 37:235–37.

Hibbard, C. W., and F. J. Hinds. 1960. A radio-carbon date for a woodland musk ox in Michigan. *Papers of the Michigan Academy of Science, Arts, and Letters* 45:103–8.

Hibbard, E. A. 1958. Occurrence of the extinct moose, *Cervalces*, in the Pleistocene of Michigan. *Papers of the Michigan Academy of Science, Arts, and Letters* 43:33–37.

Holman, J. A. 1975. Michigan's fossil vertebrates. *Publications of the Museum, Michigan State University, Educational Bulletin* 2:1–54.

———. 1976. A 25,000-year-old duck: More evidence for a Michigan Wisconsinan interstadial. *American Midland Naturalist* 96:501–3.

———. 1979. New fossil vertebrate remains from Michigan. *Michigan Academician* 11:391–97.

———. 1986. The Dansville mastodont and associated wooden specimen. *National Geographic Research* 2:416.

———. 1988a. Michigan's mastodonts and mammoths revisited. In *Wisconsinan and Holocene stratigraphy in southwestern Michigan, 1988 Midwest Friends of the Pleistocene 35th Field Conference*, edited by G. L. Larson and G. W. Monaghan, 35–41. East Lansing: Department of Geological Sciences, Michigan State University.

———.1988b. The status of Michigan's Pleistocene herpetofauna. *Michigan Academician* 20:125–32.

———. 1990a. A Late Wisconsinan woodland musk ox, *Bootherium bombifrons*, from Montcalm County, Michigan, with remarks on Michigan musk oxen. *Michigan Academician* 22:1–10.

———. 1990b. Mysteries of our past I. *Michigan Natural Resources Magazine* (March–April):26–35.

———. 1990c. Riddle of the whales. *Michigan Natural History Resources Magazine* (September–October):32–37.

———. 1991. New records of Michigan Pleistocene vertebrates with comments on the Mason-Quimby Line. *Michigan Academician* 23:273–83.

———. 1995. Issues and innovations in Pleistocene vertebrate paleontology in Michigan—the last fifty-five years. *Michigan Academician* 27:409–24.

Holman, J. A., L. M. Abraczinskas, and D. B. Westjohn. 1988. Pleistocene proboscideans and Michigan salt deposits. *National Geographic Research* 4:4–5.

Holman, J. A., and D. C. Fisher. 1993. Late Pleistocene turtle remains (Reptilia: Testudines) from southern Michigan. *Michigan Academician* 25:491–99.

Holman, J. A., D. C. Fisher, and R. O. Kapp. 1986. Recent discoveries of fossil vertebrates in the Lower Peninsula of Michigan. *Michigan Academician* 18:431–63.

Holman, J. A., and M. B. Holman. 1991. Mysteries of our past II. *Michigan Natural Resources Magazine* (September–October):16–21.

Kapp, R. O. 1970. A 24,000-year-old Jefferson mammoth from Midland County, Michigan. *Michigan Academician* 3:95–99.

———. 1978. Plant remains from a Wisconsinan interstadial dated 25,000 B.P., Muskegon County, Michigan. *American Midland Naturalist* 100:506–9.

———. 1985. Late-glacial pollen and macrofossils associated with the Rappuhn mastodont (Lapeer County, Michigan). *American Midland Naturalist* 116:368–77.

———. 1999. Michigan Late Pleistocene, Holocene, and presettlement vegetation and climate. In Retrieving Michigan's buried past: The archaeology of the Great Lakes State, edited by J. R. Halsey and M. D. Stafford. *Cranbrook Institute of Science Bulletin* 64:31–58.

Karrow, P. F., K. L. Seymour, B. B. Miller, and J. E. Mirecki. 1997. Pre-Late Wisconsinan Pleistocene biota from southeastern Michigan, U.S.A. *Palaeogeography, Palaeoclimatology, Palaeoecology* 133:81–101.

MacAlpin, A. 1940. A census of mastodon remains in Michigan. *Papers of the Michigan Academy of Science, Arts, and Letters* 25:481–90.

MacCurdy, H. M. 1920. Mastodon remains found in Gratiot County, Michigan. *Annual Report of the Michigan Academy of Science* 21:109–10.

Mason, R. J. 1958. Late Pleistocene geochronology and the Paleo-Indian penetration of the Lower Michigan Peninsula. *University of Michigan Museum of Archaeology Papers* 11:1–48.

Oltz, D. O., and R. O. Kapp. 1963. Plant remains associated with mastodon and mammoth remains in central Michigan. *American Midland Naturalist* 70:339–46.

Potts, R. 1959. Michigan mammoth. *Nature Magazine* 59:471–72.

Quimby, G. I. 1958. Fluted points and the geochronology of the Lake Michigan Basin. *American Antiquity* 24:424–26.

Semken, H. A., B. B. Miller, and J. B. Stevens. 1964. Late Wisconsin woodland musk oxen in association with pollen and invertebrates from Michigan. *Journal of Paleontology* 38:823–35.

Sherzer, W. H. 1927. A new find of the woolly elephant in Michigan. *Science* 65:616.

Shoshani, J. 1989. A report on the Shelton Mastodon Site and a discussion of the numbers of mastodons and mammoths in Michigan. *Michigan Academician* 21:115–32.

Shoshani, J., D. C. Fisher, J. M. Zawiskie, S. J. Thurlow, S. L. Shoshani, W. S. Benninghoff, and F. H. Zoch. 1989. The Shelton Mastodon Site: Multidisciplinary study of a Late

Pleistocene (Twocreekan) locality in southeastern Michigan. *University of Michigan Contributions from the Museum of Paleontology* 27:393–436.

Shott, M. J., and H. T. Wright. 1999. The Paleo-Indians: Michigan's first people. In Retrieving Michigan's buried past: The archaeology of the Great Lakes State, edited by J. R. Halsey and M. D. Stafford. *Cranbrook Institute of Science Bulletin* 64:59–70.

Simons, D. B., M. Shott, and H. T. Wright. 1984. The Gainey Site: Variability in a Great Lakes Paleo-Indian assemblage. *Archaeology of Eastern North America* 12: 66–79.

———. 1987. Paleoindian research in Michigan: Current status of the Gainey and Leavitt projects. *Current Research in the Pleistocene* 4:27–30.

Skeels, M. A. 1962. The mastodons and mammoths of Michigan. *Papers of the Michigan Academy of Science, Arts, and Letters* 47:101–33.

Stoermer, E. F., J. P. Kociolek, J. Shoshani, and C. Frisch. 1988. Diatoms from the Shelton Mastodont Site. *Journal of Paleolimnology* 1:193–99.

Stoutamire, W. P., and W. S. Benninghoff. 1964. Biotic assemblage associated with a mastodon skull from Oakland County, Michigan. *Papers of the Michigan Academy of Science, Arts, and Letters* 49:47–60.

Weston, D., and K. McMillion. 1973. Elephant hunting in Michigan. *Michigan Natural Resources* 42:18–21.

Wilson, R. L. 1967. The Pleistocene vertebrates of Michigan. *Papers of the Michigan Academy of Science, Arts, and Letters* 52:197–257.

Winchell, A. 1864. Notice of the remains of a mastodon recently discovered in Michigan. *American Journal of Science* 38:223–24.

———. 1888. Extinct peccary in Michigan. *American Geologist, Personal and Scientific News* 1:67.

Wittry, W. L. 1965. The institute digs a mastodon. *Cranbrook Institute of Science Newsletter* 35:14–19.

Wood, N. A. 1914. Two undescribed specimens of *Castoroides ohioensis* from Michigan. *Science* 39:758–59.

Ohio

Dyer, D. L., C. R. Harington, R. L. Fernandez, and M. C. Hansen. 1986. A Pleistocene stag-moose (*Cervalces scotti*) from Licking County, Ohio. *Ohio Journal of Science* 86:7 (abstract).

Falquet, R. A., and W. C. Hanebert. 1978. The Willard mastodon: Evidence of human predation. *Ohio Archaeologist* 28:17.

Fisher, D. C., B. T. Lepper, and P. E. Hooge. 1994. Evidence for butchery of the Burning Tree mastodon. In *The first discovery of America: Archaeological evidence of the early inhabitants of the Ohio area*, edited by W. S. Dancey, 43–57. Columbus: Ohio Archaeological Council.

Ford, K. M., III. 1994. Faunal list of Sheriden Pit local fauna Indian Trails Caverns. Privately published.

Ford, K. M., III, A. R. Bair, and J. A. Holman. 1996. Late Pleistocene fishes from Sheriden Pit, northwestern Ohio. *Michigan Academician* 28:135–45.

Forsyth, J. L. 1963. Ice Age census. *Ohio Conservation Bulletin* 27:16, 19, 31, 33.

Guilday, J. E. 1966. *Rangifer* antler from an Ohio bog. *Journal of Mammalogy* 47:325–26.

Hansen, M. C. 1992. Bestiary of Pleistocene vertebrates of Ohio. *Ohio Geology* 1:3–6.

Hansen, M. C., D. Davids, F. Erwin, G. R. Haver, J. E. Smith, and R. P. Wright. 1978. A radiocarbon-dated mammoth site, Marion County, Ohio. *Ohio Journal of Science* 78:103–5.

Hansen, M. C., and M. T. Sturgeon. 1984. A mastodon from Late Wisconsin lake silts, Athens County, Ohio. *Compass* 61:59–64.

Hoare, R. D., J. R. Coash, C. Innis, and T. Hole. 1964. Pleistocene peccary *Platygonus compressus* LeConte from Sandusky County, Ohio. *Ohio Journal of Science* 64: 207–14.

Holman, J. A. 1986. Turtles from the Late Wisconsinan of west-central Ohio. *American Midland Naturalist* 116:213–14.

———. 1997. Amphibians and reptiles from the Pleistocene (Late Wisconsinan) of Sheriden Pit Cave, northwestern Ohio. *Michigan Academician* 29:1–20.

Lepper, B. T. 1983a. A preliminary report of a mastodont tooth find and a Paleo-Indian site in Hardon County, Ohio. *Ohio Archaeologist* 33:10–13.

———. 1983b. Reflection on the distribution of fluted points in Ohio: Why we may not know what we think we know. *Ohio Archaeologist* 33:32–35.

———. 1989. Lithic resource procurement and early Paleoindian land-use patterns in the Appalachian Plateau of Ohio. In *Eastern Paleoindian lithic resource use*, edited by C. J. Ellis and J. C. Lothrop, 239–57. Boulder, Col.: Westview Press.

———. 1990. The first Ohioans: Licking County's Paleoindian pioneers. *Historical Times Newsletter of the Granville Ohio Historical Society* 4:1–5.

———. 1991. Licking County's oldest living residents: "Rip Van Winkle" bacteria more than 11,000 years old. *Licking County Historical Society Quarterly* 1:1–2, 5.

Lepper, B. T., T. A. Frolking, D. C. Fisher, G. Goldstein, J. E. Sanger, D. A. Wymer, J. G. Ogden III, and P. E. Hooge. 1991. Intestinal contents of a Late Pleistocene mastodon from midcontinental North America. *Quaternary Research* 36:120–25.

Lepper, B. T., and J. B. Gill. 1991. Recent excavations at the Munson Springs Site, a Paleoindian base camp in central Ohio. *Current Research in the Pleistocene* 8:39–41.

McDonald, H. G. 1992a. A Late Pleistocene fauna from the

Sheriden Pit, Indian Trail Caverns, Wyandot County, Ohio, and its relationships to the Late Wisconsinan deglaciation of Ohio. *Forty-fifth Annual Meeting, Rocky Mountain Section, Geological Society of America, Abstracts with Programs* 24:52.

———. 1992b. New records of the Elk-moose *Cervalces scotti* from Ohio. *American Midland Naturalist* 122:349–56.

———. 1994. The Late Pleistocene vertebrate fauna in Ohio: Coinhabitants with Ohio's Paleoindians. In *The first discovery of America: Archaeological evidence of the early inhabitants of the Ohio Area*, edited by W. S. Dancey, 23–41. Columbus: Ohio Archaeological Council.

McDonald, H. G., and R. A. Davis. 1989. Fossil muskoxen of Ohio. *Canadian Journal of Zoology* 67:1159–66.

Miller, G. S., Jr. 1899. A new fossil bear from Ohio. *Proceedings of the Biological Society of Washington* 13:53–56.

Mills, R. S. 1975. A ground sloth, *Megalonyx*, from a Pleistocene site in Darke County, Ohio. *Ohio Journal of Science* 75:147–55.

Mills, R. S., and J. E. Guilday. 1972. First record of *Cervalces scotti* Lydekker from the Pleistocene of Ohio. *American Midland Naturalist* 88:255.

Murphy, J. L. 1983. The Seeley mastodon: A Paleo-Indian kill? *Ohio Archaeologist* 33:12–13.

Prufer, O. H. 1971. Survey of Paleo-Indian remains in Walhonding and Tuscarawas Valleys, Ohio. *Ohio Archaeologist* 21:309–10.

Prufer, O. H., and R. S. Baby. 1963. *The Paleo-Indians of Ohio*. Columbus: Ohio Historical Society.

Sears, P. B., and K. H. Clisby. 1952. Pollen spectra associated with the Orleton Farms Mastodon Site. *Ohio Journal of Science* 5:9–10.

Seeman, M. F., and O. H. Prufer. 1982. An updated discussion of Ohio fluted points. *Midcontinental Journal of Archaeology* 7:155–69.

Taggart, R. E., and A. T. Cross. 1983. Indications of temperate deciduous forest vegetation in association with mastodont remains from Athens County, Ohio. *Ohio Journal of Science* 83:26 (abstract).

Thomas, E. S. 1952. The Orleton Farms mastodon. *Ohio Journal of Science* 5:1–5.

Todd, T. N. 1973. A Pleistocene record of North American mud minnow, *Umbra*. *Copeia* 1973:587–88.

Wood, A. E. 1952. Tooth-marks on bones of the Orleton Farms mastodon. *Ohio Journal of Science* 52:27–28.

Indiana

Cahn, A. R. 1932. Records and distribution of the fossil beaver, *Castoroides ohioensis*. *Journal of Mammalogy* 13:229–41.

Collett, J. 1881. The mammoth and mastodon: Remains in Indiana and Illinois. *Second Annual Report of the Bureau of Statistics and Geology for 1980*, 384–86.

Cope, E. D., and J. L. Wortman. 1885. Post-Pliocene vertebrates of Indiana. *Indiana Department of Geology and Natural History 14th Annual Report* 1884:4–62.

Dorwin, J. T. 1966. *Fluted points and Late Pleistocene geochronology in Indiana*. Indianapolis: Indiana Historical Society.

Ellis, G. D. 1981. A Hoosier mastodon is recovered. *Outdoor Indiana* 46:11–15.

———. 1982. The Kolarik Mastodon Site. *Proceedings of the Indiana Academy of Science* 91:346.

Engels, W. L. 1931. Two new records of the Pleistocene beaver *Castoroides ohioensis*. *American Midland Naturalist* 12:529–31.

Farlow, J. O., T. J. McNitt, and D. E. Benyon. 1986. Two occurrences of the extinct moose *Cervalces scotti* from the Quaternary of northeastern Indiana. *American Midland Naturalist* 115:407–12.

Ganz, R. A. 1925. Finds of the American mastodon (*Mammut americanum*) in Delaware County, Indiana. *Proceedings of the Indiana Academy of Science* 34:393.

Gazin, C. L. 1938. A cranium of the extinct moose, *Cervalces*, from the Quaternary of northern Indiana. *American Midland Naturalist* 19:740–41.

Gooding, A. M., and J. G. Ogden III. 1965. A radiocarbon dated pollen sequence from the Wells Mastodon Site near Rochester, Indiana. *Ohio Journal of Science* 65:1–11.

Graham, R. W., J. A. Holman, and P. W. Parmalee. 1983. Taphonomy and paleoecology of the Christensen Bog bone bed, Hancock County, Indiana. *Illinois State Museum Reports of Investigations* 38:1–29.

Graham, R. W., and E. L. Lundelius Jr. 1994. Faunmap: A database documenting Late Quaternary distribution of mammal species of the United States. 2 vols. *Illinois State Museum Scientific Papers* 25:1–690.

Hay, O. P. 1912. The Pleistocene period and its vertebrata. *Indiana Department of Geology and Natural Resources 36th Annual Report* 1911:541–782.

———. 1923. The Pleistocene of North America and its vertebrated animals from the states east of the Mississippi River and from the Canadian provinces east of longitude 95 degrees. *Carnegie Institution of Washington Publication* 322A:1–499.

Holman, J. A. 1992. Late Quaternary herpetofauna of the central Great Lakes region, U.S.A.: Zoogeographical and paleogeographical implications. *Quaternary Research Reviews* 11:345–51.

Holman, J. A., and R. L. Richards. 1981. Late Pleistocene occurrences in southern Indiana of the smooth green snake, *Opheodrys vernalis*. *Journal of Herpetology* 15:123–25.

———.1993. Herpetofauna of the Prairie Creek Site, Daviess County, Indiana. *Proceedings of the Indiana Academy of Science* 102:115–31.

Hunt, L. L., and R. L Richards. 1992. The Lewis mastodon

(*Mammut americanum*) locality, Wabash County, north-central Indiana. *Proceedings of the Indiana Academy of Science* 101:221–27.

Kintner, E. 1930. Notes on unearthing parts of a mastodon skeleton. *Proceedings of the Indiana Academy of Science* 39:237–39.

Leidy, J. 1884–85. Notice of some fossil bones discovered by Mr. Francis A. Lincke, in the banks of the Ohio River, Indiana. *Proceedings of the Academy of Natural Sciences, Philadelphia* 7:199–201.

Lyon, M. W., Jr. 1926. A specimen of the extinct musk-ox *Symbos cavifrons* (Leidy) from North Liberty, Indiana. *Proceedings of the Indiana Academy of Science* 35:321–24.

———. 1931. A small collection of Pleistocene mammals from Laporte County, Indiana. *American Midland Naturalist* 12:406–10.

———. 1939. Indiana mastodons. *Proceedings of the Indiana Academy of Science* 48:246–47.

Melhorn, W. N. 1960. The Parrish and Glasford mastodons. *Proceedings of the Indiana Academy of Science* 69:189–92.

Middleton, W. G., and J. Moore. 1900. Skull of fossil bison. *Proceedings of the Indiana Academy of Science for 1899*, 178–81.

Moodie, R. L. 1929. The geological history of the vertebrates of Indiana. *Indiana Department of Conservation Publication* 90:1–115.

Moore, J. 1890. Concerning the skeleton of the great fossil beaver, *Castoroides ohioensis*. *Journal of the Cincinnati Society of Natural History* 13:138–69.

———. 1893. The recently found *Castoroides* in Randolph County, Indiana. *American Geologist* 12:67–74.

———. 1897. The Randolph mastodon. *Proceedings of the Indiana Academy of Science for 1896*, 277–79.

———. 1900. A cranium of *Castoroides* found at Greenfield, Indiana. *Proceedings of the Indiana Academy of Science for 1899*, 171–73.

Munson, P. J., and R. W. Graham. Forthcoming 2001. Additional records of Pleistocene muskoxen from Indiana. *Proceedings of the Indiana Academy of Science.*

Munson, P. J., P. W. Parmalee, and J. E. Guilday. 1980. Additional comments on the Pleistocene mammalian fauna of Harrodsburg Crevice, Monroe County, Indiana. *National Speleological Society Bulletin* 42:78–79.

Oldham, C. E. 1977. Preserving mammoth bones. *Outdoor Indiana* 42:1, 7.

Pace, R. E. 1976. Haley Mammoth Site, Vigo County: A preliminary report. *Proceedings of the Indiana Academy of Science* 85:63.

Parmalee, P. W., P. J. Munson, and J. E. Guilday. 1978. The Pleistocene mammalian fauna of Harrodsburg Crevice, Monroe County, Indiana. *National Spelological Society Bulletin* 40:64–75.

Ray, C. E., and A. E. Sanders. 1984. Pleistocene tapirs in the eastern United States. *Special Publication of the Carnegie Museum of Natural History* 8:283–315.

Reynolds, A. E. 1961. Evidence of the mastodon in Hendricks County. *Proceedings of the Indiana Academy of Science* 71:407–11.

———. 1966. Two elephantine teeth from the Mill Creek drainage area. *Proceedings of the Indiana Academy of Science* 75:293–98.

Richards, R. L. 1980. Rice rat (*Oryzomys* cf. *palustris*) remains from southern Indiana caves. *Proceedings of the Indiana Academy of Science* 89:425–31.

———. 1981. Vertebrate remains from Carcass Crypt Cave, Lawrence County, Indiana. *Proceedings of the Indiana Academy of Science* 80:442.

———. 1982a. Hairy-tailed mole (*Parascalops breweri*) remains from south-central Indiana caves. *Proceedings of the Indiana Academy of Science* 91:613–17.

———. 1982b. Hunting Indiana bears. *Outdoor Indiana* 47:16–18.

———. 1983a. Getting down to the bear bones. *Outdoor Indiana* 48:32–34.

———. 1983b. Quaternary records of the pygmy and smoky shrews from south-central Indiana caves. *Proceedings of the Indiana Academy of Science* 92:507–21.

———. 1984a. It's the pits. *Outdoor Indiana* 49:25–27.

———. 1984b. The Pleistocene vertebrate collection of the Indiana State Museum with a list of the extinct and extralocal Pleistocene vertebrates of Indiana. *Proceedings of the Indiana Academy of Science* 93:483–504.

———. 1985. Quaternary remains of the spotted skunk, *Spilogale putorius*, in Indiana. *Proceedings of the Indiana Academy of Science* 94:657–65.

———. 1986. Late Pleistocene occurrences of boreal voles (genera *Phenacomys* and *Clethrionomys*) from southern Indiana caves. *Proceedings of the Indiana Academy of Science* 95:537–46.

———. 1987. The Quaternary distribution of the eastern woodrat, *Neotoma floridana*, in Indiana. *Proceedings of the Indiana Academy of Science* 96:513–21.

———. 1988a. Cave graves. *Outdoor Indiana* 7:4–7.

———. 1988b. *Microtus xanthognathus* and *Synaptomys borealis* in the Late Pleistocene of southern Indiana. *Proceedings of the Indiana Academy of Science* 98:561–70.

———. 1988c. Quaternary occurrence of the fisher, *Martes pennanti*, in Indiana. *Proceedings of the Indiana Academy of Science* 98:571–80.

———. 1990. Quaternary distribution of the timber rattlesnake (*Crotalus horridus*) in southern Indiana. *Proceedings of the Indiana Academy of Science* 99:113–22.

———. 1991. The Alton mammoth (*Mammuthus columbi jeffersonii*) locality, Crawford County, south-central Indiana. *Proceedings of the Indiana Academy of Science* 100:77–89.

—————. 1992a. The Lewis mastodont (*Mammut americanum*) locality, Wabash County, north-central Indiana. *Proceedings of the Indiana Academy of Science* 101:221–27.

—————. 1992b. New records of the stag moose (*Cervalces scotti*) from the Late Pleistocene of Indiana. *Proceedings of the Indiana Academy of Science* 101:83–94.

—————. 1992c. Small mammals of the Prairie Creek Site, Daviess County, Indiana. *Proceedings of the Indiana Academy of Science* 101:245–78.

Richards, R. L., and J. N. McDonald. 1991. New records of Harlan's musk ox (*Bootherium bombifrons*) and associated fauna from the Late Pleistocene of Indiana. *Proceedings of the Indiana Academy of Science* 99:211–28.

Richards, R. L., and P. J. Munson. 1988. Flat-headed peccary (*Platygonus*) and recovered Quaternary vertebrate fauna of Indun Rockshelter, Monroe County, Indiana. *National Speleological Society Bulletin* 50:64–71.

Richards, R. L., and W. D. Turnbull. 1995. Giant short-faced bear (*Arctodus simus yukonensis*) remains from Fulton County, northern Indiana. *Fieldiana Geology Publication* 1465:1–34.

Richards, R. L., and W. R. Wepler. 1985. Extinct woodland muskox, *Symbos cavifrons*, cranium from Miami County, north-central Indiana. *Proceedings of the Indiana Academy of Science* 94:667–71.

Richards, R. L., and J. O. Whitaker Jr. 1997. Indiana's vertebrate fauna: Origins and changes. In *The natural heritage of Indiana*, edited by M. T. Jackson, 144–56. Bloomington and Indianapolis: Indiana University Press.

Richards, R. L., D. R. Whitehead, and D. R. Cochran. 1987. The Dollens mastodon (*Mammut americanum*) locality, Madison County, east central Indiana. *Proceedings of the Indiana Academy of Science* 97:571–81.

Riggs, E. S. 1936. Occurrence of the moose, *Cervalces*, in Indiana and Illinois. *American Midland Naturalist* 17:664.

Simpson, P. F. 1934. The Garrett mastodon. *Proceedings of the Indiana Academy of Science* 43:154–55.

Thompson, M. 1886. Fossil mammals of the post-Pliocene in Indiana. *Indiana Department of Geology and Natural History 15th Annual Report for 1885*, 383–468.

Tomak, C. H. 1975. Prairie Creek: A stratified site in southwestern Indiana. *Proceedings of the Indiana Academy of Science* 84:65–68.

—————. 1982. *Dasypus bellus* and other extinct mammals from the Prairie Creek Site. *Journal of Mammalogy* 63:158–60.

Turnbull, W. D. 1958. Notice of a Late Wisconsin mastodon. *Journal of Geology* 66:96–97.

Volz, S. A. 1977. Preliminary report on a Late Pleistocene deathtrap fauna from Monroe County, Indiana. *Proceedings of the Indiana Academy of Science* 86:293–307.

Wayne, W. J. 1960. The Darrow mastodon. *Proceedings of the Indiana Academy of Science* 69:182.

Whitehead, D. R., S. T. Jackson, M. C. Sheehan, and B. W.

Leyden. 1982. Late-glacial vegetation associated with caribou and mastodon in central Indiana. *Quaternary Research* 17:241–57.

Wylie, T. A. 1859. Teeth and bones of *Elephas primigenius*, lately found near the western fork of White River in Monroe County, Indiana. *American Journal of Science, Arts Series* 228:283–84.

Illinois

Allen, J. A. 1876. Description of some remains of an extinct species of wolf and an extinct species of deer from the lead region of the upper Mississippi. *American Journal of Science* 11:47–51.

Anderson, N. C. 1905. A preliminary list of fossil mastodon and mammoth remains in Illinois and Iowa. Rock Island, Ill.: Book Concern Printers.

Bader, R. S., and D. Techter 1959. A list and bibliography of the fossil mammals of Illinois. *Natural History Miscellanea* 172:1–8.

Bagg, R. M., Jr. 1909. Notes on the distribution of the mastodon in Illinois. *University Studies, University of Illinois Bulletin* 6:46–57.

Baker, F. C. 1920. The life of the Pleistocene or glacial period. *University of Illinois Bulletin* 17:1–476.

Churcher, C. S., and J. D. Pinsof. 1987. Variations in the antlers of North American *Cervalces* (Mammalia: Cervidae): Review of new and previously recorded specimens. *Journal of Vertebrate Paleontology* 7:373–97.

Cole, F. C., and T. Deuel, eds. 1937. *Rediscovering Illinois: Archaeological explorations in and around Fulton County*. Chicago: University of Chicago Press.

Collett, J. 1881. The mammoth and mastodon: Remains in Indiana and Illinois. *Second Annual Report of the Bureau of Statistics and Geology for 1980*, 384–86.

Crook, A. R. 1927. *Elephas primigenius boreas* Hay at Golconda, Illinois. *Transactions of the Illinois State Academy of Science* 19:288–99.

Galbreath, E. C. 1938. Post-glacial fossil vertebrates from east-central Illinois. *Field Museum of Natural History Geological Series* 6:303–13.

—————. 1939. A second record of *Cervalces* from east central Illinois. *Journal of Mammalogy* 20:507–8.

—————. 1944. *Grus canadensis* from the Pleistocene of Illinois. *Condor* 46:35

—————. 1962. A Late Pleistocene musk-ox from east-central Illinois. *Transactions of the Illinois State Academy of Science* 55:209–10.

—————. 1974. A cranium of *Symbos cavifrons* (Mammalia) from the Mississippi River between southern Illinois and Missouri. *Transactions of the Illinois State Academy of Science* 67:393–96.

Graham, M. A., and R. W. Graham. 1986. A model of paleobiological site location probability. In *Siting the super-*

conducting super collider in Illinois: An overview and predictive model of cultural and paleobiological recourses in the SSC study area, northern Illinois, edited by C. R. McGimsey, M. A. Graham, E. K. Schroeder, R. W. Graham, M. D. Wiant, and R. Durst, 75–86. Report on file, Illinois State Museum, Springfield.

Graham, R. W., and M. A. Graham. 1989. Taphonomy and paleoecology of stag-moose (Cervalces scotti) from thermokarst deposits, Tonica, north-central Illinois. Paper delivered at the Geological Society of America twenty-third annual meeting, North-Central Section, University of Notre Dame, South Bend, Indiana (cited in Graham and Lundelius 1994).

Griffen, J. B. 1968. Observations on Illinois prehistory in Late Pleistocene and early recent times. In The Quaternary of Illinois, edited by R. E. Bergstrom. University of Illinois College of Agriculture Special Publication 14:123–37.

Gruger, E. 1972a. Late Quaternary vegetational development in south-central Illinois. Quaternary Research 2:217–31.

———.1972b. Pollen and seed studies of Wisconsinan vegetation in Illinois. U.S.A. Bulletin of the Geological Society of America 83:2715–34.

Harrison, W. F., C. Karch, and J. W. Springer. 1977. Morphology and distribution of Paleo-Indian points from the Kishwaukee basin, DeKalb County, Illinois. Wisconsin Archeologist 59:33–58.

Hay, O. P. 1921. Descriptions of some Pleistocene vertebrates found in the United States. Proceedings of the United States National Museum 58:83–46.

Holman, J. A. 1966. Some Pleistocene turtles from Illinois. Transactions of the Illinois State Academy of Science 59:214–16.

———. 1995a. Ancient life of the Great Lakes Basin. Ann Arbor: University of Michigan Press.

———. 1995b. Pleistocene amphibians and reptiles in North America. New York: Oxford University Press.

King, J. E. 1981. Late Quaternary vegetational history of Illinois. Ecological Monographs 51:43–62.

King, J. E., and J. J. Saunders. 1986. Geochelone in Illinois and the Illinoian-Sangamonian vegetation of the type region. Quaternary Research 25:89–99.

Kurtén, B., and E. Anderson. 1980. Pleistocene mammals of North America. New York: Columbia University Press.

Le Conte, J. L. 1848. Notice of five new species of fossil mammalia from Illinois. Proceedings of the Association of American Geologists and Naturalists 2:102–6.

Leidy, J. 1869. The extinct mammalian fauna of Dakota and Nebraska, including an account of some allied forms from other localities, together with a synopsis of the mammalian remains of North America. Journal of the Academy of Natural Sciences of Philadelphia, Series 27:i–vii, 8–472.

Miller, B. B., R. W. Graham, A. V. Morgan, N. G. Miller, W. D. McCoy, D. F. Palmer, A. J. Smith, and J. J. Pilny. 1994. A biota associated with Matuyama-age sediments in west-central Illinois. Quaternary Research 41:350–65.

Milstead, W. W. 1967. Fossil box turtles (Terrapene) from central North America, and box turtles of eastern Mexico. Copeia 1967:168–79.

Neumann, G. K. 1937. Faunal remains from Fulton County sites. In Rediscovering Illinois, edited by C. Cole and T. Deuel, 255–68. Chicago: University of Chicago Press.

Parmalee, P. W. 1967. Castoroides and Cervalces from central Illinois. Transactions of the Illinois State Academy of Science 60:127–30.

Ray, C. E., and A. E. Sanders. 1984. Pleistocene tapirs in the eastern United States. Special Publication Carnegie Museum of Natural History 8:283–315.

Ray, C. E., D. L. Wills, and J. C. Palmquist. 1968. Fossil muskoxen of Illinois. Transactions of the Illinois State Academy of Science 61:282–92.

Riggs, E. S. 1936. Occurrence of the extinct moose, Cervalces, in Indiana and Illinois. American Midland Naturalist 17:664.

———. 1938. A Pleistocene bog deposit and its fauna. Transactions of the Illinois State Academy of Science 29:186–89.

Smith, C. R. 1935. Mastodon and other finds in Aurora. Transactions of the Illinois State Academy of Science 28:195–96.

———. 1960. Elephants at Crystal Lake. Earth Science 13:63–64.

Strode, C. W. 1976. Extinct moose of Fulton County. Fulton County Historical and Genealogical Society Newsletter 7:1–2, 8, 14–15.

Techter, D. 1961. A list and bibliography of the fossil amphibians, reptiles, and birds of Illinois. Natural History Miscellanea 176:1–6.

Wakefield, B. 1996. Mammoth and mastodon sites of northern Illinois. Private publication of Bill Wakefield, Dekalb, Illinois.

Warren, R. E., and R. W. Graham. 1988. Cervalces: An Ice Age discovery. Living Museum 50:38–41.

Wetmore, A. 1935. A record of the trumpeter swan from the Late Pleistocene of Illinois. Wilson Bulletin 42:237.

———. 1948. A Pleistocene record for Mergus merganser in Illinois. Wilson Bulletin 60:240.

Willman, H. B., and J. C. Frye. 1970. Pleistocene stratigraphy of Illinois. Urbana: Illinois Geological Survey.

Wisconsin

Black, R. F. 1976. Quaternary geology of Wisconsin. In Quaternary stratigraphy of North America, edited by W. C. Mahaney, 93–117. Stroudsberg, Pa.: Dowden, Hutchison and Ross.

Byers, D. S. 1942. Fluted points from Wisconsin. *American Antiquity* 7:400.

Dallman, J. E. 1968. Mastodons in Dane County. *Wisconsin Academy Review* 15:9–13.

———. 1969. Giant beaver from a post-Woodfordian lake near Madison. *Journal of Mammalogy* 50:826–30.

Foley, R. L. 1984. Late Pleistocene (Woodfordian) vertebrates from the driftless area of southwestern Wisconsin. *Illinois State Museum Reports of Investigations* 39:1–50.

Huber, J. K., and Overstreet, D. F. 1990. Late glacial and early post-glacial nonsilicious algae sequences for southeast Wisconsin. *Current Research in the Pleistocene* 7:93–95.

Hussakof, L. 1918. The discovery of the great lake trout, *Cristivomer namaycush*, in the Late Pleistocene of Wisconsin. *Journal of Geology* 24:685–89.

Joyce, D. 1995. "Mammoth and man." Exhibit brochure on Schaefer Farm Mammoth Kill Site, Kenosha County, Wisconsin, Kenosha Public Museum.

Long, C. A. 1986. Pleistocene caribou in central Wisconsin. *Transactions of the Wisconsin Academy of Science, Arts, and Letters* 74:12–13.

Mason, R. J. 1986. The Paleo-Indian tradition. *Wisconsin Archeologist* 67:181–206.

———. 1988. Preliminary report on the fluted point component at the Aebischer Site (47 Ct 30) in Calumet County, Wisconsin. *Wisconsin Archeologist* 59:211–26.

———. 1995. Review of *Chesrow: A Paleoindian complex in the southern Lake Michigan Basin* by D. F. Overstreet. *Wisconsin Archeologist* 79:239–42.

Overstreet, D. F. 1993. *Chesrow: A Paleoindian complex in the southern Lake Michigan Basin.* Milwaukee: Great Lakes Archaeological Press.

———. 1996. Still more on cultural contexts of mammoth and mastodont in the southwestern Michigan basin. *Current Research in the Pleistocene* 13:36–38.

———. 1998. Late Pleistocene geochronology and the Paleoindian penetration of the southwestern Michigan basin. In From the northern tier: Papers in honor of Ronald J. Mason, edited by C. E. Cleland and R. A. Birmingham. *Wisconsin Archeologist* 79:28–52.

Overstreet, D. F., and J. K. Huber. 1990. A late glacial and early postglacial pollen record from Kenosha County, southeast Wisconsin. *Current Research in the Pleistocene* 7:98–100.

Overstreet, D. F., D. J. Joyce, K. Hallin, and D. Wasion. 1993. Cultural contexts of mammoth and mastodont in the southwestern Michigan basin. *Current Research in the Pleistocene* 10:75–77.

Overstreet, D. F., D. J. Joyce, and D. Wasion. 1995. More on cultural contexts of mammoth and mastodont in the southwestern Michigan basin. *Current Research in the Pleistocene* 12:40–42.

Palmer, H. A. 1954. A review of the Interstate Park, Wiscon-sin, bison find. *Proceedings of the Iowa Academy of Science* 61:313–19.

———. 1974. Implications of an extinct peccary—early archaic artifact association from a Wisconsin cave. *Wisconsin Archeologist* 55:218–30.

Palmer, H. A., and J. B. Stoltman. 1976. The Boaz mastodon: A possible association of man and mastodon in Wisconsin. *Midcontinental Journal of Archaeology* 1:163–77.

Parmalee, P. W. 1959. Animal remains from Raddatz Rock Shelter, Sk5, Wisconsin. *Wisconsin Archeologist* 40:83–90.

Rasmusen, D. L. 1971. Microvertebrates from a fissure deposit in the driftless area of southwestern Wisconsin. *Abstract with Programs, North-Central Section, Geological Society of America, Fifth Annual Meeting*, 275–76.

Ritzenthaler, R. 1966. The Kouba Site: Paleoindians in Wisconsin. *Wisconsin Archeologist* 47:171–87.

———. 1967. A cache of Paleo-Indian gravers from the Kouba Site. *Wisconsin Archeologist* 48:261–62.

Schneider, A. F. 1983. Wisconsinan stratigraphy and glacial sequence in southeastern Wisconsin. *Geoscience Wisconsin* 7:59–85.

Stoltman, J. B. 1998. Paleoindian adaptive strategies in Wisconsin during Late Pleistocene times. In From the northern tier: Papers in honor of Ronald J. Mason, edited by C. E. Cleland and R. A. Birmingham. *Wisconsin Archeologist* 79:53–67.

Stoltman, J. B., and K. Workman. 1969. A preliminary study of Wisconsin fluted points. *Wisconsin Archeologist* 50:189–214.

Wendt, D. 1985. Paleo-Indian site distribution in the Yahara River basin of Wisconsin. *Wisconsin Archeologist* 66:243–64.

West, R. M. 1978. Late Pleistocene (Wisconsinan) caribou from southeastern Wisconsin. *Transactions of the Wisconsin Academy of Science, Arts, and Letters* 68:50–53.

West, R. M., and J. E. Dallman. 1980. A Late Pleistocene and Holocene vertebrate fossil record of Wisconsin. *Geoscience Wisconsin* 4:25–45.

CHAPTER 9: INTERPRETATION OF THE FAUNA

Axelrod, D. I. 1967. Quaternary extinctions of large mammals. *University of California Publications in Geological Sciences* 74:1–42.

Brown, J., and C. Cleland. 1969. The late glacial and early postglacial faunal resources in midwestern biomes newly opened to human adaptation. In The Quaternary of Illinois, edited by R. E. Bergstrom. *University of Illinois College of Agriculture Special Publication* 14:114–22.

Churcher, C. S. 1980. Did the North American mammoth migrate? *Canadian Journal of Anthropology* 1980:103–5.

Cleland, C. E., M. B. Holman, and J. A. Holman. 1998. The Mason-Quimby Line revisited. In From the northern tier: Papers in honor of Ronald J. Mason, edited by C. E. Cleland and R. A. Birmingham. *Wisconsin Archeologist* 79:8–27.

De Vos, A. 1964. Range changes of mammals in the Great Lakes region. *American Midland Naturalist* 71:210–31.

Dillon, L. S. 1956. Wisconsin climate and life zones in North America. *Science* 123:167–76.

Dreimanis, A. 1967. Mastodons, their geologic age, and extinction in Ontario. *Canadian Journal of Earth Sciences* 4:663–75.

———. 1968. Extinction of mastodons in eastern North America: Testing a new climatic-environmental hypothesis. *Ohio Journal of Science* 68:257–72.

Eltringham, S. K. 1982. *Elephants.* Poole, Dorset: Blanford Press.

Fisher, D. C. 1984a. Mastodon butchery by North American Paleo-Indians. *Nature* 308:271–72.

———. 1984b. Taphonomic analysis of Late Pleistocene mastodon occurrences: Evidence of butchery by North American Paleo-Indians. *Paleobiology* 10:338–57.

———. 1987. Mastodont procurement by Paleoindians of the Great Lakes region: Hunting or scavenging? In *The evolution of human hunting,* edited by M. N. Nitecki and D. V. Nitecki, 309–421. New York: Plenum.

———. 1989. Meat caches and clastic anchors: The cryptic record of Paleoindian subsistence in the Great Lakes region. *Geological Society of America Abstracts with Programs* 21(6):A234.

———. 1990. Age, sex, and season of death of the Grandville mastodont. In Tribute to Richard E. Flanders, part 1, edited by T. J. Martin and C. E. Cleland. *Michigan Archaeologist* 36:141–60.

———. 1996. How to date a mastodon. *Michigan History* (January/February):33.

Futuyma, D. J., and M. Slatkin. 1983. *Coevolution.* Sunderland, Mass.: Sinauer Associates.

Graham, R. W. 1976. Late Wisconsin mammalian fauna and environmental gradients. *Paleobiology* 2:343–50.

Graham, R. W., and E. L. Lundelius Jr. 1984. Coevolutionary disequilibrium and Pleistocene extinctions. In *Quaternary extinctions—a prehistoric revolution,* edited by P. S. Martin and R. G. Klein, 223–49. Tucson: University of Arizona Press.

Guthrie, R. D. 1984. Mosaics, allochemics, and nutrients. In *Quaternary extinctions—a prehistoric revolution,* P. S. Martin and R. G. Klein, 259–98. Tucson: University of Arizona Press.

Haynes, C. C. 1969. The earliest Americans. *Science* 166:709–15.

Haynes, G. 1991. *Mammoths, mastodonts, and elephants: Biology, behavior, and the fossil record.* Cambridge: Cambridge University Press.

Hibbard, C. W. 1960. An interpretation of Pliocene and Pleistocene climates in North America. *Annual Report of the Michigan Academy of Science, Arts, and Letters* 62: 5–30.

Hinsdale, W. B. 1925. Primitive man in Michigan. *University of Michigan University Museum, Michigan Handbook Series* 1:1–194.

Holman, J. A. 1959. Amphibians and reptiles from the Pleistocene (Illinoian) of Williston, Florida. *Copeia* 1959: 96–102.

———. 1976. Paleoclimatic implications of "ecologically incompatible" herpetological species (Late Pleistocene: southeastern United States). *Herpetologca* 32:290–95.

———. 1988. Michigan's mastodonts and mammoths revisited. In *Wisconsinan and Holocene stratigraphy in southwestern Michigan, 1988 Midwest Friends of the Pleistocene 35th Field Conference,* edited by G. L. Larson and G. W. Monaghan, 35–41. East Lansing: Department of Geological Sciences, Michigan State University.

———. 1990. Vertebrates from the Harper Site and rapid climatic warming in Michigan. *Michigan Academician* 22:205–17.

———. 1991. North American Pleistocene herpetofaunal stability and its impact on the interpretation of modern herpetofaunas: An overview. In *Beamers, bobwhites, and blue-points: Tributes to the career of Paul W. Parmalee,* edited by J. R. Purdue and W. E. Klippel, 227–35. Springfield: Illinois State Museum.

———. 1992. Patterns of re-occupation of post-glacial Michigan: Amphibians and reptiles come home. *Michigan Academician* 24:453–66.

———. 1995a. *Amphibians and reptiles in the Pleistocene of North America.* New York: Oxford University Press.

———. 1995b. *Ancient life of the Great Lakes Basin.* Ann Arbor: University of Michigan Press.

———. 1995c. Issues and innovations in Pleistocene vertebrate paleontology in Michigan: The last fifty-five years. *Michigan Academician* 27:409–24.

Holman, J. A., L. M Abraczinskas, and D. B. Westjohn. 1988. Pleistocene proboscideans and Michigan salt deposits. *National Geographic Research* 2:416.

Holman, J. A., and M. B. Holman. 1991. Mysteries of our past II. *Michigan Natural Resources Magazine* (September–October):16–21.

Huckell, B. B. 1979. Of chipped stone tools, elephants, and Clovis hunters: An experiment. *Plains Anthropologist* 24:177–89.

Kapp, R. O. 1999. Michigan Late Pleistocene, Holocene, and presettlement vegetation. In Retrieving Michigan's buried past: The archaeology of the Great Lakes State, edited by J. R. Halsey and M. D. Stafford. *Cranbrook Institute of Science Bulletin* 64:31–58.

Kiltie, R. A. 1984. Seasonality, gestation time, and large mammal extinctions. In *Quaternary extinctions—a prehistoric revolution*, edited by P. S. Martin and R. G. Klein, 299–314. Tucson: University of Arizona Press.

King, J. E., and J. J. Saunders. 1984. Environmental insularity and the extinction of the American mastodont. In *Quaternary extinctions—a prehistoric revolution*, edited by P. S. Martin and R. G. Klein, 315–39. Tucson: University of Arizona Press.

Lepper, B. T. 1986. The Mason-Quimby Line: Paleoindian frontier or methodological illusion? *Chesopiean* 24:2–9.

Lundelius, E. L., Jr., R. W. Graham, E. Anderson, J. Guilday, J. A. Holman, D. W. Steadman, and S. D. Webb. 1983. Terrestrial vertebrates. In *Late Quaternary environments of the United States*. Vol. 1, *The Late Pleistocene*, edited by H. E. Wright Jr., 311–53. Minneapolis: University of Minnesota Press.

Martin, P. S. 1967. Pleistocene extinctions. In *Pleistocene extinctions: The search for a cause*, edited by P. S. Martin and H. E. Wright Jr., 75–120. New Haven, Conn.: Yale University Press.

———. 1973. The discovery of America. *Science* 179: 969–74.

———. 1984. Prehistoric overkill: the global model. In *Quaternary extinctions—a prehistoric revolution*, edited by P. S. Martin and R. G. Klein, 354–403. Tucson: University of Arizona Press.

Martin, P. S., and R. G. Klein, eds. 1984. *Quaternary extinctions—a prehistoric revolution*. Tucson: University of Arizona Press.

Mason, R. J. 1958. Late Pleistocene geochronology and the Paleo-Indian penetration of the Lower Michigan Peninsula. *University of Michigan, Anthropological Papers*, 11:1–48.

———. 1981. *Great Lakes archaeology*. New York: Academic Press.

McAndrews, J. H. 1981. Late Quaternary climate of Ontario: Temperature trends from the fossil record. In *Quaternary Paleoclimate*, edited by W. C. Mahaney, 319–33. Norwich, England: Geo Abstracts.

Meltzer, D. J., and J. I. Mead. 1983. The timing of Late Pleistocene mammalian extinctions in North America. *Quaternary Research* 19:130–35.

Moisimann, J. E., and P. S. Martin. 1975. Simulating overkill by Paleo-Indians. *American Scientist* 6:304–13.

Morgan, A. V., and A. Morgan. 1981. Paleontological methods of reconstructing paleoclimates with reference to interglacial and interstadial insect faunas of southern Ontario. In *Quaternary paleoclimate*, edited by W. C. Mahaney, 173–92. Norwich, England: Geo Abstracts.

Osborne, H. F. 1936. *Proboscidea: A monograph of the diversity, evolution, migration, and extinction of the mastodons and elephants of the world*. Vol. 1. New York: American Museum of Natural History.

———. 1942. *Proboscidea: A monograph of the diversity, evolution, migration, and extinction of the mastodons and elephants of the New World*. Vol. 2. New York: American Museum of Natural History.

Owen-Smith, N. 1987. Pleistocene extinctions: The pivotal role of megaherbivores. *Paleobiology* 13:351–62.

Pilny, J. J., A. V. Morgan, and A. Morgan. 1987. Paleoclimatic implications of a Late Wisconsinan insect assemblage from Rostock, southwestern Ontario. *Canadian Journal of Earth Sciences* 24:617–30.

Quimby, G. I. 1958. Fluted point of the Lake Michigan Basin. *American Antiquity* 23:247–54.

———. 1960. *Indian life in the upper Great Lakes*. Chicago: University of Chicago Press.

Saunders, J. J. 1980. A model for man-mammoth relationships in the Late Pleistocene of North America. *Canadian Journal of Anthropology* 1:87–98.

Shane, L. C. K. 1987. Late-glacial vegetational and climatic history of the Allegheny Plateau and the till plains of Ohio and Indiana, U.S.A. *Boreas* 16:1–20.

Shipman, P., D. C. Fisher, and J. J. Rose. 1984. Mastodon butchery: Microscopic evidence of carcass processing and bone tool use. *Paleobiology* 10:358–65.

Slaughter, B. H. 1967. Animal ranges as a clue to Late-Pleistocene extinction. In *Pleistocene extinctions: The search for a cause*, edited by P. S. Martin and H. E. Wright Jr., 155–68. New Haven, Conn.: Yale University Press.

———. 1975. Ecological interpretation of the Brown Sand Wedge local fauna. In Late Pleistocene environments of the southern high plains, edited by F. Wendorf and J. J. Hester, 179–92. *Fort Burgwin Research Center Publication 9*.

Taggart, R. E., and A. T. Cross. 1983. Indications of temperate deciduous forest vegetation in association with mastodont remains from Athens County, Ohio. *Ohio Journal of Science* 83:26 (abstract).

Vangengeim, E. A. 1967. The effects of the Bering land bridge on the mammalian faunas of Siberia and North America. In *The Bering land bridge*, edited by D. M. Hopkins, 281–87. Stanford, Calif.: Stanford University Press.

Wier, J. S. 1972. Spatial distribution of elephants in an African national park in relation to environmental sodium. *Oikos* 23:1–13.

Illustration Credits

INDIVIDUALS

Theodore Allen: 64

Rosemarie Attilio: 38, 39, 58

Walter Auffenberg: 69

Merald Clark: 37, 72, 127

Harry Clench: 71

Mary Ann Graham: 164

Dirk Gringhuis: 115–16, 126, 151

Barbara Gudgeon: 13, 59, 66–67, 93, 103–4, 107–8, 111–12, 118–19, 125, 145–46, 148, 150

Bonnie Hall: 105 (in part)

Lisa Hallock: 40 (in part)

James Harding: 46, 47 (in part), 48 (in part), 49 (in part), 50 (in part), 51 (in part), 53 (in part), 54 (in part), 56 (in part)

Donna Holman: 55, 57

J. Alan Holman: 1–2, 4–5, 7–12, 14 (in part),15 (in part), 16, 36, 45, 68, 70, 73, 78–79, 83–84, 95, 98, 101, 109, 113, 117, 120–21, 123, 128, 133, 136–38, 143, 152, 155, 158–59

Jane Kaminski: 48 (in part), 60, 89, 99, 163

Kathleen McNulty: 122 (in part)

George E. Miller: 92

Teresa Petersen: 3, 6, 14 (in part), 15 (in part), 17–35, 40 (in part), 41–44, 49 (in part), 62–63, 65, 86, 88 (in part), 94, 100, 129, 134, 142, 168

Irene Rinchetti: 47 (in part), 49 (in part), 50 (in part), 51 (in part), 52, 53 (in part), 54 (in part), 56 (in part), 61

Robin Ross: 110, 122 (in part), 124, 144

Julianne Snider: 88 (in part), 139

Margaret Skeels Stevenson: 102, 147

Mary Tanner: 90, 114

Cameron Wood: 149

JOURNALS AND INSTITUTIONS

Figure 6. From Holman (1997), courtesy of the *Michigan Academician*

Figure 13. From Holman (1975), courtesy of the Michigan State University Museum

Figures 23, 24 (upper), 25 (upper), 27 (upper), and 159. From Ford, Bair, and Holman (1996), courtesy of the *Michigan Academician*

Figure 39. From Holman (1984), courtesy of the Carnegie Museum of Natural History

Figure 40 (upper). From Holman (1988), courtesy of the *Michigan Academician*

Figures 41, 43 (upper), and 44. From Holman (1997), courtesy of the *Michigan Academician*

Figure 46. From Holman and Harding (1977), courtesy of the Michigan State University Museum

Figure 47A. From Holman and Richards (1993), courtesy of the Indiana Academy of Science

Figure 47B and C. From Holman and Harding (1977), courtesy of the Michigan State University Museum

Figure 48A. From Graham, Holman, and Parmalee (1983), courtesy of the Illinois State Museum

Figure 48B and C. From Holman and Harding (1977), courtesy of the Michigan State University Museum

Figure 49A. From Holman and Richards (1993), courtesy of the Indiana Academy of Science

Figure 49B. From Holman and Fisher (1993), courtesy of the *Michigan Academician*

Figure 49C. From Holman and Harding (1977), courtesy of the Michigan State University Museum

Figure 50A and B. From Holman and Richards (1993), courtesy of the Indiana Academy of Science

Figure 50C. From Holman and Harding (1977), courtesy of the Michigan State University Museum

Figure 51 (upper). From Holman and Richards (1993), courtesy of the Indiana Academy of Science

Figure 51 (lower). From Holman and Harding (1977), courtesy of the Michigan State University Museum

Figure 52. From Holman and Richards (1993), courtesy of the Indiana Academy of Science

Figure 53A. From Holman and Richards (1993), courtesy of the Indiana Academy of Science

Figure 53B and C. From Holman and Harding (1977), courtesy of the Michigan State University Museum

Figure 54 (upper). From Holman and Richards (1993), courtesy of the Indiana Academy of Science

Figure 54 (lower). From Holman and Harding (1977), courtesy of the Michigan State University Museum

Figure 56 (upper). From Holman and Richards (1993), courtesy of the Indiana Academy of Science

Figure 56 (lower). From Holman and Harding (1977), courtesy of the Michigan State University Museum

Figure 58. From Holman and Winkler (1987), courtesy of the Texas Memorial Museum

Figure 59. From Holman (1976), courtesy of the *Michigan Academician*

Figure 61. From Holman and Richards (1993), courtesy of the Indiana Academy of Science

Figures 62 and 63. From Holman (1997), courtesy of the *Michigan Academician*

Figure 65. From Holman (1990), courtesy of the *Michigan Academician*

Figure 66. From Holman (1975), courtesy of the Michigan State University Museum

Figure 67. From Holman (1976), courtesy of the *American Midland Naturalist*

Figure 69. From Auffenberg (1957), courtesy of the *Florida Academy of Sciences*

Figure 71. From Guilday (1971), courtesy of *Research Division Monographs*, Virginia Polytechnic Institute and State University, Blacksburg, Virginia

Figures 74, 75, and 77. From Repenning (1967), courtesy of U. S. Geological Survey, Washington, D.C.

Figure 87. From Graham (1991), courtesy of the Illinois State Museum.

Figure 88A and B. From Graham and Graham (1990), courtesy of the *American Midland Naturalist*

Figure 90 from Schultz, Martin, and Schultz (1985), courtesy of the Nebraska Academy of Sciences

Figure 92. From Slaughter (1963), courtesy of the *Texas Journal of Science*

Figure 93. From Holman (1975), courtesy of the Michigan State University Museum

Figure 96. From Martin and Neuner (1978), courtesy of the *Nebraska Academy of Sciences*

Figure 102. From Semken and Griggs (1965), courtesy of the *Papers of the Michigan Academy of Science, Arts, and Letters*

Figures 103 and 104. From Holman (1975), courtesy of the Michigan State University Museum

Figure 105A. From Martin and Neuner (1978), courtesy of the *Nebraska Academy of Sciences*

Figure 105B. From Hibbard and Dalquest (1962), courtesy of the *Papers of the Michigan Academy of Science, Arts, and Letters*

Figures 107, 108, 111, and 112. From Holman (1975), courtesy of the Michigan State University Museum

Figure 114. From Schultz and Hillerud (1977), courtesy of the Nebraska Academy of Science

Figures 118 and 119. From Holman (1975), courtesy of the Michigan State University Museum

Figure 122B and C. From Slaughter (1966), courtesy of the *Journal of Paleontology*

Figures 125 and 127. From Holman (1975), courtesy of the Michigan State University Museum

Figure 132A. From Slaughter and McClure (1965), courtesy of the Texas Journal of Science

Figure 132 B and E. From Guilday, Parmalee and Hamilton (1977), courtesy of the Carnegie Museum of Natural History

Figure 132C. From Martin (1991), courtesy of the Illinois State Museum

Figure 132D. From Martin (1968), courtesy of the *Journal of Mammalogy*

Figure 135A, B, D, and E. From Guilday, Parmalee, and Hamilton (1977), courtesy of the Carnegie Museum of Natural History

Figure 135C. From Martin (1979), courtesy of the *Journal of Mammalogy*

Figures 139. From Foley (1984), courtesy of the Illinois State Museum

Figure 140B. From Martin (1989), courtesy of the *Journal of Vertebrate Paleontology*

Figure 141. From Hibbard and Mooser (1963), courtesy of *Contributions from the Museum of Paleontology of the University of Michigan*

Figure 145. From Holman (1975), courtesy of the Michigan State University Museum

Figure 146. From Holman (1975), courtesy of the Michigan State University Museum

Figure 147. From Skeels (1962), courtesy of the *Papers of the Michigan Academy of Science, Arts, and Letters.*

Figure 148. From Holman (1975), courtesy of the Michigan State University Museum

Figure 150. From Holman (1975), courtesy of the Michigan State University Museum

Figure 153. From Kapp (1970), courtesy of the *Michigan Academician*

Figure 154. From DeFauw and Shoshani (1991), courtesy of the *Michigan Academician*

Figure 155. From Holman (1979), courtesy of the *Michigan Academician*

Figure 156. From Garland and Cogswell (1985), courtesy of the *Michigan Archaeologist*

Figure 157. From Abraczinskas (1993), courtesy of the *Michigan Academician*

Figure 159. From Ford, Bair, and Holman (1996), courtesy of the *Michigan Academician*

Figure 160. From McDonald (1989), courtesy of the *American Midland Naturalist*

Figure 161. From Fisher, Lepper, and Hoodge (1992), courtesy of the Ohio Archaeological Council, Inc.

Figure 162. From Richards (1991), courtesy of the Indiana Academy of Science.

Figures 163 and 164. From Graham, Holman, and Parmalee (1983), courtesy of the Illinois State Museum

Figure 165. From Richards, Whitehead, and Cochran (1987), courtesy of the Indiana Academy of Science

Figures 166 and 167. From Overstreet (1998), courtesy of the *Wisconsin Archeologist*

Figure 168. From Holman (1990), courtesy of the *Michigan Academican*

General Index

Note: The material in chapters 7 and 8 is covered in the index to common and scientific names and in the site index.

Aboral axis, as it relates to symmetry in animals, 6
Absolute chronology, in geological time, 4
Adaptive radiation, in evolution, 2
Africa: megaherbivores, 173; modern grassland communities of, 173; Pleistocene extinction of vertebrates in, 188
Agassiz, Louis, 12
Alton Mammoth site, Indiana, 187
American lion, 189
Amphibian and reptile recolonization of post-glacial Michigan, 184–86
Amphioxus, 5
Analogy, in evolution, 2–3
Asphalt pits, 8. *See also* Tar Pits
Asymmetry in the animal body, 5–6

Bilateral symmetry in animals, 5–6
Block, 28–29
Bone: discussion of, 6; inorganic components of, 6; organic components of, 6
Bone apatite, 30–31
Box turtles, ability to freeze, 192
Breathing tissues of vertebrates, discussion of, 7
Bronze Age, 12

Capybara, 189
Carbon-14 dates, 31
Carnivores: definition of, 175; of terrestrial vertebrate fauna of Great Lakes Pleistocene, 175–76; of the Late Wisconsinan of Subregion I, 178; of the Late Wisconsin of Subregion II, 179
Carter Bog site, Ohio, 183–84
Cartilage of vertebrates, discussion of, 6
Caves: formation of, 23–24; intrusive vertebrates in sediments of, 24; inverted time sequence in, 23

Celom of vertebrates, discussion of, 7
Cementing agents in fossil collecting, discussion of, 28
Cementum, as a component of vertebrate teeth, 6
Cephalization, 5
Champlain Sea, 24
Cherry Tree substage, 18
Chorus frogs, ability to freeze, 192
Christensen Bog site, Indiana, 183
Circulatory system of vertebrates, discussion of, 7
Classification of species, discussion of, 9
Clovis points, 194–95
Co-evolutionary disequilibrium model, 190
Collagen, in bone, 6, 30
Collecting Pleistocene vertebrate fossils: ethics involved in, 25–26; general discussion of, 25; in caves, 27–28; in kettle bogs and shallow basins, 26–27; in lake and stream deposits, 28; screening methods, 27–28
Communities: as dominated by large herbivorous dinosaurs, 172–173; as dominated by large mammalian herbivores, 173
Community, biological, discussion of, 8–9
Consumers, 9
Continental ice sheets, modern, 12
Convergence, in evolution, 2
Coprolites, 8
Cores, sea bottom, 10
Cuvier, Georges, 5

Darwin, Charles, 2
Decomposers, in the biological community, 9
Dentine, 6
Deposition, 3
Digestive system of vertebrates, discussion of, 7
Dire wolves, 189
Disconformities, 3
Disharmonious communities, 181
Domino effect, 173
Draught hypothesis, 190
Drift, glacial, 16–17
Drumlins, 17

Earliest mammals, 173
Ecological niches, 2
Ecosystem, definition of, 8
Enamel, 6
Eratics, glacial, 12
Erosion, 3
Eskers, 17
"Everything wrong at once," hypothesis of Pleistocene extinction, 195–96
Evolution, discussion of, 2–3
Extinction in Pleistocene of Great Lakes region: amphibians, 191; birds, 191; fishes, 190–91; large marine mammals, 191; large land mammals, 191; reptiles, 191; small land mammals, 191

Faunal succession, as it relates to geological time scale, 4
Fluted projectile points, 194
Folsom points, 194
Food web, in ecological communities, 9
Fossil, definition of, 5

Gainey points, 194
Genetic material, molecular nature of, 2
Geological cycle, 3
Geological formation, 4
Geology, definition of, 3
Giant condors, 189
Giant ground sloth, 189
Giant land tortoises, 189
Glacial advance, possibility in future, 1
Glacial features, discussion of, 16–17
Glacial Lake Chicago, 17
Glacial Lake Iroquois sites, Ontario, 183
Glacial stages, 10
Glacially derived kettles and basins: as mastodont and mammoth traps, 20–23; discussion of, 20; zones in, 20–21
Glyptodonts, 189
Gray tree frogs, ability to freeze, 192
Great Lakes, origin of, discussion of, 17
Great Pleistocene extinction, general discussion of, 188–89
Great Slave Lake, Canada, 182

Guilday, John, 77

Hard tissues of vertebrates, 6
Harper Holocene site, Michigan, 197
Harrodsburg Crevice, Indiana, 187–88
Hemoglobin, as component of vertebrate blood, 7
Herbivores: definition of, 175; of terrestrial vertebrate fauna of Great Lakes Pleistocene, 175; of the Late Wisconsinan of Subregion I, 177–78; of the Late Wisconsinan of Subregion II, 178–79
Hibbard, Claude W., 61, 174
Hibbard's Rule, 61
Homology, in evolution, 2–3
Hopwood Farm site, Illinois, 188
Horses, 189
Humans as agents in extinction, 194–95
Hydroxyapatite, as part of bone, 6
Hypsithermal warm spell, 197

Ice sheet, deepest penetration in North America, 12
Ice sheet effects: climatic changes, 10, 13; community destruction, 13; reorganization of biotic communities, 14; topographic changes, 13; vegetational changes, 13
Interglacial stages, 10
Interstadials, Wisconsinan, 15
Invasion of marine vertebrates, Late Pleistocene of Ontario, 186–88
Invertebrate paleontology, definition of, 5
Irvingtonian land mammal age, 15

Jaguar, 189
Jefferson, Thomas, 75, 144

Kalkaska County, Michigan, 18
Kames, 17
Karst topography, 23
Kettle holes, 17
Kidneys in vertebrates, discussion of 7
King Leo Pit Cave, Indiana, 187

Index to Common and Taxonomic Names